Statistics for Biology and Health

Series Editors

K. Dietz, M. Gail, K. Krickeberg, J. Samet, A. Tsiatis

Statistics for Biology and Health

Giovanni Parmigiani
Elizabeth S. Garrett
Rafael A. Irizarry
Scott L. Zeger
Editors

The Analysis of
Gene Expression Data
Methods and Software

With 113 Figures, Including 37 Color Plates

 Springer

Giovanni Parmigiani
Departments of Oncology,
    Biostatistics, and Pathology
Elizabeth S. Garrett
Departments of Oncology
    and Biostatistics
Johns Hopkins University
Baltimore, MD 21205-2013
USA

Rafael A. Irizarry
Department of Biostatistics
Scott L. Zeger
Departments of Biostatistics
    and Epidemiology
Johns Hopkins University
Baltimore, MD 21205-2013
USA

Series Editors
K. Dietz
Institut für Medizinische
    Biometrie
Universität Tübingen
Westbahnhofstrasse 55
D-72070 Tübingen
Germany

M. Gail
National Cancer Institute
Rockville, MD 20892
USA

K. Krickeberg
Le Chatelet
F-63270 Manglieu
France

J. Samet
Department of Epidemiology
School of Public Health
Johns Hopkins University
615 Wolfe Street
Baltimore, MD 21205-2103
USA

A. Tsiatis
Department of Statistics
North Carolina State University
Raleigh, NC 27695
USA

Library of Congress Cataloging-in-Publication Data
    The analysis of gene expression data : methods and software / editors, Giovanni
Parmigiani . . . [et al.].
        p.   cm. — (Statistics for biology and health)
    Includes bibliographical references and index.
    ISBN 0-387-95577-1 (alk. paper)
    1. DNA microarrays. 2. Gene expression—Research—Methodology. 3. Gene
expression—Data processing.   I. Parmigiani, G. (Giovanni). II. Series.
    QP624.5.D726 A53   2003
    572′.8′65–dc21   2002030572

ISBN 0-387-95577-1   Printed on acid-free paper.

Printed in the United States of America.

9 8 7 6 5 4 3

springer.com

Preface

The development of technologies for high–throughput measurement of gene expression in biological system is providing powerful new tools for investigating the transcriptome on a genomic scale, and across diverse biological systems and experimental designs. This technological transformation is generating an increasing demand for data analysis in biological investigations of gene expression. This book focuses on data analysis of gene expression microarrays. The goal is to provide guidance to practitioners in deciding which statistical approaches and packages may be indicated for their projects, in choosing among the various options provided by those packages, and in correctly interpreting the results.

The book is a collection of chapters written by authors of statistical software for microarray data analysis. Each chapter describes the conceptual and methodological underpinning of data analysis tools as well as their software implementation, and will enable readers to both understand and implement an analysis approach. Methods touch on all aspects of statistical analysis of microarrays, from annotation and filtering to clustering and classification. All software packages described are free to academic users.

The materials presented cover a range of software tools designed for varied audiences. Some chapters describe simple menu-driven software in a user-friendly fashion and are designed to be accessible to microarray data analysts without formal quantitative training. Most chapters are directed at microarray data analysts with master's-level training in computer science, biostatistics, or bioinformatics. A minority of more advanced chapters are intended for doctoral students and researchers.

Baltimore, Maryland

Giovanni Parmigiani
Elizabeth S. Garrett
Rafael A. Irizarry
Scott L. Zeger

Contents

7   **An S-PLUS Library for the Analysis and Visualization**
     of Differential Expression   **163**
     Jae K. Lee and Michael O'Connell

8   **DRAGON and DRAGON View: Methods for the**
     Annotation, Analysis, and Visualization of Large-Scale
     Gene Expression Data   **185**
     Christopher M.L.S. Bouton, George Henry, Carlo Colantuoni,
     and Jonathan Pevsner

12 SAM Thresholding and False Discovery Rates for Detecting Differential Gene Expression in DNA Microarrays   272
John D. Storey and Robert Tibshirani

13 Adaptive Gene Picking with Microarray Data: Detecting Important Low Abundance Signals   291
Yi Lin, Samuel T. Nadler, Hong Lan, Alan D. Attie, and Brian S. Yandell

14 **MAANOVA: A Software Package for the Analysis of
Spotted cDNA Microarray Experiments**   **313**
*Hao Wu, M. Kathleen Kerr, Xiangqin Cui,
and Gary A. Churchill*

15 **GeneClust**   **342**
Kim-Anh Do, Bradley Broom, and Sijin Wen

Contributors

Christopher M.L.S. Bouton, LION Bioscience Research, Inc., Cambridge, MA 20139, USA

Alvis Brazma, European Bioinformatics Institute, Wellcome Trust Genome Campus, Hinxton, Cambridge, England CB10 1SD

Allan D. Attie, Department of Biochemistry, University of Wisconsin-Madison, Madison, WI 53706, USA

Bradley Broom, Department of Computer Science, Rice University, Houston, TX 77005-1892, USA

Atul J. Butte, Division of Endocrinology and Informatics Program, Children's Hospital, Boston, MA 02115, USA

Vincent Carey, Department of Biostatistics, Harvard Medical School, Boston, MA 02115, USA

Gary A. Churchill, The Jackson Laboratory, Bar Harbor, ME 04609, USA

Carlo Colantuoni, Department of Biostatistics, Johns Hopkins Bloomberg School of Public Health, Baltimore, MD 21205; Clinical Brain Disorders Branch, National Institute of Mental Health, Bethesda, MD 20892, USA

Leslie M. Cope, Department of Mathematical Sciences, Johns Hopkins University, Baltimore, MD 21205, USA

Xiangquin Cui, The Jackson Laboratory, Bar Harbor, ME 04609, USA

Kim-Anh Do, Department of Biostatistics, University of Texas MD Anderson Cancer Center, Houston, TX, 77030-4095, USA

Sandrine Dudoit, Division of Biostatistics, School of Public Health, University of California, Berkeley, CA 94720-7360, USA

Jai Evans, Division of Enterprise and Custom Applications, Center for Information Technology, National Institute of Health, Bethesda, MD 20892, USA

Elizabeth S. Garrett, Departments of Oncology and Biostatistics, Johns Hopkins University, Baltimore, MD 21205-2013, USA

Laurcut Gautier, Center for Biological Sequences, Denmark Technical University, Lyngby, DK-2100, Denmark

Robert Gentleman, Department of Biostatistics, Harvard School of Public Health, Boston, MA 02155, USA

George Henry, Wadham College, Oxford University, Oxford, OX1 3PN England

Rafael Irizarry, Department of Biostatistics, Johns Hopkins University, Baltimore, MD 21205-2013, USA

Misha Kapushesky, European Bioinformatics Institute, Wellcome Trust Genome Campus, Hinxton, Cambridge, England CB10 1SD

Patrick Kemmeren, European Bioinformatics Institute, Wellcome Trust Genome Campus, Hinxton, Cambridge, England CB10 1SD; Genomics Lab, Division of Biomedical Genetics, 3508 AB Utrecht, The Netherlands

Christina Kendziorski, Department of Biostatistics and Medical Informatics, University of Wisconsin-Madison, Madison, WI 53706-1685, USA

M. Kathleen Kerr, The University of Washington, Seattle, WA 98195, USA

Isaac S. Kohane, Division of Endocrinology and Informatics Program, Children's Hospital, Boston, MA 02115, USA

Hong Lan, Department of Biochemistry, University of Wisconsin-Madison, Madison, WI 53706, USA

Jae K. Lee, Department of Health Evaluation Sciences, University of Virginia School of Medicine, Charlottesville, VA 22908-0717, USA

Peter Lemkin, Laboratory of Experimental and Computational Biology, CCR, National Cancer Institute, FCRDC, Frederick, MD 21702, USA

Cheng Li, Department of Biostatistics, Dana-Farber Cancer Institute, Boston, MA 02115, USA

Yi Lin, Department of Statistics, University of Wisconsin-Madison, Madison, WI 53706, USA

Samuel T. Nadler, Department of Biochemistry, University of Wisconsin-Madison, Madison, WI 53706, USA

Michael A. Newton, Department of Statistics, University of Wisconsin-Madison, Madison, WI 53706-1685, USA

Michael F. Ochs, Department of Bioinformatics, Fox Chase Cancer Center, Philadelphia, PA 19111, USA

Michael O'Connell, Director, BioPharm Solutions, Insightful Corporation, Seattle, WA 98109, USA

Giovanni Parmigiani, Departments of Oncology, Biostatistics, and Pathology, Johns Hopkins University, Baltimore, MD 21205-2013, USA

Jonathan Pevsner, Department of Neurology, Kennedy Krieger Institute, and Department of Neuroscience, Johns Hopkins University School of Medicine, Baltimore, MD 21205, USA

Marco Ramoni, Children's Hospital Informatics Program, Harvard Medical School, Boston, MA 02115, USA

Ugis Sarkans, European Bioinformatics Institute, Wellcome Trust Genome Campus, Hinxton, Cambridge, England CB10 1SD

Paola Sebastiani, Department of Mathematics and Statistics, University of Massachusetts, Amherst, MA 01003, USA

John D. Storey, Department of Statistics, University of California, Berkeley, CA 94720-3860, USA

Gregory C. Thornwall, SAIC, National Cancer Institute, Frederick, MD; DECA, CIT, National Institutes of Health, Bethesda, MD 20892, USA

Robert Tibshirani, Department of Health Research & Policy and Department of Statistics, Stanford University, Stanford, CA 94305, USA

Jaak Vilo, European Bioinformatics Institute, Wellcome Trust Genome Campus, Hinxton, Cambridge, England CB10 1SD

Sijin Wen, Department of Statistics, University of Texas M. D. Anderson Cancer Center, Houston, TX 77030-4095, USA

Wing Hung Wong, Department of Statistics, Harvard University, Cambridge, MA 02138, USA

Hao Wu, The Jackson Laboratory, Bar Harbor, ME 04609, USA

Brian Yandell, Department of Statistics and Department of Horticulture, University of Wisconsin-Madison, Madison, WI 53706, USA

Jean Yee Hwa Yang, Division of Biostatistics, University of California, San Francisco, CA 94143-0560, USA

Scott L. Zeger, Departments of Biostatistics and Epidemiology, Johns Hopkins University, Baltimore, MD 21205-2013, USA

1

The Analysis of Gene Expression Data: An Overview of Methods and Software

Giovanni Parmigiani
Elizabeth S. Garrett
Rafael A. Irizarry
Scott L. Zeger

Abstract

This chapter is a rough map of the book. It provides a concise overview of data-analytic tasks associated with microarray studies, pointers to chapters that can help perform these tasks, and connections with selected data-analytic tools not covered in any of the chapters. We wish to give a general orientation before moving to the detailed discussion provided by individual chapters. A comprehensive review of microarray data analysis methods is beyond the scope of this introduction.

1.1 Measuring Gene Expression Using Microarrays

1.1.1 Microarray Technologies

Proteins are the structural components of cells and tissues and perform many key functions of biological systems. The production of proteins is controlled by genes, which are coded in deoxyribonucleic acid (DNA), common to all cells in one being, and mostly static over one's lifetime. Protein production from genes involves two principal stages, known as transcription and translation, as illustrated in the schematic of Figure 1.1. During transcription, a single strand of messenger ribonucleic acid, or mRNA, is copied from the DNA segment coding the gene. After transcription, mRNA is used as a template to assemble a chain of amino acids to form the protein. Gene expression investigations study the amount of transcribed mRNA in a biological system. Although most proteins undergo modification after translation and before becoming functional, most changes in the state of a cell are related to changes in mRNA levels for some genes, making the transcriptome worthy of systematic measurement. Basic biochemistry and

1

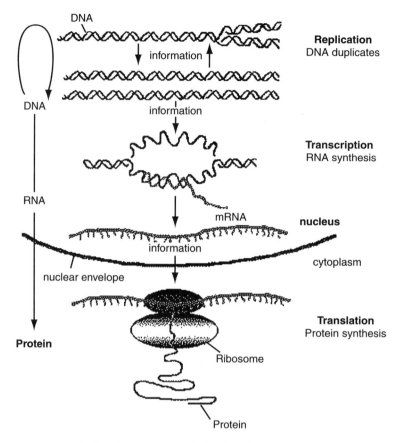

The Central Dogma of Molecular Biology

FIGURE 1.1. A schematic of the role of RNA in gene expression and protein production. Graphics from http://www.accessexcellence.org.

molecular biology textbooks, such as those of Bolsover et al. (1997) and Garrett and Grisham (2002) provide background on gene expression and its biological significance.

Several techniques are available for measuring gene expression, including serial analysis of gene expression (SAGE), cDNA library sequencing, differential display, cDNA subtraction, multiplex quantitative RT-PCR, and gene expression microarrays. Microarrays quantify gene expression by measuring the hybridization, or matching, of DNA immobilized on a small glass, plastic, or nylon matrix to mRNA representation from the sample under study. A separate experiment takes place in each of many individual spots, arranged as a regular pattern on the matrix, whence the name array. Arrays can currently have hundreds of thousands of spots. Such ability to measure simultaneously a large proportion of the genes on a genome opens the door to the investigation of the interactions among the genes on a large scale, the discovery of the role of the vast number of genes whose function is not adequately understood, and the characterization of how metabolic pathways are changed under varying conditions. Duggan et al. (1999) review the use of microarrays in genomic investigations and the impressive spectrum of biological applications. A broad range of topics regarding microarray analysis is reviewed by Collins (1999) (and in other articles in the same Supplement to *Nature Genetics*), and Kohane et al. (2002).

There are several microarray technologies. Currently, two approaches are prevalent: cDNA arrays and oligonucleotide arrays. Both are briefly described below. Although they both exploit hybridization, they differ in how DNA sequences are laid on the array and in the length of these sequences. Schena (2000) reviews in detail the technical aspects of current microarray technologies. For a brief overview, see Southern (2001) or Hardiman (2002).

In spotted DNA arrays, mRNA from two different biological samples is reverse-transcribed into cDNA, labeled with dyes of different colors, and hybridized to DNA sequences, each of which is spotted on a small region, or spot, on a glass slide. After hybridization, a laser scanner measures dye fluorescence of each color at a fine grid of pixels. Higher fluorescence indicates higher amounts of hybridized cDNA, which in turn indicates higher gene expression in the sample. A spot typically consists of a number of pixels. Image analysis algorithms either assign pixels to a spot or not and produce summaries of fluorescence at each spot as well as summaries of fluorescence in the surrounding unspotted areas (background). cDNA microarrays are described in Schena et al. (1995) and DeRisi et al. (1997).

For each location on the array, a typical output consists of at least four quantities, one of each color for both the spot and the background. Sometimes these are accompanied by measures of quality of the spot, to flag technical problems, or by measures of the pixel intensity variability. It is conventional to refer to the two colors as red and green. Throughout the book, we will denote the "red" intensities by R and the "green" intensities by G. The use of two channels allows for measurement of relative gene expression across two sources of cDNA, controlling for the amount of spotted DNA, which can be variable, as well as other experimental variation. This had led to emphasis on ratios of intensities at each spot. Although this ratio is critical, there is relevant information in all four of the quantities above.

The second common approach involves the use of high-density oligonucleotide arrays. This is an area of active technological development. As we write, the most widely used oligonucleotide array type is the Affymetrix GeneChip (for brevity Affy). In Affy arrays, expression of each gene is measured by comparing hybridization of the sample mRNA to a set of probes, composed of 11-20 pairs of oligonucleotides, each of length 25 base pairs. The first type of probe in each pair is known as perfect match (*PM* throughout the book) and is taken from the gene sequence. The second type is known as mismatch (*MM* throughout the book) and is created by changing the middle (13th) base of the *PM* sequence to reduce the rate of specific binding of mRNA for that gene. The goal of *MM*s is controlling for experimental variation and nonspecific binding of mRNA from other parts of the genome.

An RNA sample is prepared, labeled with a fluorescent dye, and hybridized to an array. Unlike in two-channel arrays, a single sample is hybridized on a given array. Arrays are then scanned, and images are produced and analyzed to obtain a fluorescence intensity value for each probe, measuring hybridization for the corresponding oligonucleotide. For each gene, or probe set, the typical output consists of two vectors of intensity readings, one for *PM*s and one for *MM*s. Chapters 4 and 5 discuss data analysis approaches for deriving probe-set summaries that best reflect the level of expression of the corresponding gene. Oligonucleotide arrays are discussed by Lockhart et al. (1996); details on Affy arrays can be found in Affymetrix (1999). Some oligonucleotide approaches are compatible with two-sample hybridization. For example, Agilent chips use 60-nucleotide probes and measure two channels, as in the cDNA arrays described above.

1.1.2   Sources of Variation in Gene Expression Measurements Using Microarrays

Gene expression microarrays are powerful, but variability arising throughout the measurement process can obscure the biological signals of interest. It is useful to classify sources of error into five phases of data acquisition: microarray manufacturing, preparation of mRNA from biological samples, hybridization, scanning, and imaging. Each of these phases can introduce an amount of artifactual variation and/or bias that complicates the estimation of expression levels as well as the comparison of expression changes between arrays. To convey a sense for the multiplicity of sources of errors and the importance of quality control, we list some examples.

Manufacturing errors are specific to the technology. In cDNA microarrays, variability arises in the amplification, purification, and concentration of DNA clones for spotting, in the amount of material spotted, in the ability of the spotted material to bind to the array, and in the shape of the deposited spot. Systematic variation can be determined by microscopic defects in the print tip of the robotic equipment used for spotting.

During the preparation of samples, sources of variability depend on the protocol and the platform used. Important examples include labeling procedures, RNA extraction, and amplification. In cDNA arrays, dye biases can arise from different physical properties of the dyes, such as decay rates, or from differential ability of the dyes to incorporate into the samples. There are examples of RNA sequences for which only one of the dyes functions efficiently.

During hybridization, variability arises from ambient conditions such as temperature and humidity, from edge effects (that is, effects seen only at the genes spotted near the edges of the array), from slight inhomogeneity of the hybridization solution, from extraneous molecules or dust binding to the array, from cross-hybridization of molecules with high sequence identity, and from washing of nonhybridized materials from the array.

During scanning, natural fluorescence and binding of genetic material to the array in unspotted regions can introduce a nontrivial, spatially varying background noise. Scanning requires separating the fluorescent label from the biological material and capturing it with sensors; both phases involve randomness, and rescanned slides usually give slightly different results. Scanning intensity is an important factor, as higher intensity improves the quality of the signal but increases the risk of saturation caused by a ceiling occurring when a channel reaches maximum intensity.

In the imaging step, some technologies require human intervention for the initialization of the imaging algorithms or the alignment of the image to a grid. Different imaging algorithms and options within these algorithms also typically lead to varying fluorescence quantifications.

Although many of these errors are relatively small, the compounding of their effects can be significant. As a result, we can generally expect variation in the expression of a given gene across different hybridizations using the same RNA sample. Also, when a sequence is spotted in multiple locations on an array, there is usually variation in the amount of hybridization measured across locations. In cDNA arrays, many sources of noise can be quantified in the aggregate by a self-versus-self hybridization in which two subsamples from the same pool of RNA are labeled with different dyes and then hybridized to the same array.

The sources of variation described so far arise from limitation of the current techniques and will be referred to as "technological." In most microarray experiments we will also need to consider the usual sources of variation that arise in selection of biological samples. For example, in comparing the gene expression of tumor tissue to normal tissue in a patient, we need to consider that the overall expression profile of the tissue may have been different if another patient had been analyzed and may also have been different if a different portion of the tissue from the patient under considereation had been sampled. In this regard, microarray experiments are the same as any other investigation of biological variation.

1.1.3   Phases of Microarray Data Analysis

The high cost, high volume, and complex experimental artifacts associated with microarray data collection have emphasized the need for statistical and data-analytic techniques at all stages of experimentation. The resulting explosion of literature on microarray data analysis can be monitored, for example, from microarray journal watch Web Sites such as "Y. F. Leung's

Functional Genomics page" (`http://genomicshome.com/`) or "Microarray & Data Analysis" (`http://linkage.rockefeller.edu/wli/microarray/`).

The most important step in a microarray experiment, like any other experiment, is the identification of biological questions of interest. The degree of specificity of the question can range from a precisely defined hypothesis about two groups –say, the study of the effect on a particular organ of a toxic compound in a population of genetically identical laboratory animals to a much more broadly defined one –for example, the search for novel hypotheses about yet-unidentified subtypes of lung cancer. The questions chosen give rise to a design of experiment, which in turn leads to data. In view of the multiple sources of errors described above, a substantial effort is typically necessary to extract from the data a reliable signal (that is, a reliable representation of gene expression under the various experimental conditions). Gene expression measures are subsequently used to address the biological questions. Because of the high potential for false-positive findings, there is wide agreement that results should be validated on a smaller scale using alternative assays, such as RNA blotting, or RT-PCR.

Ideally, the signal extraction, data analysis, and validation stages should be integrated and uncertainties propagated across stages. Because of the complexity and novelty of the tasks, only preliminary progress has been made toward this integration. The more typical approach is to perform normalization and probe-level analysis on spot-level summaries that are taken at face value and then perform data analysis on normalized gene expression measures that are also taken at face value. The prevalent strategy for dealing with uncertainty in individual spot measurements is to exclude spots for which uncertainty is considered too high to be acceptable.

Table 1.1 gives a schematic view of the phases of microarray experimentation that involve data-analytic steps. The list of tasks is not exhaustive, but can serve as a rough map. Chapters of this book cover software tools to perform all tasks in the table with the exception of image analysis and sample size calculations.

A broad spectrum of biological investigations is made possible by microarray technologies. On the one hand of this spectrum, we have highly controlled comparisons (for example, between treated and control groups of genetically identical mice). In such applications, the signal-to-noise ratios are relatively favorable and statistical questions, albeit hard to address, are generally better defined. Discussion of statistical approaches is in Section 1.5. At the opposite end, we have what we could describe as "genome biometry," that is, the description of the variability of genomic information in biological populations. In the latter, signal-to-noise ratios are less favorable, and the more exploratory nature of the biological investigation makes the statistical questions less well defined. Statistical methods, although critical, cannot alone take one from data to answers and are often used to support data exploration rather than completely automate it.

TABLE 1.1. Phases of microarray experimentation with examples of related data-analytic tasks.

Experimental Design

Choice of sample size
Assignment of experimental conditions to arrays

Signal Extraction

Image analysis
Gene filtering
Probe level analysis of oligonucleotide arrays
Normalization and removal of artifacts for comparisons across arrays

Data Analysis

Selection of genes that are differentially expressed across experimental
   conditions
Clustering and classification of biological samples
Clustering and classification of genes

Validation and Interpretation

Comparisons across platform
Use of multiple independent datasets

Table 1.2 reviews a core set of symbols that are used, with the same meaning across the book unless otherwise noted.

1.2   Design of Microarray Experiments

1.2.1   Replication and Sample Size Considerations

It is now becoming widely accepted that microarray experiments need to be replicated (Lee et al., 2000). The presence of internal controls in both two-channel cDNA arrays and Affy chips is very helpful but not sufficient to eliminate or evaluate the numerous sources of experimental error. We can distinguish two broad types of replicate experiments: "biological replicates" refer broadly to analysis of RNA of the same type from different subjects (for example, muscle tissue treated with the same drug in different mice); "technical replicates" refer to multiple-array analysis performed using the same RNA (for example, multiple samples from the same tissue).

Depending on the experimental setting, one or both of these types of replicates need to be considered. In controlled experiments, replicates are generally used to increase the reliability of conclusions. In more complex or more exploratory experiments, where biological variability is likely to ex-

TABLE 1.2. Common core of symbols used throughout the book.

	Variables
y	phenotype, sample characteristic (e.g., cancer type)
x	measure of gene expression, absolute
z	design variables for controlled experiments (e.g., drug A, drug B)
d	measure of differential gene expression across conditions in pairwise comparisons (e.g., difference of logs in cDNA arrays); differential gene expression between an individual and a population average; and others. Used for both ratios and differences.
a	overall abundance for a gene (e.g., average of logs in cDNA arrays); population average expression.

	Subscripts
$i = 1, \ldots, I$	individual, sample
$j = 1, \ldots, J$	gene
$k = 1, \ldots, K$	"technical" replicate (see Subsection 1.2.1). May denote the same sequence on multiple spots on the same array or the same RNA on multiple hybridizations.
$t = 1, \ldots, T$	time for time course experiments

ceed technical errors, obtaining biological replicates is more critical. Simon et al. (2002) provide a discussion of the relevant trade-offs.

In controlled experiments comparing gene expression in two or a small number of conditions, the goal of a microarray study can be often described as identifying as many genes that are differentially expressed across conditions as possible while keeping the probability of making false declarations of expression acceptably low. When this is the goal, we can address the question of how many replicates are required using well-developed hypothesis-testing ideas. Reviews of standard statistical approaches are provided by Adcock (1997) and Desu and Raghavarao (1990). In general, the answer depends on several factors: the signal-to-noise ratios (that is, the likely magnitudes of expression changes compared to technological error), the desired sensitivity in detecting changes, and the tolerance for false findings. In the context of microarray experiments, Pan et al. (2002) discuss how to calculate the number of replicates in the context of applying a normal-mixture-model approach to detect changes in gene expression. Sample S-PLUS code is available from the author's Web site.

Determining the number of replicates for genome biometry investigations such as cluster analysis is more complex. Because most clustering methods depend on distance matrices, Bryan and van der Laan (2001) developed a sample size formula that aims at controlling the accuracy of the estimation of the sample mean and sample covariance matrix of the array.

1.2.2   Design of Two-Channel Arrays

In two-channel arrays, investigators have the option of hybridizing two types of samples to the same array. This enables a direct comparison of those two samples, while controlling for array-specific variation. A standard approach is to compare all samples of interest to the same reference. Although practical, this strategy is such that all comparisons of biological interest are indirect, or across arrays. In a series of seminal papers, Kerr, Churchill and their colleagues (Kerr et al., 2000; Kerr and Churchill, 2001b and c) investigated the use of classical experimental designs in the context of spotted cDNA arrays and suggested general designs that estimate the baseline error variance in an array experiment and obtain normalization of results across multiple chips. A key observation they make is that the paired comparisons in two-dye systems correspond to a so-called incomplete block structure on the experiment. Chapter 14 describes methods for the analysis of data obtained from reference designs as well as more complex array designs. The basics of experimental design are found in Box et al. (1978). A review of design issues for cDNA microarray experiments is provided by Yang and Speed (2002).

1.3   Data Storage

1.3.1   Databases

The information associated with a microarray experiment has four important components:

1. A table of numbers representing absolute or relative expression values at the gene or spot level. The emerging standard is to have rows represent genes and columns represent samples/arrays.
2. A table of covariates associated with the samples. These may include information on the samples' phenotypes (e.g., cancer type) as well as design variables for controlled experiments (e.g., drug treatment). For these tables, it is more common to have rows represent samples and columns represent covariates, in which case the columns of the expression table correspond to the rows of the covariate table.
3. Detailed descriptions of the genes represented by the row of the expression data and of the phenotypic variables related to the column.
4. Information about the experiment itself. This could include identification number in a database, experimental protocols used, preprocessing information, and so forth.

This complexity creates substantial informatics challenges.

For researchers conducting numerous microarray experiments it is crucial that this information be stored efficiently. Databases make it possible to efficiently store and access complex datasets and facilitate combining information from multiple microarray experiments. Gardiner-Garden and Littlejohn (2001) give a survey and comparative analysis of microarray databases as of 2001. The Web sites `http://www.biologie.ens.fr/en/genetiqu/puces/bddeng.html`, `http://genome-www5.stanford.edu/MicroArray/SMD/databases.html`, and `http://www.wehi.edu.au/bioWeb/Suzanne/databases.html` maintain lists of microarray databases. There are several public and commercial products, fulfilling different roles and serving varying categories of users.

Various open source database servers are freely available for custom database development. Two examples are MySQL and postgresql. Free and open source scripting languages such as Perl, pyhthon, and php can be used to programmatically add and extract information to and from a database. The Bioperl Project (`http://www.bioperl.org`) is an international association of developers of open source Perl tools for bioinformatics, genomics and life science research. An example of software that makes use of these tools is DRAGON, discussed in Chapter 8. Furthermore, a common set of classes and methods for interfacing R to a relational database is becoming available; for example, the *RmySQL* library provides an interface to MySQL.

1.3.2   *Standards*

The proliferation of microarray databases and the growing appreciation for both the importance of analyses across experiments and the need for well-documented repositories have stimulated work toward the development of standards. The Microarray Gene Expression Database (MGED) group (`http://www.mged.org`) is a grass-roots movement to promote the adoption of standards in microarray experiments and data. Among its goals are establishing gene expression databases, improving comparability of microarray data from different sources, and interoperability of different functional genomics databases and data analysis software. MGED developed requirements for the minimum information about a microarray-experiment (MIAME) required to interpret and verify the results (Brazma et al., 2001). Scientific journals are beginning to require compliance with MIAME standards for data made available as supplementary information in microarray-based papers.

As we write, MGED is developing a data exchange format (MAGE-ML) and an object model (MAGE-OM) for information about microarray experiments. MAGE is a language designed to describe and communicate information about microarray-based experiments. MAGE-ML is based on XML and can be used to describe microarray designs, microarray manufacturing information, microarray experiment setup and execution information, gene expression data, and data analysis results. The materials at the MGED Web site describe the basic MAGE-ML structure and its main concepts. Additional efforts are under way to develop ontologies for microarray experiment description and biological material annotation and to develop recommendations regarding experimental controls and data normalization.

Microarray analysis environments are increasingly using MIAME and MAGE standards. For example, ArrayExpress is a public repository for microarray data based on MAGE objects and hosted at the EBI (see Chapter 6 for details). The Expression Profiler environment discussed in detail in Chapter 6 is able to import data directly from ArrayExpress. As another example, MADAM (MicroArray DAta Manager, available at `http://www.tigr.org/software`) is a tool that leads users through the microarray process, facilitating data entry into a MIAME-compliant MySQL, or other ANSI-SQL, database.

1.3.3 Statistical Analysis Languages

Proper storage and access to data is critical but not sufficient to microarray analysis. A database also needs to interact efficiently with statistical analysis languages and environments. With regard to this aspect, microarray analysis tools have been developed following three approaches: as parts of comprehensive databases, as standalone packages, and as libraries within well-established programming and analysis languages, such as S-PLUS, R, SAS, Excel, or Matlab. The first approach addresses the storage/analysis interface seamlessly but often requires new development for statistical analysis software. The second approach leaves the interface to the user. The third approach permits exploiting the array of data manipulation and statistical analysis available in these environments and facilitates tailoring analysis methods to the particular question and data collected to address it. Substantial progress is being made toward improving interfaces between statistical programming environments and databases. This book covers examples of all three kinds of analysis tools.

A special focus is on tools developed in the S language. S is an intuitive, highly functional, and extensible programming language that forms the basis of the Open Source R Project (Ihaka and Gentleman, 1996) and the commercial software system S-PLUS. We refer those interested in the S language to Becker and Chambers (1984), Chambers (1998) and Venables and Ripley (2000). Extensive support for microarray analysis is available as part of the Bioconductor Project (`http://www.bioconductor.org`), whose core R packages are discussed in Chapters 2, 3, and 4. Other R libraries for microarray analysis are discussed in Chapters 11, 13, and 16; S-PLUS libraries are in Chapters 6 and 15.

R provides users with a comprehensive, state-of-the-art microarray analysis toolbox and built-in statistical functionality based entirely on free and open source code. R may be downloaded from the Comprehensive R Archive

Network (CRAN) at `http://cran.r-project.org`, which also includes tutorial documentation. An active community of users is continuiung to contribute statistical free-ware both at the experimental/developmental level and at the data management/database interface level.

An important aspect of data managment of genomic data is the ability to program using an object-oriented language, which simplifies writing, maintaining and interpreting complicated systems. Object-oriented programming can be implemented in many languages. The reader may refer to books dedicated to object-oriented programming to understand the general principles. For example, Chambers (1998) reviews these concepts in the context of the S language. Illustration are in several chapters: Chapter 2 describes storage of information related to a microarray experiment using objects in R. Chapter 3 describes R objects specific to the analysis of cDNA arrays and Chapter 4 describes R objects for data from Affy chips.

1.4   Preprocessing

1.4.1   Image Analysis

In microarray experiments, the first quantified values are contained in the image files produced by the scanner. The pixel intensities, stored in these files, can be thought of as the raw data. Image analysis tools are then used for segmentation (that is, defininition of the areas in these images that are representing expression information) and summarization of the pixel-level data (for example, by computation of the average or median intensity of the pixels in a spot). Various methods exist for segmentation of the arrays and summarization of pixel intensities; choice of methodology can have a substantial impact on results, especially with cDNA arrays. We do not discuss imaging techniques in this book. Instead, the reader is referred to recent reviews of image analysis techniques and software by Brown et al. (2001), Yang et al. (2001) and Jain et al. (2002).

1.4.2   Visualizations for Quality Control

In most microarray technologies, it is essential to visually inspect data from every array to diagnose the presence of possible artifacts. This book includes various examples of how exploratory data analysis can contribute to quality control by revealing artifacts worthy of further investigations and facilitate investigating relationships between measured intensity and potential sources of bias. For example, Chapter 3 describes boxplots for diagnosing the presence of print tip effects. For most visualizations, logarithmic transformations of the data are recommended because of the marked differences (usually orders of magnitude) of expression data values that result.

Evaluation of spatially varying bias is also critical. One can look at the original images or, more conveniently, at images of the processed absolute or relative expression values arranged by their location (see Chapter 3), or at "gradient plots" graphing intensity versus one of the two spatial coordinates (See Chapters 4 and 5).

When interest is in comparing two sets of expression values (say, the two channels of a cDNA array or the expression levels across two arrays), it is useful to look at scatterplots of the two intensities or, as is more commonly done, to examine plots of differential expression versus overall intensity. These have been named in many ways, but perhaps the most common name is MA plot. MA plots are used throughout the book (e.g., Chapters 3, 7, 8, 9, and 14). A gentle introduction is Figure 9.2, in Chapter 9. An MA plot is a 45 degree rotation (with a re-scaling of the x-axis) of the scatterplot of the intensities. The latter are usually expressed in the $\log_2$ scale. The MA has the quantity that is usually of most interest, differential expression, on the y-axis and another important quantity, related to absolute intensity, on the x-axis. These plots have been useful, for example, for finding intensity related biases and variances, saturation effects, and other artifacts.

1.4.3 Background Subtraction

The scanning of arrays results in optical or background noise affecting pixel intensities. Images obtained from spotted arrays contain specific information on this background noise from the pixels not associated with spotted regions (Chen et al., 1997). High-density oligonucleotide arrays have minimal space between the segments of the array where probes are attached, known as cells; therefore, background information is difficult to obtain and not commonly used.

Typically, image processing software will produce an absolute expression measure X and a background measurement B for each spot or cell. If, as is likely, X is the result of signal and additional background noise, then it is a biased estimate of the true hybridization that we intend to measure (that is, it is likely to be systematically too high). To obtain an unbiased measure of expression, conventional wisdom is to subtract the background by considering $X - B$. If both X and B are unbiased and background adds to the signal, then $X - B$ is unbiased. Even in these circumstances, however, there are important trade-offs to be evaluated in deciding whether and how to subtract background noise. Because both X and B are estimates, the variability of $X - B$ is larger than that of X alone; thus, subtracting background adds variance. This is especially problematic in the low-intensity range, where the variance of B can be of the same order of magnitude as X. Generally, the assumptions of unbiasedness and additivity are far too optimistic. Also, some researchers have found that the background estimates produced by some of the most popular image-processing algorithms are not sufficiently reliable (Yang et al., 2001).

One alternative is to avoid background subtraction altogether and only use X to estimate the expression level. This avoids introducing the additional variance from inaccurately estimated background and is generally conservative in making declarations of differential expression and practical. To see this, say that the true expressions in two samples being compared are e_1 and e_2. We observe $X_1 = e_1 + B_1 + \epsilon_1$ and $X_1 = e_2 + B_2 + \epsilon_2$, where ϵ_1 and ϵ_2 are errors in measurement of the true signal. Because B's are positive, the log ratio of the nonbackground-corrected raw expression values X_1/X_2 is likely to be closer to 1 than the true ratio e_1/e_2. This bias toward one is stronger for low intensity genes. In summary, not subtracting background can be an attractive alternative, as it does not rely on potentially problematic background estimates and loses sensitivity mostly for low intensity genes; the exceptions are experiments with major spatial artifacts affecting only one channel.

In practice, decisions about background subtraction need to be made based on careful visualization of the data. In our experience, the following rules of thumb are helpful for cDNA arrays:

- Inspect images of background alone (see also Chapter 3 for methods and software), and focus on major spatial artifacts affecting only one channel. If those are present, then background subtraction is critical.

- Avoid generating negative values and zeroes after background subtraction; these indicate that the error in the measurement of background is greater than the signal in the spot. The spot does not necessarily need to be discarded. Also avoid infinitesimal values. The resulting estimated ratio is both unreliable and likely to generate extreme ratios.

- Inspect MA plots, discussed in Subsection 1.4.2. Major "fishtail effects" at low intensities often indicate that spot-level background subtraction cannot be carried out reliably. Consider alternatives such as global background subtraction, subtraction of the faintest background, or no background subtraction.

1.4.4   Probe-level Analysis of Oligonucleotide Arrays

High-density oligonucleotide arrays pose the challenge of summarizing data from a probe set into a single measure, which estimates the level of expression of the gene of interest. Affy software provides default approaches for this step. Two important issues suggest that probe-level data should be considered an integral part of Affy data analysis. First, visualization of probe-level data can help identify artifacts on Affy chips. Second, there is evidence that alternative summarizations to the defaults currently implemented by Affy may provide improved ability to detect biological signal. Chapter 3 presents R tools for the analysis of probe-level data arising from Affy chips, including data entry, annotation, object management, and visualization. Chapter 5 reviews the Li-Wong model (Li and Wong, 2001) for analysis of problem-level oligonucleotide data and the associated dChip software, which performs probe-level analysis as well as other analyses and visualizations. This is an active field of methodological research. Recent contributions to the literature can be surveyed from the Web page of the recent "Genelogic Workshop on Low Level Analysis of Affymetrix Genechip data" at `http://stat-www.berkeley.edu/users/terry/zarray/Affy/GL_Workshop/genelogic2001.html`

1.4.5   Within-Array Normalization of cDNA Arrays

In two-channel cDNA arrays, the two signals allow for internal correction of a number of commonly occurring artifacts. Evidence is emerging to support the notion that the normalization process is difficult to automate (Tseng et al., 2001; Yang et al., 2002). Normalization is best understood as an iterative process of visualization, identification of likely artifacts; and removal of artifacts when feasible. Examples of artifacts that can be at least partially removed include differential nonlinear response of the two channels to hybridization intensity, biases in DNA spotting from defective print tips and the fainting of the signal in large regions of an array. Chapters 3 and 9 present motivation, visualization tools, and examples of normalization.

1.4.6   Normalization Across Arrays

Most applications of microarrays require comparisons of gene expression measures across arrays. Variation across arrays will reflect the genetic, experimental, and environmental differences under study but will also include variation introduced during the sample preparation, during the manufacture of the arrays, and during the processing of the arrays (labeling, hybridization, and scanning). When different arrays are used to represent the different samples, it is important to normalize across arrays. As in within-array normalization, visualization and artifact detection are critical components of normalization. A useful visualization approach is to first create an artificial reference array (for example, by considering the median gene expression across arrays) and visualize each array against this reference (Parmigiani et al., 2002). In Affy arrays and other platforms providing a single reading per spot, one can then construct MA plots for each array versus the reference.

Initial approaches for normalization across arrays focused on standardizing overall intensity. This is useful but often inadequate, as one commonly encounters systematic nonlinear distortions. Vertical residuals from robust regression of MA plots against a common reference, as described above, can provide normalized values that account for nonlinear effects (see Chapter 7). A popular alternative is quantile normalization (Bolstad et al., 2002). This is based on transforming each of the array-specific distributions of intensities so that they all have the same values at specified quantiles. A simple example is to center arrays on their medians. Another example is to register all distributions so that they have the same deciles, rescaling observations linearly within deciles. Increasing the number of quantiles improves comparability, but beyond a certain point it starts to obscure the biological signal. For this reason, quantile normalization at the gene level is appropriate when most of the genes are believed to be unaltered across conditions. In oligonucleotide arrays, quantile normalization can be carried out very effectively at the probe level (see Chapter 4) simply by imposing that all probe-level distributions be the same. One implementation is to map each quantile to the average quantile across arrays and then derive gene indices from the quantile-normalized arrays.

Discussion of normalization approaches across arrays is included, for example, in Chapters 3, 4, 5, 7, and 14. In addition, many of the environments for microarray analysis include normalization tools. Of interest are the normalization methods described in Tseng et al. (2001), and implemented in the Windows-based tool cDNA Microarray Analysis `http://biosun1.harvard.edu/~ctseng/download.html`.

1.5   Screening for Differentially Expressed Genes

1.5.1   Estimation or Selection?

The goal of many controlled microarray experiments is to identify genes that are regulated by modifying conditions of interest. For example one may wish to compare wild type to knockout laboratory animals or alternative drug treatments. The goal of these experiments is generally that of identifying as many of the genes as possible that are differentially expressed across the conditions compared. Gene expression can often be thought of as the response variable in statistical models.

There are two broad categories of situations that it is useful to keep in mind when choosing a statistical approach. In the first situation, we are comparing samples where the majority of genes are expected to show some difference, albeit in varying degrees. An example would be cells at different stages of development. This problem is best approached by estimating the differences or ratios of expression across conditions for each gene, and by reporting interval estimates across the genome. Other useful presentations include aggregate changes of expression over functional classes of genes.

In the second situation, we compare samples where a relatively small fraction of genes are altered. For example, we may be studying alterations in expression caused by loss of a gene active in a specific pathway. In this type of application, microarrays are often used to screen genes for further analyses by more reliable assays, and the data analysis is best approached by ranking genes and/or by selecting a subset of genes for further validation.

In addition, analyses should provide an indication of the degree to which the reported subset is believed to represent the subset of truly altered genes. A comparative review of statistical methods for identifying differentially expressed genes is given by Pan (2002).

1.5.2   One Problem or Many?

Perhaps the simplest screening approach is to select genes based on average change in expression (say, difference in mean log expression across groups). One problem with this is that it ignores variation in how reliably each gene is measured. The problem is partly mitigated by careful normalization. For example, the approaches described in Chapter 9 attempt to correct systematic patterns across intensities in both the fold change and the gene-specific variance by making use of information from the ensemble of genes. Even after normaliztion, however, within-group variation of expression will be highly gene-dependent. This calls for the use of statistical methods that weigh differences across groups against variation within groups.

Comparing gene expression across two conditions for a single gene is an instance of one of the most basic statistical questions: a two-sample comparison. Estimation and testing tools in this case are very well developed. In genomics applications, however, there is increasing consensus that it is not efficient to consider each gene in isolation, and gains can be made by considering the ensemble of gene expression measures at once. This occurs for at least two reasons: first, genes measured on the same array type in the same laboratory are all affected by a number of common sources of noise; second, changes in expression are all part of the same biological mechanism, and their magnitudes, although different, are not completely unrelated. For example, suppose that the comparison of interest is truly altering 100 genes on an array, and imagine that an oracle revealed to us 99 of the fold changes. Generally, we should have a better guess at the fold change of the 100th regulated gene than if this revelation had not been made.

Joint estimation of many related quantities is a time-honored problem in statistics, dating back at least to the pioneering work of Stein and colleagues (James and Stein, 1961) and continuing with empirical Bayes approaches (Efron and Morris, 1973) and hierarchical Bayesian multilevel modeling (Lindley and Smith, 1972). The idea behind the multilevel models and the associated empirical Bayes and hierarchical Bayes estimation techniques is to proceed in two stages. The first defines some useful summaries at the gene level (for example, test statistics, or parameter estimates of fold change and noise for parametric models). These describe the variability of samples for each gene. The second stage posits a distribution for these gene-level summaries. This approach has benefits in both estimation and selection. In the case of microarrays, examples of implementations are provided by Efron et al. (2001), Lönnstedt and Speed (2002) for selection and Ibrahim et al. (2002) and Newton et al. (2001) for both selection and estimation. In this book, formal multilevel models are utilized in Chapters 11 and 16, while analysis strategies that, broadly speaking, use information from the full genomic distribution are also featured in other chapters, such as 5, 12, and 13.

Most microarray analysis packages include basic one-gene tools for testing the null hypothesis of no differential expression, such as the t-test or its nonparametric counterparts. Cyber-T provides R-based software for regularized t-tests that offer the opportunity to incorporate genome-wide information via a prior distribution (Baldi and Long, 2001). The SAM software (Tusher et al., 2001) described in Chapter 12 uses genome-wide information to transform the signal-to-noise ratio, making it roughly independent of noise. The type of transformation used by SAM is designed to protect against false discoveries generated by very small denominators in the t-ratios. The denominators are highly likely to be small only by chance because genes share many sources of variability and a certain amount of variation is to be expected from all of them. An example of implementation of SAM is given by Xu et al. (2002). Chapter 6 describes a local pooled error (LPE) method to develop z-statistics for differential expression tests by pooling variance estimates for genes with similar expression intensities and constructing a smooth variance function in terms of average expression intensity. The LPE approach is designed for microarray experiments with limited number of replicates. In this case, error estimates constructed solely using gene-specific information can be inefficient. Pickgene, described in Chapter 14, implements a robust approach that accounts for the ensemble of measurement by tailoring the analysis to the level of intensity. EBAR-RAY, described in Chapter 11, performs empirical Bayes analysis of various multilevel models, addressing both estimation and testing.

1.5.3 Selection and False Discovery Rates

In analyses aimed at selecting differentially expressed genes, there are several approaches for reporting the degree of reliability of results. Conventional approaches based on gene-specific p-values are generally criticized on the grounds of the multiplicity of comparisons involved. Several proposals exist for adjusting p-values (see, for example, Dudoit et al., 2002 and references therein) to account for multiplicities. A second approach is to compute the a posteriori probability that a gene is differentially expressed, as discussed, for example, in Chapters 11 and 16. For a discussion of the differences between the two approaches, see Berger and Delampady (1987).

A third and perhaps currently most popular approach is to report the false discovery rate, or FDR (Benjamini and Hochberg, 1995), for a group of genes or for a specified cutoff value of a statistic of choice. Assuming that the population of genes truly divides into two groups, the altered and unaltered genes, and that a statistical approach selects a set of "significant genes," the FDR is an estimate of the fraction of truly altered genes among the genes declared significant. This approach often reflects appropriately the fact that array experiments are performed to guide future validation work on individual genes, which is usually expensive and time-consuming. It is also directly interpretable as the probability that a gene in the list of selected genes is differentially expressed if one takes the set of genes on the array as the population of reference for the calculation of the probability. Additional discussion is in Storey (2001); Genovese and Wasserman (2002); Storey (2002) Applications to microarrays and software for computing FDR within the SAM approach are discussed in Chapter 12.

1.5.4   Beyond Two Groups

Methods for identifying genes that are differentially expressed across two experimental conditions can be extended to more general settings in which multiple conditions are considered. Extensions include time course experiments where conditions correspond to multiple time points, factorial designs, in which the effects of multiple factors and their interaction are explored simultaneously and so forth. The empirical Bayes methods discussed in Chapter 11 can be used to assess differential expression among multiple conditions. A 4 group analysis is presented. In general, multilevel models can proceed by first specifying a statistical model for expression as a function of condition (for example, a time series model in a time course experiment, as shown in Chapter 18, or an ANOVA model for a factorial design, as shown in Chapter 14). The gene-to-gene variability of the parameters describing this model can be represented by additional distributions as discussed earlier.

Standard ideas from analysis of variance can be generalized to this context to incorporate randomness in the spots and in the arrays (see Chapter 14). Software for linear models with random effects for arrays is also available in SAS (Wolfinger et al., 2001).

1.6   Challenges of Genome Biometry Analyses

Important microarray studies also take place outside experimental settings. These observational and more exploratory studies describe the variation of genomic information in biological populations, with broadly ranging goals including refinement of current taxonomies, identification of genome-phenotype relationships, classification and annotation of genes, and exploration of unknown pathways. The statistical tools brought to bear in these investigations cover the full range of traditional multivariate analysis, cluster analysis, and classification. A thorough exposition of the field can be gained from Ripley (1996). In addition, novel and more specific tools are being developed.

A challenge for the application of these analysis tools to genome-wide studies of gene expression comes from the large number of genes that are studied simultaneously and the high gene-to-sample ratio. Having many more genes or ESTs than biological replicates makes possible a number of strategies for analysis. These can be categorized into three broad categories: "summarize then analyze" (STA); "analyze then summarize" (ATS), and "summarize while analyzing" (SWA). In the first category, STA, a multivariate procedure such as cluster analysis or multidimensional scaling is used to reduce the large number of genes to a smaller number of summary variables or "profiles." These are then taken as outcome variables in, say, a regression analysis of expression on experimental conditions or as predictor variables in a model with a health outcome. For example, Rosenwald et al. (2002) use hierarchical cluster analysis to identify three summary variables that are used to predict survival in a population of patients who have undergone chemotherapy for large B-cell lymphoma.

In the analyze then summarize (ATS) approach, the modeling is conducted for each gene, producing an estimate of a statistic of scientific interest (say a regression coefficient) and its standard error. These gene-specific coefficients are then summarized; for example, by identifying those that are largest or most statistically significant using methods such as those covered in Section 1.5. These methods may ignore important relationships between genes, suggesting that it may be more efficient to identify summaries of gene expression and conduct classification analyses in a single procedure (SWA). Prediction Analysis of Microarrays, or PAM (Tibshirani et al., 2002), nonparametric regression methods such as CART (Breiman et al., 1984), and logic regression (Ruczinski et al., 2002) are examples of methods that identify a small subset of genes to optimally predict a phenotype.

High-dimensional responses are not unique to gene expression studies. Sparse data analysis deals in general with the problem of detecting faint signal buried in high noise in very high dimensional data (Abramovich, Benjamini, Donaho, and Johnstone, 2000). Specific application areas with potential synergies with microarrays include statistical image analysis (Friston et al., 1995; Worsley et al., 2002), item-response data analysis from psychometric tests and surveys, and chemometrics (Sundberg, 1999). The dimensionality of the image analysis problem is even more challenging than the microarray's because it is common to have 100,000 or more voxels (volume elements) and a time series of activation levels at each. These analyses must contend with autocorrelated noise over time at each location and spatial correlation among both the noise and the signal across neighboring voxels. SPM is a leading image analysis software package, written as a suite of Matlab functions, that uses the STA approach and incorporates multiple comparison methods for dealing with the multiplicity issue. See the Web site http://www.fil.ion.ucl.ac.uk/spm for more details.

A second challenge and opportunity for the analysis of gene expression data is the incorporation of prior biological information. Genes need not be treated as exchangeable in statistical analyses. There is considerable biological data available about the genes, the proteins they encode and the functional groups to which they may belong. This information is readily available using modern bioinformatics tools such as DRAGON, described in Chapter 8. The challenge is to develop methods that include relevant biology as another way to address the dimensionality hurdle.

The next two sections consider examples of statistical tools for genome biometry investigations. The main goal of the sections is to provide enough background to place the contributions of each chapter in context. There is no attempt at being exhaustive. For convenience of exposition, we separated the discussion into supervised and unsupervised methods, even though the distinction is somewhat artificial and hybrid approaches can be useful. We also point to R libraries that provide free tools to implment the methods cited. In most instances, similar libraries are also available in S-PLUS as a well as other commercial languages and statistical analysis systems.

1.7   Visualization and Unsupervised Analyses

1.7.1   Profile Visualization

As with quality control and signal extraction, multivariate analysis of microarrays relies substantially on visualization. The most commonly used tool is a color map of either relative or absolute hybridization intensities after proper normalization. This type of map was introduced by Eisen et al. (1998), along with the popular software packages Cluster and TreeView, available from `http://rana.lbl.gov`. In this book, expression maps are illustrated, for example, in Chapter 6. An expression map is arranged as a matrix in which each row corresponds to a gene and each column to an array, and the color intensity at each point represents the expression level. Rows and columns are often sorted in a way that facilitates visualization; for example, genes may be grouped by functional classes, samples may be ordered by time, or either genes or samples may be sorted by hierarchical clustering (see below). The choice of colors for the map can be critical. An in-depth illustration is given in Figure 6.2 in Chapter 6.

In this book, many of the chapters make use of these maps in different contexts. Special attention is given to them in Chapters 6, 7, 8, 10, 15, 16, and 18. Many free packages provide functionality to make these maps. An example is MeV (Multiple Experiment Viewer), a tool for data mining and visualization of arrays that was developed at TIGR (`http://www.tigr.org/software`).

Maps arrange genes along a single dimension and are often inadequate for investigating networks of gene relationshisps. Alternative methods include the so-called relevance networks, discussed in Chapter 19. MeV also performs relevance networks calculations and visualizations. In some applications, additional insight can be gained by clustering concomitantly both the genes and the samples, a technique sometimes called biclustering and illustrated in the work of Getz et al. (2000) and Lazzeroni and Owen (2002), who also provide software.

1.7.2   Why Clustering?

Clustering algorithms divide a set of objects (genes or samples) into groups so that gene expression patterns within a group are more alike than patterns across groups. There are two broad approaches to clustering techniques. Hierarchical techniques provide a series of successively nested clusters, and their result resembles a phylogenetic classification. Nonhierarchical techniques generally find a single partition, with no nesting. Both are used extensively in microarray analysis for two main goals. The first is representing distances among high-dimensional expression profiles in a concise, visually effective way, such as a tree or dendrogram. The second is to identify candidate subgroups in complex data. The two tasks can be closely tied, as a concise lower-dimensional representation of the objects studied often simplifies manual identification of subgroups. Clustering techniques have been successful in supporting visualization and as a method for generating hypotheses about the existence of groups of genes or samples with similar behavior. An example of the latter is the identification of novel subtypes in cancer. Oftentimes, even when phenotype information is available, unsupervised cluster analyses are used to explore gene expression data and form gene groups that are then correlated to phenotype.

Some of the most commonly used clustering techniques are briefly reviewed in the remainder of this section. Quackenbush (2001) and Tibshirani et al. (1999) provide excellent reviews of the use of cluster analysis in microarrays. Comprehensive treatments of clustering include Everitt (1980) and Kaufmann and Rousseeuw (1990). Most statistical analysis environments offer ample choices of clustering methods. The Classification Society of North America (`http://www.pitt.edu/~csna/software.html`) maintains a list of software for clustering and multivariate analysis. In R and S-PLUS, for example, the library *cluster* (Rousseeuw et al., 1996) includes a variety of algorithms and visualization tools, while the library *mclust* performs model-based cluster analysis and numerous others perform more specialized clustering tasks. Most microarray analysis environments include at least some clustering tools. In this book, clustering and related mining tools are discussed in Chapters 5, 6, 7, 10, 15, and 16.

Some general comments apply to the use of clustering techniques in genomic analysis. First, these techniques are exploratory: their strength is in providing rough maps and suggesting directions for further study. In good studies, context and meaning for groups found by automated algorithms is provided by substantial additional work either in the lab or on the Web.

The outcome of a clustering procedure is therefore the beginning, rather than the end, of a genome biometry analysis.

Second, clustering results are sensitive to a variety of user-specified inputs. The clustering of a large and complex set of objects, such as a genome, is akin to arranging books in a library: it can be done sensibly in many different ways, depending on the goals. From this perspective, good clustering tools are responsive to users' choices, not insensitive to them, and sensitivity to input is a necessity of cluster analysis rather than a weakness. This also means, however, that use of a clustering algorithm without a thorough understanding of its workings, the meaning of inputs, and their relationship to the biological questions of interest is likely to yield misleading results.

Third, clustering results are generally sensitive, sometimes extremely so, to small variations in the samples and the genes chosen and to outlying observations. This means that a number of the data-analytic decisions made during normalization, filtering, data transformations, and so forth will have an effect on clustering. In this context, it is challenging, but all the more important, to provide accurate assessments of the uncertainty that should be associated with the clusters found. Uncertainty from sampling and outliers can be addressed within model-based clustering (see below) or alternatively using resampling techniques (Kerr and Churchill, 2001a; McShane et al., 2001). Effects of choosing among reasonable alternative transformations, normalizations, and filtering should be addressed by sensitivity analysis (that is, by repeating the analysis in various sensible ways and reporting conclusions that are consistent across analyses).

1.7.3 Hierarchical Clustering

Hierarchical clustering consists of a series of partitions of the data extending from a single cluster to single member clusters. To produce a hierarchical cluster, a similarity or distance matrix between the M objects to be clustered (in this case, either genes or arrays) must first be chosen and calculated. There is a wide assortment of metrics (Gordon, 1999) reflecting the heterogeneity of users' needs. For example, distance measures based on correlations (e.g., $1 - |\rho|$) cluster together objects whose patterns are in synch in deviating from their own average, while Euclidean distance clusters together objects whose patterns are in synch with each other. The resulting clusters are typically radically different. Additionally, a strategy for deciding how to measure similarity between clusters of objects needs to be chosen. This is referred to as the "linkage" method. Some commonly used linkage methods are complete, average, and connected.

There are two kinds of hierarchical approaches: agglomerative and divisive clustering. Agglomerative clustering algorithms (bottom-up algorithms) begin with every object belonging to its own cluster. At the first stage, the two most similar objects will be joined. Then, the next most similar objects and/or clusters will be joined. This process is repeated until there is only one cluster containing all of the M objects. Divisive clustering algorithms instead proceed from the top down. Initially, all objects belong to the same cluster. At the first stage, the objects are broken into two groups. This continues until there are M clusters, each containing only one object. It is usually the case that agglomerative and divisive methods produce different hierarchies. In gene expression analysis, divisive algorithms can be more stable because the focus is on large clusters, and fewer steps in the divisive algorithms occur to achieve the large clusters compared to the agglomerative algorithms. More detailed descriptions of the algorithms, choice of metrics, and choice of linkage can be found in Gordon (1999) and Quackenbush (2001).

This scheme of hierarchical classifications leads to a graphical representation known as a dendrogram, which has been used effectively in identifying and displaying patterns in gene expression data (Eisen et al., 1998). Two dendrograms are shown in Figure 1.2, where 30 tissue samples (10 normal, 10 stage 1–2 cancer, and 10 stage 3–4 cancer) have been clustered based on simulated data for 50 genes. Dendrograms arrange the clustered objects in terms of their similarity to one another. Objects that are most similar are joined at low heights and those that are dissimilar are joined at larger heights. In each of the dendrograms in Figure 1.2, the three tissue types

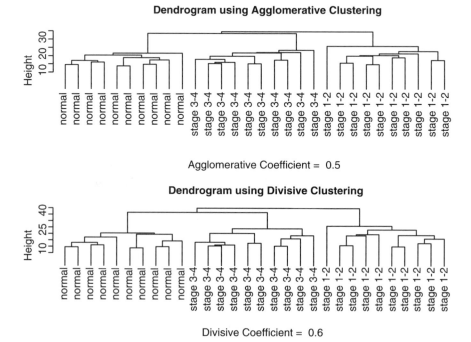

FIGURE 1.2. Examples of dendrograms.

appear to be clearly distinguished based on these 50 genes. However, when phenotypic data are not available, assessing the validity and strength of clustering is less obvious. See Yeung et al. (2001) for approaches to this issue. Applications of hierarchical clustering to gene expression analysis are now commonplace, following Eisen et al. (1998). For recent examples, see Blader et al. (2001) and Chen et al. (2002).

An important caveat about the use of dendrograms concerns the interpretation of the order of objects at the bottom. At each split, it is arbitrary which branch is drawn on the left versus the right. As a result, a multitude of dendrograms and orderings are consistent with a given hierarchical classification. Closeness of objects should be judged based on the length of the path that connects them and not on their distance in the ordering.

1.7.4 k-Means Clustering and Self-Organizing Maps

Nonhierarchical techniques find a partition of the objects. Commonly used examples are k-means clustering (Hartigan and Wong, 1979) and self-organizing maps, or SOMs (Kohonen, 1982; Kohonen, 1995). Both require the user to define ahead of time the number of clusters (or partitions) into which the objects should be divided. In k-means clustering, the goal is to break objects into groups that have low variance within clusters and large variance across clusters. The k-means algorithm begins by randomly assigning each object to one of the k clusters. Distances between clusters are calculated based on the mean vector within each cluster. In the case of clustering samples, the mean vector is the mean expression vector within a cluster. Using an iterative process, objects are retained or moved to other clusters to increase across-cluster variance and decrease within-cluster variance. The algorithm stops when no more improvements can be made. One difficulty with k-means clustering is choosing k. However, other methods, such as PCA (described below), can help in choosing the number of clusters (Quackenbush, 2001).

SOMs accomplish two goals: they reduce dimensionality by producing a map that highlights the similarities among objects, and they group similar objects together. SOMs were originally developed in machine learning, building on artificial neural networks. Ripley (1996) describes the connection between SOMs and k-means clustering. In R, the library *GeneSOM* implements self-organizing maps. See Gordon (1999) and Kohonen (1989) for a more detailed description of SOM and Tamayo et al. (1999) for a discussion of SOMs as applied to gene expression data. SOM classifications lend themselves to interesting visualization techniques such as the Ultsch representation (Ultsch, 1993), which can be especially helpful when the number of clusters is moderate or large. The software package GeneCluster, among others, implements a version of SOM tailored to gene expression data (Tamayo et al., 1999).

1.7.5   Model-Based Clustering

Model-based clustering is a powerful approach that is based on assuming a mixture model for the multivariate distribution of the expression values. Mixture components represent classes. Model-based clustering provides approaches for choosing the number of clusters and optionally allows for noise and outliers (Yeung et al., 2001). Tools are available in the library `mclust` in S-PLUS (Banfield and Raftery, 1993) and R. Another example of how formal statistical modeling can be used to develop clustering algorithms for microarray data analysis is CAGED, discussed in Chapter 18.

1.7.6   Principal Components Analysis

Principal components analysis (PCA) is a method for creating a small number of summary variables (called principal components) from a much larger set. These summary variables can be used for visualization or for more complex statistical modeling. There are two aspects to the procedure. The first aspect is the creation of the components, which are weighted averages of the original variables, constructed to be uncorrelated with each other and to capture as much of the original variability as possible. This is accomplished via a rotation of the original coordinates. The second aspect is the selection of the most representative (or principal) components based on the fraction of variability that is retained using them only.

As in cluster analysis, a principal components analysis is based on a similarity (or distance) matrix between objects. In gene expression, it is common to use the correlation matrix. In that case, the first principal component is the direction along which there is the greatest variation in the data. If we have J genes, the first principal component is described by a vector of "loadings" of length J, consisting of weights to be applied to each gene's expression. A loading close to 0 implies that a gene does not vary much across arrays. Large negative and large positive loadings imply that a gene shows considerable variation across arrays in the dataset. The second principal component is created so that it is orthogonal to the first principal component. It can be interpreted as the weighted average across the genes that explains the largest amount of variation in the data after controlling for the variation explained by the first principal component.

A strength of PCA is that redundant information (e.g., genes showing similar expression patterns across samples) can be represented in a single variable. A drawback is that the summary variables do not necessarily have a clear biological interpretation. PCAs are sometimes used to visually identify clusters. This is sometimes successful, but there is in general no guarantee that the data will cluster along the dimensions identified by the principal component.

There is a large body of literature on principal components analysis and the associated mathematical details, which are too complex to describe here. Yeung and Ruzzo (2001) provide a discussion of PCA as applied to gene expression data. The data transformation implicit in PCA can be used in more elaborate modeling schemes, as, for example, in West et al. (2001). Useful statistical references include a conceptual description by Kachigan (1991) and more technical descriptions by Dunteman (1989) and Everitt (2001). An introduction of PCA for biologists can be found in Knudsen (2002). Quackenbush (2001) discusses the use of PCA in combination with k-means clustering for analyzing gene expression data. For an example of principal components applied to gene expression data, see Raychaudhuri et al. (2000), Granucci et al. (2001) and Alter et al. (2000). Most statistical analysis environments include convenient PCA tools. In R, see the functions *princomp* and *prcomp*.

1.7.7   Multidimensional Scaling

Multidimensional scaling (Kruskal, 1964), or MDS, is another technique for representing high-dimensional data in a small number of dimensions. The goal of multidimensional scaling is to identify variables (say, coordinates in a map) that are as consistent as possible with the observed distance matrix. This results in a graphical representation that portrays the objects in a 2D or 3D figure. As in other analyses discussed above, user-specified options will affect the resulting representation. The most common method for MDS is metric MDS or principal coordinates analysis, but a nonmetric method is sometimes used. For more detail on this approach and other specifics of MDS, including three-dimensional representations, see Gordon (1999) and Everitt (2001). Khan et al. (1998) and Bittner et al. (2000) provide examples of applications of MDS to cancer classification using microarrays. In R, the library *mva*, a comprehensive multivariate analysis library, includes the function *cmdscale* for metric scaling, while the library *MASS* provides for nonmetric scaling.

1.7.8   Identifying Novel Molecular Subclasses

One of the most highly specialized areas of unsupervised microarray analysis is the search for novel cancer subtypes for individualized prognosis and therapy. This requires a complex interaction between elicitation of biological information and use of bioinformatics and biostatistics techniques. Alizadeh et al. (2000) and Bittner et al. (2000) provide examples of complex analyses using a combination of formal and informal steps. A successful classification does not necessarily need to assign all samples to a subtype but only to identify interesting subgroups for further analysis. Also, molecular classes need to be interpretable and lend themselves to further biological analysis. Eventually, subtypes will need to be identified in clinical settings using less expensive and more accurate assays. For this reason, it is not necessary, nor generally useful, to use all genes on the array to define a subtype. Subtypes based on a small number of genes can be more efficiently implemented and validated.

The "not-all-genes" and "not-all-samples" features of molecular classification of cancer distinguish it from most clustering and classification approaches. Methodologies for unsupervised classification of microarray data are growing rapidly. See Herrero et al. (2001) or Segal et al. (2001) and references therein for recent developments. Most approaches pursue some dimension reduction, recognizing that the gene-to-sample ratios make it difficult to develop empirical classifications that make full use of the genomic dimensionality. One approach, prevalent in the statistical literature, is to generate low-dimensional summaries of the full gene expression information, such as distance matrices, or lower-dimensional projections such as principal components. An alternative approach, prevalent in successful contributions to the cancer literature, is to identify, via visualization, expert opinion, or more formal tools, a manageable number of genes. One problem with the first approach is that molecular profiles are defined in a way that depends on the entire set of genes on the array and do not easily transport across platforms.

Chapter 15 covers the so-called "gene shaving" algorithm Hastie et al. (2000). Gene shaving searches for clusters of genes showing high variation across the samples, high correlation across the genes within a cluster, and high diversity of gene expression from cluster to cluster. In applications, clusters are often selected for further analysis based on expert elicitation. Cluster averages are used for classification. The same gene can belong to more than one cluster. An alternative approach for molecular classification is based on using latent categorical variables for gene expression, as described in Chapter 16.

1.7.9 Time Series Analysis

Time series analysis is different in relation to other gene expression analyses in which arrays that are being compared are treated as unrelated to one another. In time series analysis, the goal is often to identify genes that show similar trends over time within the same organism or sample type and to identify samples that are differentiated by such patterns. These analyses are often performed using regression, where time is the primary predictor variable and gene expression is the outcome. Chapters 17 and 18 describe methods for analysis of time series data. Zhao et al. (2001) and Spellman et al. (1998) provide other examples of time series analysis of gene expression data. Background on time series models can be found in Diggle et al. (1994).

1.8   Prediction

1.8.1   Prediction Tools

Another important task in microarray analysis is class prediction, or the classification of samples based on gene expression patterns into known categories based on morphology, known biological features, clinical outcomes, and so on. Examples include prediction of subtypes of leukemia (Golub et al., 1999) or identification of patterns that are associated with specific cancer sites, as in the NCI60 dataset (Ross et al., 2000) available at `http://genome-www.stanford.edu/nci60`. Depending on the goals of the analysis, the richness of background knowledge, and the number of tumors available, class prediction can range from an exploratory exercise, directed mostly at suggesting interesting directions for future investigations, to the full-fledged development of prognostic models based on molecular characterizations. A similar set of quantitative techniques can be applied throughout this range, provided that the interpretation of the significance of the results is placed in the appropriate context. As in Section 1.5, one wishes to identify genes that are differentially expressed across groups, but here gene expression is the predictor rather than the response, and the focus is often on finding a small, rather than a large group of genes and in finding a group that reliably generalizes beyond the sample analyzed.

The complexity of gene expression array analysis is stimulating the development of novel and specific statistical modeling tools for this purpose. However, the existing body of pattern recognition and prediction algorithms developed in computer science and statistics can provide an excellent starting point for prediction in molecular problems for the immediate future. National Research Council; Panel on Discriminant Analysis Classification and Clustering (1988) offers a concise and authoritative review of the field from its inception to the late 1980s. Dudoit et al. (2002) offer a practical comparison of discrimination methods for the classification of tumors using gene expression data. Relevant tools from the statistical modeling tradition include: discriminant analysis (Gnanadesikan, 1977), including linear, logistic, and more flexible discrimination techniques; tree-based algorithms, such as classification and regression trees (CART) by Breiman et al. (1984) and variants; generalized additive models (Hastie and Tibshirani, 1990); and neural networks (Neal, 1996; Ripley, 1996; Rios Insua and Mueller, 1998). Appropriate versions of these methods can be used for both classification and prediction of quantitative responses such as continuous measures of aggressiveness. Some of these methods are reviewed in more detail later. Statistical computing environments typically offer a rich set of alternatives. In R, the library *MASS* provides tools for classification and neural networks.

1.8.2   Dimension Reduction

In many applications, a practical approach is in two stages: dimension reduction and modeling. Dimension reduction selects a subset of genes likely to be predictive; modeling defines more carefully the relationship between the predictors (genes) and the outcome (patient's response, morphological categories, and so forth). Promising models need to undergo a subsequent stage of independent validation, discussed later. Two of the most useful dimension reduction approaches are screening of variables to reduce the number of candidate predictors by eliminating uninteresting ones and substitution of gene expression with a more parsimonious representation of the data. Some screening is almost always appropriate. For example, genes with constant expressions can contribute no discrimination ability. Additional screening can be based on explanatory power, measures of marginal association, such as the ratio of within-group variation to between-group variation, or the measure used in Slonim et al. (1999).

Parsimonious representations of the data may be identified when there is biological knowledge about a pathway; the presence of a pathway (say, overexpression of gene 1 causing underexpression of gene 2, causing in turn overexpression of gene 3) can then be used to construct new and more highly explanatory variables. Normally, such knowledge is not available and we need to proceed by applying the discovery techniques described earlier; for example, the centroids of clusters or the variables identified by projection pursuit can be used as predictors. A critical aspect in this is the ability to create new variables that are easily measurable and interpretable in terms of the original gene expression.

1.8.3   Evaluation of Classifiers

Modeling of gene expression data aims at usable classification and prognostic and therapeutic models. For the foreseeable future, these models will still be probabilistic (e.g., they will only predict that a certain percentage of patients in a given tumor category will relapse); therefore, statistical validation is critical before models can be employed, especially in clinical settings. A general discussion of evaluation of classifiers is in Chapter 7 of Michie et al. (1994).

The most satisfactory approach to validation is based on the use of independent data, which can often be achieved by setting aside samples for validation purposes, as illustrated by Dudoit et al. (2002). Statistical validation of probabilistic models (DeGroot and Fienberg, 1983) has two goals: assessing calibration (that is, the correspondence between the fraction predicted and the fraction observed in the validation sample) and measuring refinement (that is, the ability of the model to discriminate between classes). The latter is quantified, for example, by ROC curve analysis.

As an alternative to setting aside samples for validation, one can use cross-validation, one variant of which is splitting the data in K portions, and training the classifier K times, setting aside each portion in turn for validation. The average classification rates in the K analyses is an unbiased estimate of the correct classification rate. Toussaint (1974) provides a bibliography on estimation of misclassification rates.

A potentially serious mistake is to evaluate classifiers on the same data that were used for training. When the number of predictors is very large, a relatively large number of predictors will appear to be highly correlated with the phenotype of interest as a result of the random variation present in the data. These spurious correlations have no biological foundation and do not generally reproduce outside of the sample studied. As a result, evaluation of classifiers on training data tends to give overly optimistic assessments of the validity of a classifier. In plausible settings, classifiers can appear to have a near perfect classification ability in the training set without having any biological relation with phenotype. Radmacher et al. (2001) explore this issue and provide compelling examples.

1.8.4   Regression-Based Approaches

Linear models, generalized linear models, generalized additive models, and the associated variable selection strategies provide ways to select a useful subset of genes and develop probabilistic prediction models. An important challenge in the use of these standard statistical approaches in gene expression classification is to adapt them to situations where the number of predictors (genes) is large compared to the number of samples. A popular way of restricting such searches is based on a forward stepwise model. Recent work in data mining and in large dimensional statistical modeling problems has emphasized the limitation of traditional stepwise variable selection procedures and led to a variety of more scalable approaches. These include so-called stochastic search methods (George and McCulloch, 1993), that generate a sample of plausible subsets of explanatory variables. The selected subsets are then subjected to additional scrutiny to determine the most appropriate classification algorithm. A combination of stochastic search with principal component analysis and other orthogonalization techniques has proven very effective in high-dimensional problems (Clyde et al., 1996; Clyde and Parmigiani, 1998), and has recently been employed in microarray data analysis (West et al., 2001). Another related direction of interest is multiple shrinkage of regression coefficient, now formally implemented in the Prediction Analysis of Microarray, or PAM (Tibshirani et al., 2002), software.

1.8.5   Classification Trees

Classification trees recursively partition the space of expression profiles into subsets that are highly predictive of the phenotype of interest. They are ro-

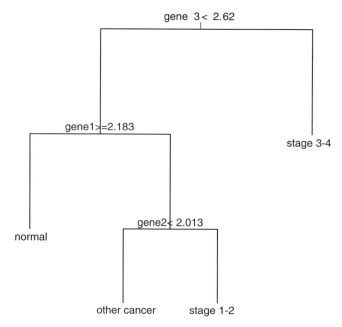

FIGURE 1.3. Example of a classification tree.

bust and easy-to-use. They can automatically sift large genomic databases, searching for and isolating significant patterns and relationships and can be applied without prescreening of the genes (Breiman et al., 1984). The resulting predictive models can be displayed using easy-to-grasp graphical representations. An example is shown in Figure 1.3. In this case, only three genes are used in the classification algorithm, although the number of candidates may have been large. Classification trees also provide summaries of precision and misclassification. An example in which classification trees have been applied to gene expression data can be found in Zhang and Yu (2002). In R, classification trees can be computed using the *rpart* library or the *tree* library.

1.8.6   *Probabilistic Model-Based Classification*

Model based classification is similar to model based clustering in that it is based on the assumption of a mixture model for the multivariate distribution of the expression values. Mixture components now represent known classes (Pavlidis et al., 2001). These approaches are computation-intensive but can provide a solid formal framework for the evaluation of many sources of uncertainty and for assessing the probability of a sample belonging to a class. Methods that assign samples or genes to specific profiles based on probabilistic modeling are also discussed in Chapter 11 and Chapter 16.

1.8.7   Discriminant Analysis

Discriminant analysis (Fisher, 1936) and its derivatives are approaches for optimally partitioning a space of expression profiles into subsets that are highly predictive of the phenotype of interest, for example by maximizing the ratio of between-classes variance to within-class variance. Ripley (1996) and Everitt (2001) give descriptions of the details of this method. Hastie et al. (1994) discuss flexible extensions of discriminant analysis (FDA). Li and Yang (2002) provide a discussion of discriminant analysis in the context of gene expression array data. In R, the functions *lda* and *qda* perform linear and quadratic discriminant analysis, while the function *mda* performs mixture and flexible discriminant analysis.

1.8.8   Nearest-Neighbor Classifiers

A simple and powerful class of algorithms for classification is that of nearest-neighbors algoritms (Cover and Hart, 1967), which classify a sample of known gene expression and unknown phenotype by comparing the given gene expression profile to those of a sample of known phenotype. A simple implementation is to choose a rule for finding the k nearest neighbors and then deciding the classification by majority vote. Nearest-neighbor classifiers are robust, simple to interpret and implement, and do not require, although they may benefit from, preliminary dimension reduction. GeneCluster and other microarray packages perform nearest-neighbor classification. In R, *knnTree* performs nearest-neighbor classification with variable selection inside leaves of a tree. Nearest-neighbor algorithms are also used in several packages for imputation of missing expression measures.

1.8.9   Support Vector Machines

Support vector machines (SVMs) (Vapnik, 1998) are a method for creating classification or general regression functions from a set of training data in which phenotype data are available. For classification, SVMs operate by finding a hypersurface in the space of gene expression profiles. As in FDA, this hypersurface will attempt to split the groups. In SVMs, however, the split will be chosen to have the largest distance from the hypersurface to the nearest of the patterns in the two groups. See Burges (1998) and Christianini and Shawe-Taylor (2000) for details of SVMs and generalizations. Lee and Lee (2002) and Brown et al. (2000) give examples of analysis of gene expression data using SVMs.

1.9   Free and Open-Source Software

All packages described in this book are either open code or free for educational and academic purposes. Links for downloading and information about platforms and languages required can be found on the book's Web site at www.arraybook.org. In this section we provide background information on other projects that have a focus similar to the chapters in this book. They range from stand-alone software for specific tasks to comprehensive systems for microarray management and data analysis. The goal of this section is to broaden the reader's view beyond the software projects covered in this book. Our list is by no means inclusive. Although there are several useful and widely used commercial systems, we do not cover them here. The site http://ihome.cuhk.edu.hk/~b400559/arraysoft.html maintains a catalog of microarray software.

1.9.1   Whitehead Institute Tools

GeneCluster is a stand-alone Java application implementing the methodology described in Tamayo et al. (1999). It provides tools to filter and preprocess data, cluster expression profiles using SOMs, classify using weighted voting and k-nearest neighbors, and visualize the results. It was developed at the Whitehead Institute Center for Genome Research http://www-genome.wi.mit.edu/cancer. Installation support is available for Microsoft Windows and Apple Macintosh.

1.9.2   Eisen Lab Tools

Cluster and TreeView http://rana.lbl.gov/EisenSoftware.htm (Eisen et al., 1998) are an integrated pair of programs for analyzing and visualizing the results of complex microarray experiments. Cluster implements a variety of cluster analysis algorightms and other types of processing on large microarray datasets. It currently includes hierarchical clustering, SOMs, k-means clustering, and principal component analysis. TreeView permits users to graphically browse results of clustering and other analyses from Cluster. It supports tree-based and image-based browsing of hierarchical trees. It is only available for Microsoft Windows95/98/NT. Combined Expression Data and Sequence Analysis (GMEP) http://rana.lbl.gov/EisenSoftware.htm (Chiang et al., 2001) is software for computing genome-mean expression profiles from expression and sequence data.

1.9.3   TIGR Tools

The Quackenbush lab at TIGR developed a suite of tools for microarray data analysis. The system, called TM4, is freely available as open-source code with a well-defined application programming interface and consists of four integrated applications. MADAM (MicroArray DAta Manager) leads users through the microarray process, facilitating data entry into a MIAME-compliant MySQL (or other ANSI-SQL) database. A series of report generators and embedded tools allow for the accurate tracking and querying of stored data. Spotfinder analyzes microarray TIFF images to detect spots and report intensity values using a dynamic thresholding algorithm. Quality-control functions assist in the assessment of hybridizations. MIDAS (MIcroarray Data Analysis System) can be used to compute normalized, trimmed, and filtered raw intensity data generated from image analysis. MIDAS offers an interface that allows users to build an extensible data pipeline by linking methods of data manipulation. Researchers can harness powerful algorithms, such as lowess. Finally, MeV (Multiple Experiment Viewer) is a tool for data mining and visualization. Correlations between patterns of expression can be investigated using a variety of techniques, including bootstrapped and jackknifed hierarchical trees, principal components analysis, template matching, and relevance networks. Clusters of interest can be tracked throughout the analyses, allowing for validation of results and fresh perspectives of the underlying expression data.

1.9.4 GeneX and CyberT

GeneX, available at `http://www.ncgr.org/genex`, is an open source database system. It consists of four parts: a curation tool, a database, an XML data exchange protocol, and ways of querying and analyzing data. Cyber-T, available at `http://genomics.biochem.uci.edu/genex/cybert` is a Web interface designed to detect changes in gene expression and operates on a set of functions written in R. The functions are based on Bayesian statistical approaches (Baldi and Long, 2001) and include regularized t-tests. Cyber-T is Web-based, but R functions can be downloaded. It is mutually supported with GeneX.

1.9.5 Projects at NCBI

Homologene is a homology resource that includes both curated and calculated orthologs and homologs for nucleotide sequences represented in UniGene and LocusLink. (`http://www.ncbi.nlm.nih.gov/HomoloGene`). GeneExpression Omnibus (GEO) is a gene expression and hybridization array data repository, as well as an online resource for the retrieval of gene expression data from any organism or artificial source (`http://www.ncbi.nlm.nih.gov/geo`).

1.9.6 BRB

BRB ArrayTools is an integrated package for the visualization and statistical analysis of DNA microarray gene expression data developed at the Biometrics Research Branch of the NCI. It was developed by statisticians experienced in the analysis of microarray data and involved in the development of improved methods for the design and analysis of microarray based experiments. The array tools package utilizes an Excel front end, but the visualization and analysis tools are developed in R, in C and Fortran programs, and in Java applications. BRB is available at `http://linus.nci.nih.gov/BRB-ArrayTools.html`.

1.9.7 The OOML library

The Object Oriented Microarray Library, developed by Coombes and colleagues at M. D. Anderson Cancer Center, is a suite of object-oriented programming modules written in S-PLUS. The library contains tools for both two-channel arrays and radioactivity-based nylon membranes. Focus is primarily on selecting genes that are statistically significantly differentially expressed betwen two groups. Information and download are available from the site `http://www3.mdanderson.org/depts/cancergenomics/OOMAL/index.html`.

1.9.8 MatArray

Venet and colleagues released a Matlab toolbox called MatArray, designed for the analysis of microarray data. It features advanced normalization schemes, including loess-like approaches, efficient implementation of hierarchical clustering, with leaf ordering and export to TreeView, and efficient implementation of k-means clustering. The toolbox is geared toward seasoned Matlab users who want to treat microarray data with efficient command-line algorithms. For more information, one can consult `http://iribhn.ulb.ac.be/microarrays/toolbox`.

1.9.9 BASE

BioArray Software Environment, or BASE (Saal et al., 2002) is a free and open source (GPL) MIAME-compliant microarray database and analysis engine for microarray laboratories. It is meant to serve many users simultaneously via the Web. Analysis modules can be added as "plug-ins" to increase the functionality of the system. Currently there are plug-ins for various types of normalization, as well as hierarchical clustering, multidimensional scaling, principal component analysis, and others. Detailed information is available at `http://base.thep.lu.se/`.

1.10 Conclusion

Free and open-code software provides efficient and rigorous approaches for the management of microarray data on scales ranging from that of a small laboratory to that of a multicenter data repository. In addition to the usual advantages, in microarrays, freeware offers the availability of constantly improving state-of-the-science statistical and bioinformatics methodology and the support of a large active and interactive community of scientists. We hope that this book will promote an increased use of these tools, contribute to a more appropriate interpretation of the results obtained using them, and facilitate their improvement by making them known to a broader community. The Web site www.arraybook.org contains links to all packages discussed and will continue to be updated after the publication of the book.

Acknowledgments. The work of Garrett and Parmigiani was partly supported by NCI grants P50CA88843, P50CA62924-05, DK-58757, 5P30 CA06973-39. The work of Irizarry, Parmigiani, and Zeger was partly supported by NIH grant HL 99-024. The work of Zeger was partly supported by NIMH grant R01-MH56639.

Christina Kendziorski, Michael O'Connell, and Paola Sebastiani gave useful comments on an earlier draft.

References

Abramovich F, Yoav Benjamini DD, Donaho D, Johnstone I (2000). Adapting to unknown sparsity by controlling the false discovery rate. Discussion paper, Department of Statistics, Stanford University.

Adcock CJ (1997). Sample size determination: A review. *The Statistician* 46:261–283.

Affymetrix (1999). *Affymetrix Microarray Suite User Guide.* Affymetrix, Santa Clara, CA.

Alizadeh AA, Eisen MB, Davis RE, Ma C, Lossos IS, Rosenwald A, Boldrick JC, Sabet H, Tran T, Yu X, Powell JI, Yang L, Marti GE, Moore T, J. Hudson Jr J, Lu L, Lewis DB, Tibshirani R, Sherlock G, Chan WC, Greiner TC, Weisenburger DD, Armitage JO, Warnke R, Levy R, Wilson W, Grever MR, Byrd JC, Botstein D, Brown PO, Staudt LM (2000). Distinct types of diffuse large B-cell lymphoma identified by gene expression profiling. *Nature* 403:503–511.

Alter O, Brown PO, Botstein D (2000). Singular value decomposition for genome-wide expression data processing and modeling. *Proceedings of the National Academy of Science, USA* 97(18):10101–10106.

Baldi P, Long AD (2001). A Bayesian framework for the analysis of microarray expression data: Regularized t–test and statistical inferences of gene changes. *Bioinformatics* 17(6):509–519.

Banfield JD, Raftery AE (1993). Model-based gaussian and non-gaussian clustering. *Biometrics* 49:803–822.

Becker RA, Chambers JM (1984). *S: an interactive environment for data analysis and graphics.* Belmont, California: Duxbury Press.

Benjamini Y, Hochberg Y (1995). Controlling the false discovery rate: A practical and powerful approach to multiple testing. *Journal of the Royal Statistical Society, Series B* 57:289–300.

Berger JO, Delampady M (1987). Testing precise hypotheses. *Statistical Science* 2:317–335.

Bittner M, Meltzer P, Chen Y, Jiang Y, Seftor E, Hendrix M, Radmacher M, Simon R, Yakhini Z, Ben-Dor A, Sampas N, Dougherty E, Wang W, Marincola F, Gooden C, Lueders J, Glatfelter A, Pollock P, Carpten J, Gillanders E, Leja D, Dietrich K, Beaudry C, Berens M, Alberts D, Sondak V, Hayward N, Trent J (2000). Molecular classification of cutaneous malignant melanoma by gene expression profiling. *Nature* 406:536–540.

Blader IJ, Manger ID, Boothroyd JC (2001). Microarray analysis reveals previously unknown changes in toxoplasma gondii-infected human cells. *Journal of Biological Chemistry* 276:24223–24231.

Bolsover SR, Hyams JS, Jones S, Shepard EA, White HA (1997). *From Genes to Cells*. New York: Wiley.

Bolstad B, Irizarry R, Åstrand M, Speed T (2002). A comparison of normalization methods for high density oligonucleotide array data based on variance and bias. Technical report, UC Berkeley.

Box GEP, Hunter WG, Hunter JS (1978). *Statistics for experiments: An introduction to design, data analysis, and model building*. New York: Wiley.

Brazma A, Hingamp P, Quackenbush J, Sherlock G, Spellman P, Stoeckert C, Aach J, Ansorge W, Ball CA, Causton HC, Gaasterland T, Glenisson P, Holstege FC, Kim IF, Markowitz V, Matese JC, Parkinson H, Robinson A, Sarkans U, Schulze-Kremer S, Stewart J, Taylor R, Vilo J, Vingron M (2001). Minimum information about a microarray experiment (MIAME)–toward standards for microarray data. *Nature Genetics* 29:365–371.

Breiman L, Friedman JH, Olshen RA, Stone CJ (1984). *Classification and Regression Trees*. Belmont, CA: Wadsworth International Group.

Brown CS, Goodwin PC, Sorger PK (2001). Image metrics in the statistical analysis of dna microarray data. *Proceedings of the National Academy of Science, USA* 98(16):8944–8949.

Brown MPS, Grundy WN, Lin D, Cristianini N, Sugnet CW, Furey TS, Ares MJ, Haussler D (2000). Knowledge-based analysis of microarray gene expression data by using support vector machines. *Proceedings of the National Academy of Science, USA* 97:262–267.

Bryan J, van der Laan M (2001). Gene expression analysis with the parametric bootstrap. *Biostatistics* 2(4):445–461.

Burges CJC (1998). A tutorial on support vector machines for pattern recognition. *Data Mining and Knowledge Discovery* 2:121–167.

Chambers JM (1998). *Programming with Data: A Guide to the S Language*. New York: Springer.

Chen X, Cheung ST, So S, Fan ST, Barry C, Higgins J, Lai KM, Ji J, Dudiot S, Ng IOL, van de Rijn M, Botstein D, Brown PO (2002). Gene expression patterns in human liver cancers. *Molecular Biology of the Cell* 13:1929–1939.

Chen Y, Dougherty E, Bittner M (1997). Ratio-based decisions and the quantitative analysis of cDNA micro-array images. *Journal of Biomedical Optics* 2:364–374.

Chiang DY, Brown PO, Eisen M (2001). Visualizing associations between genome sequence and gene expression data using genome-mean expression profiles. *Bioinformatics* 17:S49–S55.

Christianini N, Shawe-Taylor J (2000). *An Introduction to Support-Vector Machines*. Cambridge: Cambridge University Press.

Clyde MA, DeSimone H, Parmigiani G (1996). Prediction via orthogonalized model mixing. *Journal of the American Statistical Association* 91:1197–1208.

Clyde MA, Parmigiani G (1998). Bayesian variable selection and prediction with mixtures. *Journal of Biopharmaceutical Statistics* 8(3):431–443.

Collins FS (1999). Microarrays and macroconsequences. *Nature Genetics* 21S:2.

Cover TM, Hart PE (1967). Nearest neighbor pattern classification. *IEEE Transactions on Information Theory* IT-13:21–27.

DeGroot MH, Fienberg SE (1983). The comparison and evaluation of forecasters. *The Statistician* 32:12–22.

DeRisi JL, Iyer VR, Brown PO (1997). Exploring the metabolic and genetic control of gene expression on a genomic scale. *Science* 278:680–686.

Desu M, Raghavarao D (1990). *Sample Size Methodology*. New York: Academic Press.

Diggle P, Liang KY, Zeger SL (1994). *Analysis of Longitudinal Data*. Oxford: Oxford University Press.

Dudoit S, Fridlyand J, Speed TP (2002a). Comparison of discrimination methods for the classification of tumors using gene expression data. *JASA* 97:77–87.

Dudoit S, Yang YH, Callow MJ, Speed TP (2002b). Statistical methods for identifying genes with differential expression in replicated cDNA microarray experiments. *Statistica Sinica* 12:111–139.

Duggan D, Bittner M, Chen Y, Meltzer P, Trent J (1999). Expression profiling using cDNA microarrays. *Nature Genetics* 21:10–14.

Dunteman GH (1989). *Principal Components Analysis, Vol. 69*. Sage University Paper series on Quantitative Applications in the Social Sciences, series no. 07-064. Newbury Park, CA: Sage.

Efron B, Morris C (1973). Combining possibly related estimation problems (with discussion). *Journal of the Royal Statistical Society, Series B* 35:379–421.

Efron B, Tibshirani R, Storey JD, Tusher V (2001). Empirical Bayes analysis of a microarray experiment. *Journal of the American Statistical Association* 96:1151–1160.

Eisen MB, Spellman PT, Brown PO, Botstein D (1998). Cluster analysis and display of genome-wide expression patterns. *Proceedings of the National Academy of Science, USA* 95:14863–14868.

Everitt B (1980). *Cluster Analysis*. New York: Halsted.

Everitt B (2001). *Applied Multivariate Data Analysis*. Edward Arnold, London.

Fisher RA (1936). The use of multiple measurements in taxonomic problems. *Annals of Eugenics* 7(part 2):179–188.

Friston KJ, Holmes AP, Worsley KJ, Poline JB, Frith CD, Frackowiak R (1995). Statistical parametric maps in functional imaging: A general linear approach. *Human Brain Mapping* 2:189–210.

Gardiner-Garden M, Littlejohn T (2001). A comparison of microarray databases. *Briefings in Bioinformatics* 2:143–158.

Garrett RH, Grisham CM (2002). *Principles of Biochemistry*. Pacific Grove, CA: Brooks/Cole.

Genovese C, Wasserman L (2002). Operating characteristics and extensions of the false discovery rate procedure. *Journal of the Royal Statistical Society, Series B* 64:499–518.

George EI, McCulloch RE (1993). Variable selection via Gibbs sampling. *Journal of the American Statistical Association* 88:881–889.

Getz G, Levine E, Domany E (2000). Coupled two-way clustering analysis of gene microarray data. *Proceedings of the National Academy of Science, USA* 97(22):12079–12084.

Gnanadesikan R (1977). *Methods for Statistical Data Analysis of Multivariate Observations*. New York: Wiley.

Golub TR, Slonim DK, Tamayo P, Huard C, Gaasenbeek M, Mesirov JP, Coller H, Loh M, Downing JR, Caligiuri MA, Bloomfield CD, Lander ES (1999). Molecular classification of cancer: Class discovery and class prediction by gene expression monitoring. *Science* 286:531–537.

Gordon AD (1999). *Classification*. New York: Chapman and Hall/CRC.

Granucci F, Vizzardelli C, Pavelka N, Feau S, Persico M, Virzi E, Rescigno M, Moro G, Ricciardi-Castagnoli P (2001). Inducible IL-2 production by dendritic cells revealed by global gene expression analysis. *Nature Immunology* 2:882–888.

Hardiman G (2002). Microarray technologies—an overview. *Pharmacogenomics* 3(3):293–7.

Hartigan JA, Wong MA (1979). A k-means clustering algorithm. *Applied Statistics* 28:100–108.

Hastie T, Tibshirani R (1990). *Generalized Additive Models*. London: Chapman and Hall.

Hastie T, Tibshirani R, Eisen MB, Alizadeh A, Levy R, Staudt L, Chan WC, Botstein D, Brown P (2000). "Gene shaving" as a method for identifying distinct sets of genes with similar expression patterns. *Genome Biology* 1:research0003.1–research0003.21.

Hastie TJ, Tibshirani R, Buja A (1994). Flexible discriminant analysis by optimal scoring. *Journal of the American Statistical Association* 89:1255–1270.

Herrero J, Valencia A, Dopazo J (2001). A hierarchical unsupervised growing neural network for clustering gene expression patterns. *Bioinformatics* 17:126–136.

Ibrahim JG, Chen MH, Gray RJ (2002). Bayesian models for gene expression with DNA microarray data. *Journal of the American Statistical Association* 97:88–99.

Ihaka R, Gentleman R (1996). R: A language for data analysis and graphics. *Journal of Computational and Graphical Statistics* 5:299–314.

Jain AN, Tokuyasu TA, Snijders AM, Segraves R, Albertson DG, Pinkel D (2002). Fully automatic quantification of microarray image data. *Genome Research* 12(2):325–332.

James W, Stein C (1961). Estimation with quadratic loss. *Proceedings of the Fourth Berkeley Symposium on Mathematical Statististics and Probability* 1:361–380.

Kachigan SK (1991). *Multivariate Statistical Analysis: A Conceptual Introduction*. New York: Radius Press.

Kaufmann L, Rousseeuw PJ (1990). *Finding Groups in Data: An introduction to Cluster Analysis*. New York: Wiley.

Kerr MK, Churchill GA (2001a). Bootstrapping cluster analysis: Assessing the reliability of conclusions from microarray experiments. *Proceedings of the National Academy of Science, USA* 98:8961–8965.

Kerr MK, Churchill GA (2001b). Experimental design in gene expression microarrays. *Biostatistics* 2:183–201.

Kerr MK, Churchill GA (2001c). Statistical design and the analysis of gene expression microarray data. *Genetics Research* 77:123–128.

Kerr MK, Martin M, Churchill GA (2000). Analysis of variance for gene expression microarray data. *Journal of Computational Biology* 7:819–837.

Khan J, Simon R, Bittner M, Chen Y, Leighton S, Pohida T, Smith PD, Jiang Y, Gooden GC, Trent JM, Meltzer PS (1998). Gene expression profiling of alveolar rhabdomyosarcoma with cDNA microarrays. *Cancer Research* 58:5009–5013.

Knudsen S (2002). *A Biologist's Guide to Analysis of DNA Microarray Data*. New York: John Wiley and Sons.

Kohane IS, Kho A, Butte AJ (2002). *Microarrays for an Integrative Genomics*. Cambridge, MA: MIT Press.

Kohonen T (1982). Analysis of a simple self-organizing process. *Biological Cybernetics* 43:59–69.

Kohonen T (1989). *Self-Organization and Associative Memory*. Berlin: Springer-Verlag.

Kohonen T (1995). *Self Organizing Maps*. Berlin: Springer-Verlag.

Kruskal JB (1964). Multidimensional scaling by optimizing goodness of fit to a nonmetric hypothesis. *Psychometrika* 29:1–27.

Lazzeroni L, Owen AB (2002). Plaid models for gene expression data. *Statistica Sinica* 12:61–86.

Lee ML, Kuo FC, Whitmore GA, Sklar J (2000). Importance of replication in microarray gene expression studies: Statistical methods and evidence from repetitive cDNA hybridizations. *Proceedings of the National Academy of Sciences USA* 97(18):9834–9839.

Lee Y, Lee CK (2002). Classification of multiple cancer types by multicategory support vector machines using gene expression data. Technical Report 1051, University of Wisconsin, Madison, WI.

Li C, Wong W (2001). Model-based analysis of oligonucleotide arrays: Expression index computation and outlier detection. *Proceedings of the National Academy of Science, USA* 98:31–36.

Li W, Yang Y (2002). How many genes are needed for a discriminant microarray data analysis? In: SM Lin, KF Johnson (eds.), *Methods of Microarray Data Analysis*, 137–150. Dordrecht: Kluwer Academic.

Lindley DV, Smith AFM (1972). Bayes estimates for the linear model (with discussion). *Journal of the Royal Statistical Society, Series B* 34:1–41.

Lockhart DJ, Dong H, Byrne MC, Follettie MT, Gallo MV, Chee MS, Mittmann M, Wang C, Kobayashi M, Horton H, Brown EL (1996). Expression monitoring by hybridization to high-density oligonucleotide arrays. *Nature Biotechnology* 14:1675–1680.

Lönnstedt I, Speed T (2002). Replicated microarray data. *Statistica Sinica* 12(1):31–46.

McShane LM, D RM, Freidlin B, Yu R, Li MC, Simon R (2001). Methods for assessing reproducibility of clustering patterns observed in analyses of microarray data. Tech report #2, BRB, NCI, Bethesda, MD.

Michie D, Spiegelhalter DJ, Taylor CC (eds.) (1994). *Machine Learning, Neural and Statistical Classification*. New York: Ellis Horwood.

National Research Council; Panel on Discriminant Analysis Classification and Clustering (1988). *Discriminant Analysis and Clustering*. Washington, D. C.: National Academy Press.

Neal RM (1996). *Bayesian Learning for Neural Networks*. New York: Springer-Verlag.

Newton MA, Kendziorski CM, Richmond CS, Blattner FR, Tsui KW (2001). On differential variability of expression ratios: Improving statistical inference about gene expression changes from microarray data. *Journal of Computational Biology* 8:37–52.

Pan W (2002). A comparative review of statistical methods for discovering differentially expressed genes in replicated microarray experiments. *Bioinformatics* 18:546–554.

Pan W, Lin J, Le CT (2002). How many replicates of arrays are required to detect gene expression changes in microarray experiments? A mixture model approach. *Genome Biology* 3(5):research0022.1–0022.10.

Parmigiani G, Garrett ES, Anbazhagan R, Gabrielson E (2002). A statistical framework for expression-based molecular classification in cancer. *Journal of the Royal Statistical Society, Series B*, 64:717–736.

Pavlidis P, Tang C, Noble WS (2001). Classification of genes using probabilistic models of microarray expression profiles. In: MJ Zaki, H Toivonen, JTL Wang (eds.), *Proceedings of BIOKDD 2001: Workshop on Data Mining in Bioinformatics*, 15–18. New York: Association for Computing Machinery.

Quackenbush J (2001). Computational analysis of microarray data. *Nature Reviews Genetics* 2:418–427.

Radmacher MD, McShane LM, Simon R (2001). A paradigm for class prediction using gene expression profiles. Tech report #1, BRB, NCI, Bethesda, MD.

Raychaudhuri S, Stuart JM, Altman RB (2000). Principal components analysis to summarize microarray experiments: Application to sporulation time series. In: RB Altman, AK Dunker, L Hunter, K Lauderdale, TE Klein (eds.), *Fifth Pacific Symposium on Biocomputing*, 455–466.

Rios Insua D, Mueller P (1998). Feedforward neural networks for nonparametric regression. In: *Practical Nonparametric and Semiparametric Bayesian Statistics*, 181–194. New York: Springer.

Ripley BD (1996). *Pattern Recognition and Neural Networks*. Cambridge: Cambridge University Press.

Rosenwald A, Wright G, Chan W, Connors JM, Campo E, Fisher RI, Gascoyne RD, Muller-Hermelink HK, Smeland EB, Giltnane JM, Hurt EM, Zhao H, Averett L, Yang L, Wilson WH, Jaffe ES, Simon R, Klausner RD, Powell J, Duffey PL, Longo DL, Greiner TC, Weisenburger DD, Sanger WG, Dave BJ, Lynch JC, Vose J, Armitage JO, Montserrat E, Lopez-Guillermo A, Grogan TM, Miller TP, LeBlanc M, Ott G, Kvaloy S, Delabie J, Holte H, Krajci P, Stokke T, Staudt LM (2002). The use of molecular profiling to predict survival after chemotherapy for diffuse large b-cell lymphoma. *New England Journal of Medicine* 346(25):1937–1947.

Ross DT, Scherf U, Eisen MB, Perou CM, Rees C, Spellman P, Iyer V, Jeffrey SS, van de Rijn M, Waltham M, Pergamenschikov A, Lee JCF, Lashkari D, Shalon S, Myers TG, Weinstein JN, Botstein D, Brown PO (2000). Systematic variation in gene expression patterns in human cancer cell lines. *Nature Genetics* 24:227–235.

Rousseeuw P, Struyf A, Hubert M (1996). Clustering in an object-oriented environment. *Journal of Statistical Software* 1:1–30.

Ruczinski I, Kooperberg C, LeBlanc M (2003). Logic regression. Manuscript submitted for publication.

Saal LH, Troein C, Vallon-Christersson J, Gruvberger S, Borg A, Peterson C (2002). Bioarray software environment (base): a platform for comprehensive management and analysis of microarray data. *Genome Biolog* 3:software0003.10003.

Schena M (2000). *Microarray Biochip Technology*. Westborough, MA: BioTechniques Press.

Schena M, Shalon D, Davis R, Brown P (1995). Quantitative monitoring of gene expression patterns with a complementary DNA microarray. *Science* 270:467–470.

Segal E, Taskar B, Gasch A, Friedman N, Koller D (2001). Rich probabilistic models for gene expression. *Bioinformatics* 17:S243–S252.

Simon R, Radmacher MD, Dobbin K (2002). Design of studies using dna microarrays. *Genetic Epidemiology* 23:21–36.

Slonim DK, Tamayo P, Mesirov P, Golub TR, Lander ES (1999). Class prediction and discovery using gene expression data. Discussion paper, Whitehead/M.I.T. Center for Genome Research, Cambridge, MA.

Southern EM (2001). DNA microarrays. History and overview. *Methods in Molecular Biology* 170:1–15.

Spellman PT, Sherlock G, Zhang MQ, Iyer VR, Anders K, Eisen MB, Brown PO, Botstein D, Futcher B (1998). Comprehensive identification of cell cycle-regulated genes of the yeast *saccharomyces cerevisiae* by microarray hybridization. *Molecular Biology of the Cell* 9:3273–3297.

Storey JD (2001). The positive false discovery rate: A bayesian interpretation and the q-value. Discussion paper, Department of Statistics, Stanford University.

Storey JD (2002). A direct approach to false discovery rates. *Journal of the Royal Statistical Society, Series B* 64:479–498.

Sundberg R (1999). Multivatiate calibration —direct and indirect regression methodology. *Scandinavian Journal of Statistics* 26:161–207.

Tamayo P, Slonim D, Mesirov J, Zhu Q, Dmitrovsky E, Lander ES, Golub TR (1999a). Interpreting gene expression with self-organizing maps: Methods and application to hematopoietic differentiation. *Proceedings of the National Academy of Science, USA* 96:2907–2912.

Tamayo P, Slonim D, Mesirov J, Zhu Q, Kitareewan S, Dmitrovsky E, Lander ES, Golub TR (1999b). Interpreting patterns of gene expression with self-organizing maps. *Proceedings of the National Academy of Science USA* 96:2907–2912.

Tibshirani R, Hastie T, Eisen M, Ross D, Botstein D, Brown P (1999). Clustering methods for the analysis of DNA microarray data. Technical report, Department of Statistics, Stanford University, Stanford, CA.

Tibshirani R, Hastie T, Narasimhan B, Chu G (2002). Diagnosis of multiple cancer types by shrunken centroids of gene expression. *Proceedings of the National Academy of Science, USA* 99:6567–6572.

Toussaint GT (1974). Bibliography on estimation of misclassification. *IEEE Transactions on Information Theory* IT-20:472–79.

Tseng GC, Oh MK, Rohlin L, Liao J, Wong W (2001). Issues in cDNA microarray analysis: quality filtering, channel normalization, models of variations and assessment of gene effects. *Nucleic Acids Research* 29:2549–2557.

Tusher V, Tibshirani R, Chu G (2001). Significance analysis of microarrays applied to the ionizing radiation response. *Proceedings of the National Academy of Science, USA* 98:5116–5121.

Ultsch A (1993). Self-organizing neural network for visualization and classification. In: O Opitz, B Lausen, R Klar (eds.), *Information and Classification*, 307–313. Springer.

Vapnik V (1998). *Statistical Learning Theory*. New York: Wiley.

Venables WN, Ripley BD (2000). *S programming*. New York: Springer.

West M, Blanchette C, Dressman H, Huang E, Ishida S, Spang R, Zuzan H, Marks JR, Nevins JR (2001). Predicting the clinical status of human breast cancer using gene expression profiles. *Proceedings of the National Academy of Science, USA* 98:11462–11467.

Wolfinger RD, Gibson G, Wolfinger E, Bennett L, Hamadeh H, Bushel P, Afshari C, Paules RS (2001). Assessing gene significance from cDNA microarray expression data via mixed models. *Journal of Computational Biology* 8:625–637.

Worsley K, Liao C, Aston J, Petre V, Duncan G, Morales F, Evans A (2002). A general statistical analysis for fMRI data. *NeuroImage* 15:1–15.

Xu Y, Selaru F, Yin J, Zou T, Shustova V, Mori Y, Sato F, Liu T, Olaru A, Wang S, Kimos M, Perry K, Desai K, Greenwald B, Krasna M, Shibata D, Abraham J, Meltzer S (2002). Artificial neural networks and gene filtering distinguish between global gene expression profiles of Barrett's esophagus and esophageal cancer. *Cancer Research* 62:3493–3497.

Yang H, Speed TP (2002). Design issues for cDNA microarray experiments. *Nature Genetics Reviews* 3:579–588.

Yang YH, Buckley MJ, Speed TP (2001). Analysis of cDNA microarray images. *Briefings in Bioinformatics* 2(4):341–349.

Yang YH, Dudoit S, Luu P, Lin DM, Peng V, Ngai J, Speed T (2002). Normalization for cDNA microarray data: A robust composite method addressing single and multiple slide systematic variation. *Nucleic Acids Research* 30(4):e15.

Yeung K, Fraley C, Murua A, Raftery A, Ruzzo W (2001a). Model-based clustering and data transformations for gene expression data. *Bioinformatics* 17:977–987.

Yeung KY, Haynor DR, Ruzzo WL (2001b). Validating clustering for gene expression data. *Bioinformatics* 4:309–318.

Yeung KY, Ruzzo WL (2001). Principal component analysis for clustering gene expression data. *Bioinformatics* 17:763–774.

Zhang H, Yu CY (2002). Tree-based analysis of microarray data for classifying breast cancer. *Frontiers in Bioscience* 7:63–67.

Zhao LP, Prentice R, Breeden L (2001). Statistical modeling of large microarray data sets to identify stimulus-response profiles. *Proceedings of the National Academy of Science, USA* 98:5631–5636.

2

Visualization and Annotation of Genomic Experiments

ROBERT GENTLEMAN
VINCENT CAREY

Abstract

We provide a framework for reducing and interpreting results of multiple microarray experiments. The basic tools are a flexible gene-filtering procedure, a dynamic and extensible annotation system, and methods for visualization. The gene-filtering procedure efficiently evaluates families of deterministic or statistical predicates on collections of expression measurements. The expression-filtering predicates may involve reference to arbitrarily complex predicates on phenotype or genotype data. The annotation system collects mappings between manufacturer-specified probe set identifiers and public use nomenclature, ontology, and bibliographic systems. Visualization tools allow the exploration of the experimental data with respect to genomic quantities such as chromosomal location or functional groupings.

2.1 Introduction

The Bioconductor project (`www.bioconductor.org`) was initiated at Dana Farber Cancer Institute to gather statisticians, software developers, and biologists interested in advancing computational biology through the design, deployment, and dissemination of open-source software. The R statistical computing environment (`www.r-project.org`) is the central computing and interaction platform for tools created in Bioconductor. Bioconductor's software offerings consist of R packages and data images to support many aspects of computing and inference in bioinformatics. Key packages that will be reviewed in this chapter include:

- `Biobase`: the basic data structures and algorithms required for storage and exploration of genomic, phenotypic, and gene expression data obtained in microarray experiments;

- **genefilter**: routines for efficient identification of genes satisfying arbitrarily complex, statistically defined conditions;
- **edd** (expression density diagnostics): algorithms for evaluating the diversity of gene expression distributions in phenotypically defined cohorts;
- **annotate**: programs and data structures for connecting genomic and phenotypic data to biological and clinical annotation and literature to facilitate interpretation and hypothesis-driven modeling;
- **geneplotter**: programs to assist in the visualization of experimental results.

Other Bioconductor/R packages are described in Dudoit and Yang (Chapter 3, this volume) and Irizarry et al. (Chapter 4, this volume).

After describing and formalizing the motivations and choices of data structures and algorithms for this project, we will illustrate the use of these software tools with two large gene expression databases. The first is the set of 47 U68 Affymetrix arrays discussed in Golub et al. (1999). The data are available in their original form at http://www-genome.wi.mit.edu. The samples were collected on individuals with acute myelogenous or acute lymphocytic leukemia. The second dataset is a collection of 89 U95 arrays from Zhang, Derdeyn, Gentleman, Leykin, Monti, Ramaswamy, Wong, Golub, Iglehart and Richardson (2002) with additional data on HER2 status provided by Dr. A. Richardson. These data are samples collected on metastatic and nonmetastatic breast cancer tumors. Both datasets and transcripts of data analysis sessions related to this chapter are available at www.bioconductor.org/Docs/Papers/2002/Springer.

2.2   Motivations for Component-Based Software

Microarray experiments are based on collections of tissue samples (usually fewer than 100) on which expression of messenger RNA (mRNA) has been measured.

For concreteness, we focus on the analysis of oligonucleotide arrays. Let $i = 1, \ldots, I$ index independent microarrays. The ith array supplies x_{ji}, $j = 1, \ldots, J$ expression measures. There is usually substantial processing of the raw experimental data needed to obtain these expression values. The steps involved in this processing are discussed in Irizarry et al. (Chapter 4, this volume) and Li and Wong (Chapter 5, this volume). We assume that the expression values are comparable across arrays although not necessarily across genes.

The large number of genes J (usually $J > 1000$ and often $J > 10,000$) places special requirements on computational and inferential tools for the analysis of microarray experiments. New methods of computation and inference are also required for the complex task of connecting numerically detected patterns of gene expression and resources, typically textual in nature, that allow biologic interpretation of these patterns. We now discuss details of some of the basic motivations for establishing flexible component-based approaches to filtering, annotation, and visualization of gene expression data.

Diversity and dynamic nature of microarray formats and outputs. A microarray generally consists of a spot (or a set of spots) on which specific portions of mRNA should cohybridize. The number of spots, the length of the sequence against which cohybridization is performed, and the nature of the sequence depend on the manufacturer of the chip. The two most popular techniques are cDNA arrays as described by Shalon et al. (1996) and Affymetrix short oligomer arrays. It is important to realize that there are limitations to the inferences that can be derived from these arrays. In both cases, we can measure the amount of mRNA that hybridizes to the spot(s) on the array. Biologic or clinical inference relies on the imperfect and evolving mapping from the EST to the gene. This mapping should be carried out on data that are as recent as possible, so software implementations must avoid "hard-coding" any but the most permanent features of this mapping. Ideally, the software will make real-time use of Web-based repositories that satisfy given requirements for currency and accountability.

Fallibility of EST construction. Most arrays use probes that are based on expressed sequence tags (ESTs). We will refer to the probes and their targets incorrectly but interchangeably as ESTs and genes. An EST is a short (typically 100–300 bp) partial cDNA sequence. cDNA is DNA synthesized by the enzyme reverse transcriptase using mRNA as a template. In practice, cellular mRNA is used together with reverse transcriptase to build cDNA. The resulting cDNA sequences are then used to build ESTs. This process is not infallible, and some proportion of the ESTs will be incorrect (or incorrectly labeled).

Need to distinguish transcript abundance from gene activity and protein abundance. The expression of the genetic information contained in DNA occurs in two stages. The first is transcription, where DNA is transcribed into RNA. The second stage is translation, where the RNA is translated into a protein. The central dogma of molecular biology is that DNA makes RNA makes protein. Microarray technologies measure levels of mRNA in different samples. Thus, they measure transcript abundance, which may or may not relate to the presence of a gene (since the gene need not be transcribed). Transcript abundance may or may not relate to the presence of a protein (since the mRNA need not be translated). Interpretation of microarray outputs may also be affected by translocations or increases in copy number that may be present in only a fraction of samples with a given phenotype. Either increased copy number with the usual expression control or increased/enhanced expression with the copy number left the same can result in increased mRNA expression.

Need to separate potentially functional ESTs from housekeeping sequences.
For any given tissue in the human body, it is estimated that about 40% of
the genome is expressed. Another portion of the genes measured will have a
relatively constant transcript abundance across samples. These genes may
be performing *housekeeping activities* or other activities that are unrelated
to the processes being studied. Therefore, in most cases there are a relatively large number of ESTs that are inherently uninteresting, and some
means of removing them from the remainder of the analysis will be helpful.
Many of the tools that we employ are computationally expensive, and any
reduction of the data will be rewarded by decreased analysis times. One
approach is to filter out the uninteresting genes so that attention can be
focused on those genes that have some potential to be interesting.

Necessity and limitations of gene-specific processing. With Affymetrix arrays, comparisons between expression levels estimated from different probe
sets should be made with caution. For these arrays, the intensity may be
affected by the probe sequences used, and two different probe sets may have
greatly different estimated expression values when in fact the abundances
of the mRNAs are quite similar. With cDNA arrays, provided all arrays
have been hybridized with a common baseline, a comparison between genes
is probably valid. Gene-at-a-time processing is a severely limited analytic
framework and is incapable of great fidelity to the biology involved. The
systems that we are studying are complex, and there are always interactions between different gene products. There are many systems that are
more complex. For example, the ratio of BAX (BCL2-associated X protein) to BCL2 determines the cell's fate. If the level of BCL2 is larger than
that of BAX, then apoptosis (programmed cell death) is suppressed. When
the level of BAX is larger than that of BAD, apoptosis is promoted (Helmreich, 2001; p. 241). Hence, we are not interested in changes in one of these
two genes but rather in changes in their ratio.

The considerations just enumerated have played a significant role in shap-
ing our design of Bioconductor software components. We put a high pre-
mium on designs that allow adaptation to new approaches and resources—
at both the biotechnologic and inferential levels—that help overcome the
limitations and ambiguities inherent in the current state of the art of microarray experimentation and interpretation. Formalization of the key resources of our approach is provided in the next section.

2.3   Formalism

Recall that $i = 1, \ldots, I$ indexes the samples or chips, and $j = 1, \ldots, J$ indexes genes that are assumed to be common across all chips. For sample i,
a q-vector y_i contains phenotypic and/or demographic data on the patients
from which samples were derived, such as age, sex, disease status, duration
of symptoms, and so on. These data are conceptually and often adminis-

tratively distinct from the expression data (which may be managed in a completely different database). Labels $s_i = s(y_i)$ are available to classify samples; for example, $s_i \in \{normal, diseased\}$. Note that we will commonly use the term *phenotypic data* very loosely to refer to any nongenomic data related to tissue samples or their donors.

An *experiment metadata structure* is a description of conditions and materials used in a biological experiment. Standard specifications of metadata structures have been proposed (e.g., MIAME `http://www.mged.org/Workgroups/MIAME/miame.html`; Brazma et al., 2001). Our framework is sufficiently broad to accommodate such annotation.

An *annotation structure* is a mapping between EST identifiers and standardized nomenclatures for associated genes. The mapping may be formalized as a set of ordered sequences with specified positions for conventional (e.g., manufacturer-defined) EST identifiers and associated standard nomenclature tokens for the genes related to that EST.

A *microarray experiment database* is the ordered quadruple $S = \langle X, Y, A, M \rangle$, where X is a $J \times N$ matrix with columns representing J-vectors of gene expression measurements on each of N tissue samples, Y is an $N \times p$ matrix with rows representing p-vectors of phenotypic information on the N tissue samples or their donors, A is an annotation structure, and M is an experiment metadata structure. The floating-point number x_{ji} is the (j, i) entry of matrix X, representing the measured expression level for gene j on tissue sample i. The datum y_{iq} is the value of phenotypic variable q on tissue sample i or its donor. x_j is the N-vector of expression values for gene j over all donors.

A *gene filter element* is a Boolean-valued function $f(x, Y)$ of an N-vector of expression values x and an $N \times p$ phenotype matrix Y. An example is a function that uses the phenotype data to divide x into two samples (e.g., diseased and nondiseased) and returns TRUE if and only if a two-sample test of location shift rejects the null hypothesis of a common location for the two samples.

A *gene filter* is a collection of gene filter elements that can be applied to a microarray experiment database. Let F denote such a collection. Then, the action of a gene filter on a microarray experiment database S is to define the index set $I_{S,F} = \{j = 1, \ldots, J | f(x_j, Y_S) = 1, \text{ all } f \in F, x_j \in X_S\}$, where X_S and Y_S are the expression and phenotype components of S.

2.4   Bioconductor Software for Filtering, Exploring, and Interpreting Microarray Experiments

2.4.1   Formal Data Structures and Methods for Multiple Microarrays

The choice and design of data structure to represent multiple microarray experiments is one of the most critical processes of software development in this domain. Regimentation of the structure is important to establish trustworthiness of related computations and to facilitate reuse of software components that effectively navigate and process elements of the structure. Flexibility of the structure is also of great importance so that changes in biotechnology and research directions do not necessitate complete redesign. Our approach to data structure design for this problem preserves the conceptual independence of genomic and phenotypic information types and exploits the *formal methods and classes* in the R package `methods` which was developed by Chambers (1998).

The `methods` package provides a substantial basis for object-oriented programming in R. Object-oriented programming is a well-established approach to dealing with complex data structures and algorithms. When we can conceive of our data as an object with a well-defined set of properties and components of various types, then an object-oriented programming system may be used to represent our data as an instance of a class of similarly structured objects. Analysis algorithms defined in the object-oriented system are freed from responsibility for checking the structure and contents of the components of objects instantiating the class. Furthermore, when proper design and deployment discipline is maintained, it is often possible to extend the structure of an object without affecting the behavior of software developed for previous versions of the object. Thus, the object-oriented approach gives access to the regimentation and flexibility required for the creation and support of durable software for bioinformatic inference.

Classes `exprSet` and `phenoData`

To coordinate access in R to the genomic and phenotypic data generated in a typical microarray experiment, we defined two classes of objects. One is called the `exprSet` class and the other the `phenoData` class.

The `phenoData` class was designed to hold the phenotypic (or sample level) data. Objects of this class have the following properties or *slots*:

pData A `data.frame` with samples as the rows and the phenotypic variables as the columns.

varLabels A list with one element per phenotypic variable. Names of list elements are the variable names; list-element values are character strings with brief textual descriptions of the associated variables.

An instance of class `exprSet` has the following slots:

exprs An array that contains the estimated expression values, with columns representing samples and rows representing genes.

se.exprs An array of the same size as `exprs` containing estimated standard errors. This may be NULL.

phenoData An instance of the `phenoData class` that contains the sample level variables for this experiment. This class is described above.

description A character string describing the experiment. This will probably change to some form of documentation object when and if we adopt a more standard mechanism for documentation.

annotation The name of the annotation data that can be used for this exprSet.

notes A character string for notes regarding the analysis or for any other purpose.

Once a class has been defined, instances of it can be created. A particular dataset stored in this format would be called an instance of the class. Although we should technically always say something like *x is an instance of the* exprSet *class*, we will often simply say that *x is an* exprSet.

Formal Methods for exprSets

A number of methods for the exprSet and phenoData classes have been implemented. The reader is directed to the documentation in our packages for definitive descriptions. Here, we illustrate the various methods with examples based on a celebrated collection of array experiments. The golubEsets package includes exprSets embodying the leukemia data of Golub et al. (1999). Upon attaching the golubTrain element of the package, we may invoke the following exprSet methods:

Show the data. The generic `show` function (invoked upon mention of an exprSet instance) gives a concise report.

```
> golubTrain
Expression Set (exprSet) with
  7129 genes
  38 samples
   phenoData object with 11 variables and 38 cases
   varLabels
    Samples: Sample index
    ALL.AML: Factor, indicating ALL or AML
    BM.PB: Factor, sample from marrow or peripheral blood
    T.B.cell: Factor, T-cell or B-cell leuk.
    FAB: Factor, FAB classification
    Date: Date sample obtained
    Sex: Factor, sex of patient
    pctBlasts: pct of cells that are blasts
    Treatment: response to treatment
```

```
PS: Prediction strength
Source: Source of sample
```

Subset of genes. Subscripting on the first index returns a matrix of expression values of the corresponding genes. Notice that the row names are Affymetrix identifiers for the ESTs used. The `annotate` package will use these identifiers to map to different quantities such as the LocusLink identifiers or chromosomal location. Affymetrix uses the convention that identifiers that begin with `AFFX` are used for quality-control purposes and are generally uninteresting for any analysis. One may want to remove these before analyzing the data.

```
> golubTrain[1:4,]
                  [,1] [,2] [,3] ...
AFFX-BioB-5_at    -214 -139  -76
AFFX-BioB-M_at    -153  -73  -49
AFFX-BioB-3_at     -58   -1 -307
AFFX-BioC-5_at      88  283  309
```

Extract phenotype vector. This method exploits a specialization of the `$` operator to deal with instances of the `exprSet` class.

```
> table(golubTrain$ALL.AML)

ALL AML
 27  11
```

Subset of patients. Subscripting on the second index returns a matrix of expression values of the corresponding tissue donors. In this example, we obtain the dimensions after restriction to patients whose `phenoData` indicates that they have acute lymphocytic leukemia.

```
> print(dim(golubTrain[
      ,golubTrain$ALL.AML=="ALL"]))

[1] 7129  27
```

Accessor functions. A number of functions are provided to allow access to more primitive representations of expression data. The `exprs` function applied to an `exprSet` returns the $J \times N$ matrix of expression results. The `pData` function returns the $N \times p$ `data.frame` of phenotypic data.

Before concluding this topic, we make some remarks on the problem of concisely specifying `exprSet` subsets. Familiar syntax has been established to allow reference to subsets of collections of arrays. A key obligation is to keep the genomic and phenotypic data correctly aligned, so, if eS is an instance of an exprSet, we may ask for eS[,1:5]. The second index in this subscripting expression refers to tissue samples. The value of this expression is an exprSet whose contents are restricted to the first five tissue samples in eS. The new exprSet's exprs element contains the first five columns of eS's expr array. The new exprSet's se.exprs array is similarly restricted. But the phenoData must also have a subset made and for it we want the first five *rows* of eS's phenoData dataframe since it is in the more standard format where columns are variables and rows are cases.

Other tools for working with exprSets are provided, including methods for iterating over genes with function application and sampling from patients in an exprSet (with or without replacement).

2.4.2   Tools for Filtering Gene Expression Data: The Closure Concept

We have seen that the golubTrain exprSet includes results for 7129 genes. The genefilter package of Bioconductor provides a collection of functions allowing computationally efficient reduction of the set of genes of interest. The user interface is the genefilter function, which accepts a $J \times N$ matrix of expression data and an instance of class filterfun. The filterfun class supports the combination of functions to define a sequence of filtering criteria. In the current formulation, the genefilter package supports *marginal* filtering: criteria are constructed and evaluated to retain or exclude a gene based solely on features of the distribution of expression values of that gene (in relation, of course, to the phenotypic features of the samples on which gene expression was measured). This is to be distinguished from *joint* filtering, in which the retention or exclusion of a gene may depend on distributions of other genes. An example of joint filtering is given in subsection 2.6.1.

We have designed the genefilter package to support very flexible specifications and combinations of filtering criteria. Suppose that we wish to restrict attention to genes in golubTrain that have expression values exceeding 100 for at least five donors and for which the coefficient of variation is at least 2. The following commands carry this out.

```
myAbsLB <- kOverA( k=5, A=100 )  # absolute lower bound
myCVspec <- cv( a=2 )            # lower bound on CV
myFilterSeq <- filterfun( myCVspec, myAbsLB )
myInds <- genefilter( exprs(golubTrain), myFilterSeq )
```

Now

```
Train2 <- golubTrain[myInds,]
```

has 1534 genes. To introduce more stringent filtration, arguments to the filter components can be altered and the process of building and applying the filter collection `myFilterSeq` can be repeated, or, new filters could be constructed and applied to `exprs(Train2)`.

The command

```
myAbsLB <- kOverA( k=5, A=100 )
```

defines a *closure*. This is an R function accompanied by an environment defining local data. The limit, `A`, and number of exceedents required, k, are bound to 5 and 100 in `myAbsLB`, which is applied to all N-vectors of expression values for all genes in `exprs(golubTrain)`. The use of closures allows us to provide a simple but very flexible interface. Arbitrarily many filters can define a filtration. Each of the filters may make use of user-specified parameters or phenotypic data structures. For more background on closures in R, see Gentleman and Ihaka (2000).

This approach can be contrasted with one where long lists of optional parameters are supplied. Our experience suggests that the latter is error-prone. Separation of the task of identifying and creating specific filters from the task of applying those filters provides a substantial simplification.

2.4.3   Expression Density Diagnostics: High-Throughput Exploratory Data Analysis for Microarrays

A basic problem in gene discovery exercises is the formulation of test procedures that powerfully discriminate distinct patterns of gene expression in phenotypically distinct tissues. In standard statistical applications, exploratory data analyses using graphics and goodness-of-fit appraisals are used to match the test procedure to the characteristics of the data. For example, data from long-tailed distributions will often be log-transformed or scrutinized for outliers. The magnitude and relative complexity of gene expression array datasets has been a hindrance to exploratory analysis and investigation of test selection procedures based on characteristics of gene expression distributions across samples. If gene expression distributions are highly diverse in shape, discovery procedures will need to adapt to the shapes present in order to possess reasonable power.

As a first step toward more adaptive gene discovery methods, the `edd` package performs a species of high-throughput exploratory data analysis. To overcome the difficulty of evaluating thousands of histograms or density smooths to assess skewness, multimodality, outlier-proneness, or other departures from Gaussianity, we compare gene- and stratum-specific empirical distribution functions for expression data to a catalog of specified reference distributions, labeling each gene with the closest matching reference distribution or "doubt." This allows the grouping of genes into broad classes to facilitate more focused diagnostic analysis or discriminative testing. We now provide some formal definitions of components of the procedure. We suppose that the underlying distributions are of continuous type for all genes. Expression distribution classification involves three steps.

Location and scale reduction. Define $m_F = F^{-1}(1/2)$ to be the median of the cumulative distribution function (cdf) F and $a_F = c \cdot \text{median}_{x \sim F}|x - m_F|$ to be the scaled median absolute deviation (MAD) of F. Scaling factor $c = 1.483$ is used to obtain consistency of MAD for σ at $N(\mu, \sigma^2)$. The centered and scaled cumulative distribution function (cdf) corresponding to an arbitrary cdf F is defined as

$$F^*(y) = \Pr(a_F^{-1}[Y - m_F] < y).$$

We will use the asterisk superscript to denote cdfs or deviates from distributions that have been subjected to this transformation. For example, $X^* \sim \text{Beta}^*(r, t)$ implies that $X^* = s^{-1}[X - m]$, where $X \sim \text{Beta}(r, t)$ and m(resp. s) are the median (resp. MAD) of the Beta distribution with parameters r and t. The function `centerScale` in `edd` can be applied to any real vector X to obtain X^*.

Formation of reference catalog. We develop the tools to simulate from and evaluate quantiles of a collection of location- and scale-reduced parametric distributions: $\text{Beta}^*(2,8)$, $\text{Beta}^*(8,2)$, χ_1^{2*}, $\text{LN}^*(0,1)$ (log normal with mean zero and standard deviation unity after log transformation), MIX_p^*, where MIX_p denotes the two-component mixture $pN(0,1) + (1-p)N(4,1)$, t_3^*, and $U^*(0,1)$. Other catalog elements can be added as needed. These candidates were chosen to represent a diversity of shapes and to allow some assessment of the possibility of confusion (χ_1^{2*} and $\text{LN}^*(0,1)$ are very similar shapes). In general, there will be R location- and scale-reduced reference distributions in the catalog, which we will denote $\mathbb{H} = \{H_1, \ldots, H_R\}$. The functions `makeCandmat.raw` and `makeCandmat.theor` create reference catalogs based on simulation and quantile calculation after location and scale reduction, respectively.

Classification. The empirical distribution function (EDF) of location- and scale-reduced expression observations on each gene is treated as an I-dimensional multivariate datum. For gene g (in a stratum s), this EDF is denoted $\hat{F}_{sg}$; the stratum index will be suppressed unless required. Three approaches to classification of gene-specific EDFs are investigated:

- *Nearest-neighbor classification.* A prespecified number c of I-dimensional reference candidates are simulated from each element of $\mathbb{H}$. This leads to $c \cdot R$ I-vectors that serve as a training set, each with known shape. k-NN classification is conducted for each $\hat{F}_g$ with parameters k (number of neighbors to be polled) and l (minimum number of concordant votes of type s required to classify the candidate into category s). The expression distribution of gene j is declared to be of shape $r \in \{1, \ldots, R\}$ if l of the k closest catalog members are simulated from H_r and is declared to be of unknown shape otherwise.

- *Test-based classification.* For each j and each $s \in \{1, \ldots, R\}$, compute the p-value p_s of the Kolmogorov–Smirnov statistic testing the hypothesis that $x_j \sim H_s$. The expression distribution of gene j is declared to be of shape r if p_r exceeds a specified lower bound and is uniquely the largest of the p_s and is declared to be of unknown shape otherwise.
- *Model-based classification.* The $c \cdot R$ I-dimensional training set described under the nearest-neighbor approach is used to construct a neural network that predicts class membership or "doubt" for each $\hat{F}_g$.

The function `edd.unsupervised` includes options for the type of classification algorithm to be used and can be applied to `exprSets`. It returns a set of J classification labels indicating for each gene in the `exprSet` the best-fitting element of the reference catalog.

2.4.4   Annotation

Relating the ESTs comprising a class of arrays to various biological data resources (e.g., genomic maps, protein function characterizations, general literature of clinical genetics) is an essential part of the analysis. If an analysis is primarily data-driven, one needs to relate the numerical patterns discovered among ESTs and phenotypic conditions to information about gene structure and function. If an analysis is hypothesis-driven, one needs to use information on gene structure and function to restrict or otherwise structure the sets of ESTs and phenotypes considered in the analysis to conform to the hypotheses of interest. These activities are problematic because biological and clinical data resources are continually changing, as is our understanding of their contents and how they are to be harvested and used. Based on these observations, we have chosen a mechanism for capturing and supplying annotation data that is flexible and easily updated.

We have divided the process into two components. One component consists of building and collating annotation data from public databases. The second component is the extraction and formatting of the collated data into a format suitable for end users. Data analysts can simply obtain a set of annotation tables for the set of ESTs that they are using. We will consider only this use of the data in this discussion. This is a rapidly changing field, so readers should consult the Bioconductor Web site for current details on the system's resources and methods.

Our aim is to provide data structures facilitating selection of genes according to certain features or a priori conventional classifications such as membership in functional groups, involvement in biological processes, or chromosomal location. Currently, annotation structures are available from the Bioconductor project in a variety of formats. Datasets have been marked up as XML files with explicit DTDs and have also been formatted as R objects suitable for loading into R using the `load` function. Internally (to R), they are managed using hash tables, and hence we require a unique key for each entry. Mapping from the EST (or manufacturer's identification) is then fairly straightforward.

Affymetrix Inc. produces several chips for the human genome. The most popular in current use are the U95v2 chips (with version A being used most often). The probes that are arrayed here have Affymetrix identifiers. We provide functions that map the Affymetrix identifiers to a number of other identifiers such as those provided by LocusLink and GenBank.

In addition, we have assembled relations between the different identifiers and their Gene Ontology (GO) values. See http://www.geneontology.org. GO is an attempt to provide standardized descriptions of the biological relevance of genes into the three categories *biological process*, *cellular component*, and *molecular function*. Within a category, the set of terms forms a directed acyclic graph. Genes typically get a specific value for each of these three categories. We trace each gene to the top (root node) and report the last three nodes in its path to the root of the tree. These top three nodes provide general groupings that may be used to examine the data for related patterns of expression. Using the GO annotation allows users to consider subsets of genes for analysis using groupings under any of these headings. syntenic region, and sets of orthologs.

Often, researchers are interested in finding out more about their genes. For example, they would like to look at the different resources at NCBI. HTML pages with active links to the different online resources can easily be produced. The function ll.htmlpage is one example of a function designed to provide links to the LocusLink Web page for a list of genes.

It is also possible using connections and the XML package to open http connections and read the data from the Web sites directly. The resulting text could then be processed using other R tools. For example, if abstracts or full-text searchable articles were available, these could be downloaded and searched for relevant terms. Examples will be provided in Section 2.6 below.

2.5   Visualization

Visualization of data is an important and potentially very powerful method for exploring and understanding data. Like all other techniques for the analysis of genomic data, visualization is in its infancy. We consider just a few plots that might be useful. We demonstrate how these plots help data analysts understand their data.

Perhaps the most used visualization aid for microarray data are the *heat maps*, or displays of genome-wide expression patterns, proposed by Eisen et al. (1998). These plots generally cluster together both genes and samples that have similar expression levels. The argument for doing this is that genes with similar levels of expression often have a similar function. Finding genes that have a similar function is often one of the things in which we are interested.

A typical rendering is as a rectangular region with rows given by ESTs and columns by samples. Each small rectangle is colored to indicate the expression level of the specific EST for the specific sample. Expression level is usually indicated by both intensity and color (often red for high and green for low, which is rather unfortunate for the color-blind). Usually, some form of hierarchical clustering is performed to arrange the rows and the columns so that there are relatively large homogeneous regions of red and green. Rather than examine these rather well-known plots—easily constructed in R using the `image` function and a suitable set of colors—we will provide a few suggestions for other types of graphics that might be interesting.

2.5.1   Chromosomes

One might be interested in locating where a gene or set of genes is located in the genome. The chromosomal locations can be obtained from the `annotate` package and can then be used to construct visual tools. In this section, we demonstrate some of the tools available in the `geneplotter` package.

Different genomes have different numbers of chromosomes, each of different length. Additionally, chromosomes are generally double stranded, and a gene can be encoded on either strand (one strand is called the plus strand and the other the minus strand). We generally give an indication of which strand a gene is on in all visualizations.

Plots of expression and its association with chromosomal location might be helpful. The function `alongChrom` in the `geneplotter` package provides plots of expression level along the chromosomes. There are several options available. One that seems initially promising is to plot cumulative expression along a chromosome or region of a chromosome. Large jumps in this plot indicate high levels of expression, whereas flat spots indicate a lack of expression.

An alternative to plotting the cumulative expression level along a chromosome is to plot individual levels at the appropriate position. There are a variety of transformations that might be useful, such as the z-transformation $(x - \bar{x})/\mathrm{sd}(x)$ or transforming to ranks.

In breast cancer, one of the important prognostic factors is the presence of the ERBB2 gene. This particular genomic region is subject to amplification through duplication. Increased expression of ERBB2 is often associated with an increased copy number, that is, a region of the chromosome containing ERBB2 is replicated. There can be many replicates of the region contained in the genome, and there can be many genes on each replicate. A region of this type is often referred to as an amplicon. The amplicon can be of a different size in different patients.

Using `annotate` the probes for ERBB2 on the HgU95A chip were identified as `1802_s_at`, `1901_s_at`, and `33218_at`. A `pairs` plot for the log of the expression data for these three probes is given in Figure 2.1.

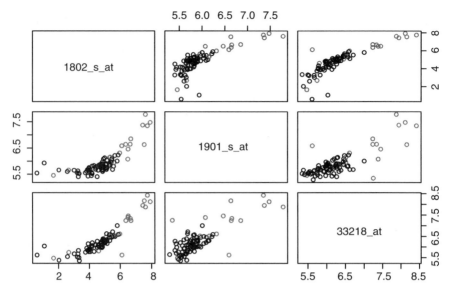

FIGURE 2.1. A pairs plot of the expression data for probes for the gene ERBB2.

We next locate ERBB2 using the data available in the annotate package. It is located (approximately) at position 41940228 on the plus strand of Chromosome 17. Thus, we next examine Chromosome 17 more closely. The first plot, Figure 2.2, is simply the cumulative expression of genes on Chromosome 17 by strand. In this case, the plot was not particularly informative, and a more detailed look at the region of the amplicon seems justified.

An examination of the relevant literature by Dr. A. Richardson implicated 32064_at, PPARBP; 37355_at, MLN64; and 1680_at, GRB7 as genes that might also be included in the amplified region, or *amplicon*. To better see their effects, we now only plot expression levels for that region of Chromosome 17. Once again, we use the alongChrom function. The plot region is restricted to the portion of Chromosome 17 between 41800000 and 42000000 bases. The cumulative expression (ERBB2-positive patients are colored red) is provided in Figure 2.3. Now we can easily see the high levels of expression of the genes (for ERBB2-positive patients) on both strands of the chromosome.

In these plots, we have chosen to plot the genes equally spaced. There are a great number of options available for alongChrom, and the reader is encouraged to explore them.

Another question that one might like to ask about these genes is whether the expression levels of different genes are correlated (across subjects). To explore this, one could look at pairs plots as we did above. However, when

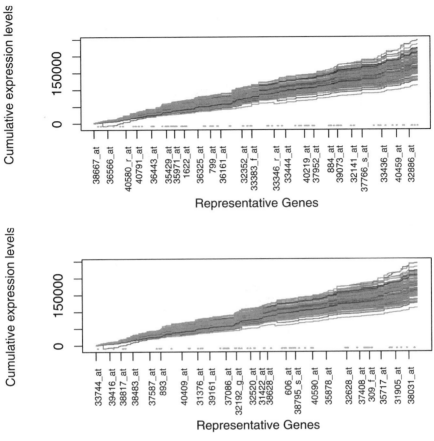

FIGURE 2.2. A plot of cumulative expression on Chromosome 17.

there are a large number of genes, this can become unwieldy. We have used the `plotcorr` function in the `ellipse` package of Murdoch (2002).

In Figures 2.4 and 2.5 we have provided these plots, one computed for each of the ERBB2-positive subset and the ERBB2-negative subset. These plots demonstrate a number of interesting features. (Note that here genes are included regardless of whether they are expressed. It might be useful to provide the expression level by coloring the ellipses.) These plots use pictographs to represent the correlation between the variables. They are symmetric.

We can see some interesting features. In both sets, we can see the clusters of highly correlated ERBB2 probe sets and TOP2A probe sets. We also see that for the ERBB2-positive patients, KIAA0130 also seems to be expressed (or at least correlated). Additionally, in this group there seems to be a high correlation between the TOP2A expression and that of SMARCE1. Further in the neighborhood of ERBB2, we can see positive correlation of ERBB2

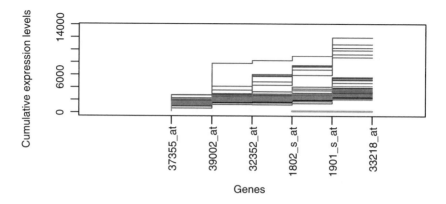

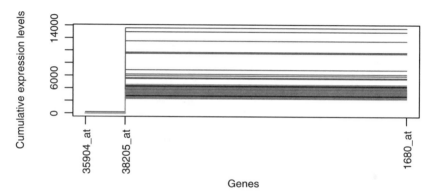

FIGURE 2.3. A plot of cumulative expression on Chromosome 17 in the region of ERBB2.

expression with GRB7 and MLN64. These relationships seem less strong in the ERBB2-negative patients.

What it means for one gene to be *near* or *close to* another gene is not yet well-understood. However, it does seem that we will want to consider concepts such as involvement in a particular pathway or process as a measure of distance as well as sharing common regulatory control as another means of finding genes that are *close to* one another. It is not entirely inconceivable that a change in regulation of one gene in these sets might be important.

Whole Genome Plots

Once chromosomal location is available for the dataset, it can be used to examine the locations of groups of genes of interest. One might be interested in whether genes group on a particular chromosome or whether they are

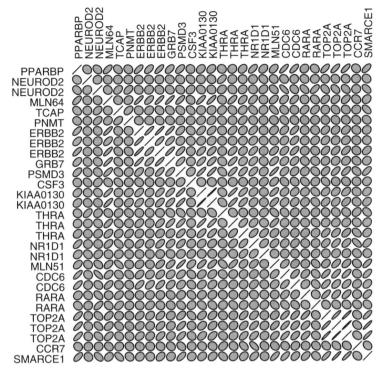

FIGURE 2.4. A plot of the correlation of expression between samples for the ERBB2-positive patients.

clustered near the ends of the chromosome or in other regions where there is some genomic instability.

We provide some tools to begin these explorations in cPlot and cColor. The first plots the location of all genes in a reference set for the genome of interest. The helper function cColor is then used to color locations differently.

Some form of interaction, such as brushing, would be extremely useful. We will be exploring ways of incorporating these plots into other software, such as GGobi (www.ggobi.org), to take advantage of the interactive facilities provided there.

Some examples of the potential use of this type of plot are:

- If gene X is of interest and we obtain data on the, say 100, ESTs with similar expression levels, then these could be colored to show where they are located.

- It is common practice to cluster genes according to expression levels subsequent to gene selection via filtering. The results of this process are

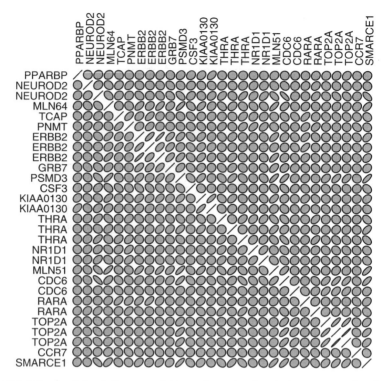

FIGURE 2.5. A plot of the correlation of expression between samples for the ERBB2-negative patients.

used to produce the usual *heat map* of genes by samples. An additional graphic showing the location of genes in different clusters may be useful.

2.6   Applications

2.6.1   A Case Study of Gene Filtering

The golubEsets package includes exprSets embodying the leukemia data reported in Golub et al. (1999). The data consist of two parts, a training set and a test set. We will show how one can perform various analyses on these data using some of the Bioconductor packages. We first load the necessary libraries and load the gene expression data objects (code not shown). For the following analyses, we will concentrate on filtering, annotation, and plotting.

The first task described in Golub et al. (1999) was the selection of the top 50 genes for predicting whether the sample was AML or ALL. We attempt a similar selection; however, rather than repeat their analysis, we will select the 50 genes with the smallest *p*-value according to a *t*-test. Note that this is a *joint filtering* exercise—retention of a gene depends on the *p*-value of other genes, so the `genefilter` package is not directly applicable. First, we construct a *t*-test builder or constructor function. It will carry out the *t*-test and return the *p*-value.

```
ttemp <- function (m, na.rm = TRUE)
{
  function(x) {
      if (na.rm) {
          drop <- is.na(x) | is.na(m)
          x <- x[!drop]
          m <- m[!drop]
      }
      t.test(x ~ m)$p.value
  }
}
tf2 <- ttemp(golubTrain$ALL.AML)
```

At this point `tf2` is a function that will carry out a *t*-test. The variable m is bound to the value golubTrain$ALL.AML.

```
pvals <- esApply(golubTrain, 1, tf2)
ord <- order(pvals)
gTr50 <- golubTrain[ord[1:50],]
```

We have the top 50 genes for discriminating between these two groups using a *t*-test. We can find out more about these genes using the `annotate` package.

```
hg68sym<- read.annotation("hgu68sym")
hg68ll <- read.annotation("hgu68ll")
syms50 <- multiget(geneNames(gTr50), hg68sym)
ll50 <- multiget(geneNames(gTr50), hg68ll)
ll.htmlpage(ll50, "GolubTop50",
          "Top 50 ESTs discriminating ALL and AML ",
          list(syms50, round(pvals[ord[1:50]], 5)))

hg68chrom <- read.annotation("hgu68chrom")
chrom50 <- multiget(geneNames(gTr50), hg68chrom)
table(unlist(chrom50))
```

```
#output
# 1 10 11 12 13 14 15 16 17 18 19 20 3 4 5 6 7 8 9 NA
# 8  2  1  3  2  3  1  2  4  1  6  1 2 2 2 4 3 1 1  1
```

The resulting Web page can be viewed (we have it permanently on the Bioconductor Web site at www.bioconductor.org/Docs/Papers/2002/

`Springer/GolubTop50`). Note that LocusLink values and symbols that could not be resolved are reported as `NA`.

We can now use these 50 genes to see how well we can classify patients in the test sample.

```
gTest50 <- golubTest[geneNames(gTr50),]
gTest50  library(class)
knn1 <- knn.cv(t(exprs(gTest50)), gTest50$ALL.AML, k=3)
table(knn1, gTest50$ALL.AML)

#knn1 ALL AML
# ALL 17   0
# AML  3  14

gTest.cent <- scale(t(exprs(gTest50)), center=FALSE)
knn2 <- knn.cv(gTest.cent, gTest50$ALL.AML, k=3) #$
table(knn2, gTest50$ALL.AML)
```

```
#knn2 ALL AML  # ALL 20  0  # AML  0 14
```

Notice that centering and scaling the genes improved the prediction on the test cases. There is some reason to believe that scaling (centering makes no difference) all genes to have variance one is generally desirable. Carrying out this part of the analysis was particularly simple due to the richness of the available software for R.

Concentrating on a Single EST

We now consider a single gene in more detail and show how the use of some additional tools together with some visualization techniques provides some insight. In our analysis, reported above, one of the genes that was important in discriminating between the two classes of leukemia was the gene HOXA9, homeo box A9. Golub et al. (1999) also reported that this gene was related to clinical outcome and found that it was overexpressed in AML patients with treatment failure. The Affymetrix identifier associated with this gene is `U82759_at`. We can examine its pattern of expression or select other genes that have patterns of expression that are highly correlated with the pattern of expression of HOXA9.

```
gTrHox <- golubTrain["U82759_at",]
gTeHox <- golubTest["U82759_at",]
```

These data are plotted in Figure 2.6, where we can see that there is some indication that HOXA9 has a higher level of expression in the samples from AML patients than in the ALL patients for both the test and the training samples.

The function `genefinder` can be used to find ESTs that have patterns of expression that are similar to that of HOXA9.

ALL/AML by Train and Test

FIGURE 2.6. A pairs plot of the expression data for probes for the gene ERBB2.

```
gStr <- genefinder(golubTrain, "U82759_at", num=100)
gSte <- genefinder(golubTest, "U82759_at", num=100)

sum(gSte[[1]]$indices %in% gStr[[1]]$indices)
# 20


##could just look at AML
gTrAML <- golubTrain[,golubTrain$ALL.AML=="AML"] #$
gTeAML <- golubTest[,golubTest$ALL.AML=="AML"] #$
gStrAML <- genefinder(exprs(gTrAML), "U82759_at", num=100)
gSteAML <- genefinder(exprs(gTeAML), "U82759_at", num=100)


##now how many in common?
sum(gStrAML[[1]]$indices %in% gSteAML[[1]]$indices)
#9
##what if we compare to gSte
sum(gSteAML[[1]]$indices %in% gSte[[1]]$indices)
#33
```

2.6.2 Application of Expression Density Diagnostics

In this example, we are concerned with understanding the diversity of expression distributions presented in two strata of 89 women with breast cancer. The women were clinically classified into 47 cases of metastasis (lymph node positive) and 42 controls (lymph node negative). Although our ultimate aim is to identify features of gene expression that are predictive of avoidance of metastasis, in this example we are concerned with the diversity of gene-specific distributions within strata.

The 89 U95v2 array results are stored in the exprSet BCes. This was filtered to eliminate those genes for which at least 23 of the samples had a measurement above 100. There were 7297 genes remaining:

```
load("BCes.rda")

kF <- kOverA(23, 100)
ff <- filterfun(kF)
wh <- genefilter(exprs(BCes), ff)

nGenes <- sum(wh)
BCok <- BCes[wh,]
```

We can view a summary of the BCok dataset.

```
Expression Set (exprSet) with
        7297 genes
          89 samples
               phenoData object with 3 variables and 89 cases
            varLabels
      Chip: chip number
      lymph.nodes: indicates whether metastasis was detected
                   in the lymph nodes
      HER2: her2 status, p-positive, lp-low positive,
                                    n-negative
```

Two exprSets were then extracted corresponding to cases and controls.

```
CasES <- BCok[ ,BCok$lymph=="positive"]
ConES <- BCok[ ,BCok$lymph=="negative"]
```

The function edd.unsupervised was applied to each set, using the
"nnet" shape-matching method. An example call for the metastatic samples is

```
library(edd)
set.seed(12345)
# needed because nnet has random initialization
CasNN <- edd.unsupervised(CasES, "nnet")
ConNN <- edd.unsupervised(ConES, "nnet")
```

Running with both metastatic and nonmetastatic samples, we find

```
   table(CasNN)
CasNN
 b28 b82 csq1   ln mix1 mix2  n01   t3   u
2682  70   55 1059  396   38 1422 1421 154
   table(ConNN)
ConNN
 b28 b82 csq1   ln mix1 mix2  n01   t3   u
2633 140  142 1505  398   33 1109 1155 182
```

Here, the reference catalog of distributional shapes consists of $\beta(2,8)$, $\beta(8,2)$, χ_1^2, standard lognormal, $.75N(0,1) + .25N(4,1)$ (mix1), $.25N(0,1)$ $+ .75N(4,1)$ (mix2), $N(0,1)$, t_3, and $U(0,1)$. We see that the majority of genes in both cases and controls are found to have a shape resembling $\beta(2,8)$, which is a right-skewed distribution with compact support. The prevalence of right-skewness (present in the $\beta(2,8)$ along with the χ_1^2 and lognormal shapes) is consistent with the prevailing tendency to apply log transformation. However, we note that there are nontrivial numbers of genes with distributional shape matching $N(0,1)$, thus not requiring transformation, and genes with distributions shaped like clearly multimodal mixtures. The joint classification of gene-specific distributional shapes can be tabulated easily:

```
table(CasNN,ConNN)
      ConNN
CasNN   b28 b82 csq1  ln mix1 mix2 n01  t3  u
  b28  1127  26   42 650  175    5 275 316 66
  b82    15   7    0   2    3    0  25  13  5
  csq1    6   0   22  23    2    0   0   2  0
  ln    381   3   44 425   50    0  53  94  9
  mix1  158   5    9  75   24    2  54  59 10
  mix2    7   5    0   3    0    1  14   8  0
  n01   462  53   10 103   74   14 342 315 49
  t3    417  36   14 217   61    9 311 321 35
  u      60   5    1   7    9    2  35  27  8
```

We note from this table that there are 53 genes for which the controls present an expression distribution having the shape of a standard Gaussian, while the cases present expression distribution with lognormal shape. Additionally, from the summary above, there are 434 genes for which the cases appear to have a mixture distribution.

Figure 2.7 gives substance to these distinctions by providing gene-specific density estimates for all genes in these strata defined by distributional shape pattern matching. All gene distributions were transformed on a gene-specific, stratum-specific basis to median zero and unit MAD prior to density estimation. The top panel is the set of density estimates for those genes classified as shape mix1 among cases. There is a clear tendency to present a second mode at about $x = 2$ MADs from the median. Genes for which the stratum-specific distribution appears to be multimodal may have particular interest in that they may help unearth new diagnostic categories or may be the basis for screening procedures (Pepe et al., 2001).

The middle and bottom panels of Figure 2.7 contain density estimates for the 34 genes on which controls were found to have approximately Gaussian distributions; while cases were found to have approximately lognormal shape. In general, simple t-tests to contrast case and control expression distributions in this group of genes will be suboptimal.

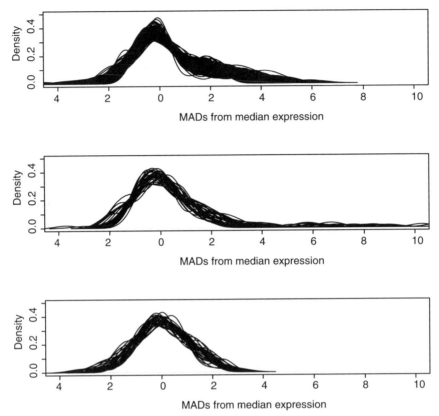

FIGURE 2.7. Superimposed default kernel density estimates for all genes possessing common distributional shape according to the edd procedure. Top panel: 396 genes from metastatic patients classified as mix1. There were 53 genes for which cases were found to have lognormal shape and controls were found to have Gaussian shape. The middle (resp. bottom) panel provides density estimates for these genes as measured on cases (resp. controls).

This application of expression density diagnostics is primarily conceptual. It exposes the existence and form of diversity of expression distributions within and across strata and genes. Work is in progress on using the results of expression diversity studies to guide the choice of discriminatory tests to increase the power of gene discovery exercises.

2.7 Conclusions

The Bioconductor project endeavors to enrich R, an interactive statistical computing and graphics environment, so that it may serve as a resource of very broad utility to bioinformatics. In this chapter, we have focused on three processes in the analysis of expression array data: filtering from very large collections of genes to more manageable sets and then annotating and visualizing the resultant data.

Filtering is supported by the adoption of a simple and extensible data structure for the collection of multiple microarrays and associated phenotypic data and by the creation of a highly flexible interface that permits concise programming of arbitrarily complex statistical predicates to define filter behavior. A tool (expression density diagnostics) that helps guide the appropriate choice of filter by characterizing gene-expression distributions has also been reviewed.

Annotation is supported in two independent ways. First, images of biological annotation data have been introduced into R to permit use of various vocabularies to identify and interpret genomic phenomena. These images may be navigated interactively or programmatically and are conveniently updated as required. Second, functions that mine the World Wide Web interactively for up-to-the-minute interpretation of detailed analytical findings have been introduced to R. Several graphical tools that help add pictorial guidance to experiment interpretation have been discussed. These graphical modules interoperate with the annotation infrastructure so that pictures are as biologically informative as possible.

Efficient progress in bioinformatics requires that barriers to entry be lowered for both statisticians and biologists. Statisticians must be able to move conveniently, with clear documentation and a rich stock of examples, into the vocabulary and experimental frameworks of high-throughput genomics. Biologists must be able to understand and apply statistically sound protocols for experimental design and inference. Again, documentation and examples are a central resource for supporting such cross-disciplinary connections. Our objective in the Bioconductor project is to provide an open-source, integrated platform that is useful to both statisticians and biologists. The efforts of both the Bioconductor and R core developer groups are gratefully acknowledged as central to the pursuit of this objective.

Acknowledgments. Robert Gentleman's work is supported by NIH/NCI Grant 2P30 CA06516-38, by the Dana-Farber/Harvard SPORE in Breast Cancer from the NCI and by the High Tech Industry Multidisciplinary Research Fund at the Dana-Farber Cancer Institute. Vincent Carey's work was supported in part by U.S. NHLBI Grant HL66795, Innate Immunity in Heart, Lung, and Blood Disease.

We would like to thank the members of R core and Bioconductor core for providing help, discussion, and interesting, useful code. We would like to thank Drs. A. Richardson, J. D. Iglehart, and S. Chiaretti for helpful discussions on the biological aspects of modeling genomic data.

References

Brazma A, Hingamp P, Quackenbush J (2001). Minimum information about a microarray experiment: Towards standards for microarray data. *Nature Genetics*, 29:365–371.

Chambers JM (1998). *Programming with Data: A Guide to the S Language.* Springer-Verlag New York.

Eisen MB, Spellman PT, Brownand PO, Botstein D (1998). Cluster analysis and display of genome-wide expression patterns. *Proceedings of the National Academy of Sciences*, 95(25):14863–14868.

Gentleman R, Ihaka R (2000). Lexical scope and statistical computing. *Journal of Computational and Graphical Statistics*, 9:491–508.

Golub TR, Slonim DK, Tamayo P, Huard C, Gaasenbeek M, Mesirov JP, Coller H, Loh M, Downing JR, Caligiuri MA, Bloomfield CD, Lander ES (1999). Molecular classification of cancer: Class discovery and class prediction by gene expression monitoring. *Science*, 286:531–537.

Helmreich, EJM (2001). *The Biochemistry of Cell Signalling.* Oxford University Press: Oxford.

Murdoch D (2002). *ellipse: Functions for drawing ellipses and ellipse-like confidence regions.* http://cran.r-project.org/.

Pepe MS, Longton G, Anderson G, Schummer M (2001). Selecting differentially expressed genes from microarray experiments. Technical report, University of Washington: Seattle.

Shalon D, Smith SJ, Brown PO (1996). A DNA microarray system for analyzing complex DNA samples using two-color fluorescent probe hybridization. *Genome Research*, 6(7):639–645.

Zhang X, Derdeyn C, Gentleman R, Leykin I, Monti S, Ramaswamy S, Wong WH, Golub TR, Iglehart JD, Richardson AL (2002). Molecular determinants of lymph node metastasis in breast cancer. Submitted for Publication.

3

Bioconductor R Packages for Exploratory Analysis and Normalization of cDNA Microarray Data

SANDRINE DUDOIT
JEAN YEE HWA YANG

Abstract

This chapter describes a collection of four R packages for exploratory analysis and normalization of two-color cDNA microarray fluorescence intensity data. R's object-oriented class/method mechanism is exploited to allow efficient and systematic representation and manipulation of large microarray datasets of multiple types. The `marrayClasses` package contains class definitions and associated methods for pre- and postnormalization intensity data for batches of arrays. The `marrayInput` package provides functions and `tcltk` widgets to automate data input and the creation of microarray-specific R objects for storing these data. Functions for diagnostic plots of microarray spot statistics, such as boxplots, scatterplots, and spatial color images, are provided in `marrayPlots`. Finally, the `marrayNorm` package implements robust adaptive location and scale normalization procedures, which correct for different types of dye biases (e.g., intensity, spatial, plate biases) and allow the use of control sequences spotted onto the array and possibly spiked into the mRNA samples. The four new packages were developed as part of the Bioconductor project, which aims more generally to produce an open-source and open-development statistical computing framework for the analysis of genomic data.

3.1 Introduction

3.1.1 Overview of Packages

Microarray experiments generate large and complex multivariate datasets. Careful statistical design and analysis are essential to improve the efficiency and reliability of microarray experiments from the early design and preprocessing stages to higher-level analyses. Access to an efficient and well-designed statistical computing environment is a related and equally critical aspect of the analysis of gene expression data. This chapter describes a collection of four R (Ihaka and Gentleman, 1996) packages for exploratory analysis and normalization of two-color cDNA microarray fluorescence intensity data. An earlier version of these packages can be found in the sma package, which was written in the Fall of 2000. The four new packages were developed as part of the Bioconductor project, which aims more generally to produce an open-source and open-development statistical computing framework for the analysis of genomic data (www.bioconductor.org). Particular emphasis is placed on facilitating access to the powerful statistical methodology implemented in the R language and on enhancing its effectiveness within a biological context. Like most Bioconductor packages, these four packages rely on R's *object-oriented class/method mechanism* (John Chambers' methods package) to allow efficient and systematic representation and manipulation of large microarray datasets of multiple types. Efforts to reduce the barrier of entry into R include providing *widgets* (i.e., small-scale graphical interfaces) for data input and basic analysis procedures. A brief description of the four marray packages is given next.

marrayClasses. This package contains class definitions and associated methods for pre- and postnormalization intensity data for batches of arrays. Methods are provided for the creation and modification of microarray objects, basic computations, printing, subsetting, and class conversions.

marrayInput. This package provides functionality for reading microarray data into R, such as intensity data from image-processing output files (e.g., .spot and .gpr files for the Spot and GenePix packages, respectively) and textual information on probes and targets (e.g., from .gal files and god lists). tcltk widgets are supplied to facilitate and automate data input and the creation of microarray-specific R objects for storing these data.

marrayPlots. This package provides functions for diagnostic plots of microarray spot statistics, such as boxplots, scatterplots, and spatial color images. Examination of diagnostic plots of intensity data is important in order to identify printing, hybridization, and scanning artifacts that can lead to biased inferences concerning gene expression.

marrayNorm. This package implements robust adaptive location and scale normalization procedures, which correct for different types of dye biases (e.g., intensity, spatial, plate biases) and allow the use of control sequences spotted onto the array and possibly spiked into the mRNA samples. Normalization is needed to ensure that observed differences in intensities are indeed due to differential expression and not experimental artifacts; fluorescence intensities should therefore be normalized before any analysis that involves comparisons among gene expression measures within or between arrays.

Bioconductor packages are distributed under an open-source license, such as GPL or LGPL, and may be downloaded from the project Web site, www.bioconductor.org, for Linux, Unix, MS Windows, and Mac OSX operating systems. Sources, binaries, and documentation for R and other R packages can be obtained from the "Comprehensive R Archive Network" (CRAN, cran.r-project.org/). As with any other Bioconductor R package, detailed information on the functions and their arguments and values can be obtained in the help files. For instance, to view the help file for the function maNorm in a browser, use help.start() followed by help(maNorm) or ?maNorm. In addition, each Bioconductor package contains step-by-step tutorials in the /doc subdirectory. These tutorials are generated using the Sweave function from the R 1.6.1 package tools (Leisch, 2002). They form integrated statistical documents intermixing text, R code, and code output (numerical, textual, and graphical). Within this framework, documents can be updated automatically whenever data or analyses are modified. The present document was also generated using Sweave; the .Rnw source file is available at www.stat.berkeley.edu/~sandrine/. In addition, a demonstration script is provided for the marrayPlots package in the /demo subdirectory and can be run using demo(marrayPlots).

The chapter is organized as follows. The remainder of this section gives background on two-color cDNA microarray experiments. Section 3.2 provides an overview of the statistical and computational methodology for exploratory analysis and normalization of cDNA microarray data. Section 3.3 describes the Swirl microarray experiment, which serves as a case study for illustrating the statistical methods and software implementation. Section 3.4 discusses the four R packages, marrayClasses, marrayInput, marrayPlots, and marrayNorm, in greater detail and provides a demonstration of package functionality using the swirl dataset. Finally, Section 3.5 summarizes our findings and outlines ongoing efforts as part of the Bioconductor project.

3.1.2 Two-Color cDNA Microarray Experiments

DNA microarrays consist of thousands of individual DNA sequences printed in a high-density array on a glass microscope slide using a robotic printer, or *arrayer*. The *relative abundance* of these spotted DNA sequences in two DNA or RNA samples may be assessed by monitoring the *differential hybridization* of the two samples to the sequences on the array. For mRNA samples, the two samples, or *targets*, are reverse-transcribed into cDNA, labeled using different fluorescent dyes (usually a red-fluorescent dye, Cyanine 5 or Cy5, and a green-fluorescent dye, Cyanine 3 or Cy3), then mixed in equal proportions and hybridized with the arrayed DNA sequences, or *probes* (following the definition of probe and target adopted in "The Chipping Forecast," a January 1999 supplement to *Nature Genetics*). After this competitive hybridization, the slides are imaged using a *scanner*, and fluorescence measurements are made separately for each dye at each spot on the array. The ratio of the red and green fluorescence intensities for each spot is indicative of the relative abundance of the corresponding DNA probe in the two nucleic acid target samples. See Brown and Botstein (1999) and Schena (2000), for a more detailed introduction to the biology and technology of cDNA microarrays.

The term *array layout* refers to the layout of DNA probe sequences on the array as determined by the printing process. In general, probe sequences are spotted on a glass microscope slide using an arrayer that has an $ngr \times ngc$ *print-head*; that is, a regular array of ngr rows and ngc columns of *print-tips*, or pins. The resulting microarrays are thus partitioned into an $ngr \times ngc$ *grid matrix*. The terms *grid, sector, pin-group*, and *print-tip-group* are used interchangeably in the microarray literature. Each grid consists of an $nsr \times nsc$ *spot matrix* that was printed with a single print-tip. DNA probes are usually printed sequentially from a collection of 384-well plates (or 96-well plates); thus, in some sense, plates are proxies for time of printing. In addition, a number of control probe sequences may be spotted on the array for normalization or other calibration purposes. The term *array batch* is used to refer to a collection of arrays with the same layout.

The *raw data* from a microarray experiment are the *image files* produced by the scanner; these are typically pairs of 16-bit tagged image file format (TIFF) files, one for each fluorescent dye (images usually range in size from a few megabytes (MB) to 10 or 20 MB for high-resolution scans). Image analysis is required to extract *foreground* and *background fluorescence intensity* measurements for each spotted DNA sequence. Image processing is beyond the scope of this chapter; the reader is referred to Yang et al. (2002a) for a detailed discussion of microarray image analysis, a description of the image-processing R package `Spot`, and additional references.

3.2 Methods

3.2.1 Standards for Microarray Data

There is still much debate regarding standards for storing and reporting microarray-based gene expression data. Significant progress toward the definition of such standards is found in the Minimum Information About a Microarray Experiment (MIAME) documents produced by the Microarray Gene Expression Database (MGED) group (Brazma et al., 2001, `www.mged.org/Workgroups/MIAME/miame.html`). The MIAME documents focus on the content and structure of the necessary information rather than the technical format for representing and storing the data. The standards apply to different microarray platforms such as cDNA spotted microarrays and Affymetrix oligonucleotide chips.

A description of a microarray experiment should contain information about the genes whose expression has been measured (*gene annotation*) and about the nature and preparation of the target samples hybridized to the arrays (*sample annotation*) in addition to the quantitative gene expression measurements. At least three levels of expression data are relevant: (i) the microarray scanned images, or raw data; (ii) the microarray image quantification data (i.e., output files from image analysis software packages); and (iii) the matrix of derived gene expression measures, where rows correspond to spots and columns to target samples. Reliability information and a detailed description of how the expression values were obtained should also be stored (e.g., record of image analysis, normalization, and quality-based filtering procedures).

Here, we begin our analysis of microarray data with the output files of image-processing packages such as `GenePix` or `Spot` (intermediate expression data, level (ii) above). In what follows, red and green background intensities for a given spot are denoted by R_b and G_b, respectively, and red and green foreground intensities by R_f and G_f, respectively. Background-corrected red and green fluorescence intensities are denoted by $R = (R_f - R_b)$ and $G = (G_f - G_b)$, respectively, and M denotes the corresponding base 2 log ratio, $M = \log_2 R/G$. We use R's object-oriented class/method mechanism for representation and manipulation of microarray data. Microarray-specific object classes were defined as described next to keep track of the three main types of microarray data (gene annotation, sample annotation, and expression measures) at different stages of the analysis process.

3.2.2 Object-Oriented Programming: Microarray Classes and Methods

Microarray experiments generate large and complex multivariate datasets, which contain textual information on probe sequences (e.g., gene identifiers, annotation, layout parameters) and mRNA target samples (e.g., description of samples, protocols, and hybridization and scanning conditions) in addition to the primary fluorescence intensity data. Efficient and coordinated access to these various types of data is an important aspect of computing with microarray data. The `marray` packages rely on the *class/method mechanism* in John Chambers' `methods` package which provides tools for *object-oriented programming* in R. To facilitate the management of microarray data at different stages of the analysis process, a collection of microarray-specific data structures, or *classes*, were defined (see also Gentleman and Carey (Chapter 2, this volume) for a discussion of the Bioconductor packages `Biobase` and `annotate`, which provide basic class definitions and methods for microarray and annotation data). Broadly speaking, classes reflect how we think of certain objects and what information these objects should contain. Classes are defined in terms of *slots* that contain the relevant data for the application at hand. *Methods* define how a particular function should behave depending on the class of its arguments and allow computations to be adapted to particular classes. For example, a microarray object should contain intensity data as well as information on the probe sequences spotted on the array and the target samples hybridized to it. Useful methods for microarray classes include specializations of printing, subsetting, and plotting functions for the types of data represented by these classes. The use of classes and methods greatly reduces the complexity of handling large and varied datasets associated with microarray experiments by automatically coordinating different sources of information.

3.2.3 Diagnostic Plots

Before proceeding to normalization or any higher-level analysis, it is instructive to look at diagnostic plots of spot statistics, such as red and green foreground and background log intensities, intensity log ratios, and quality measures (e.g., spot area). Stratifying spot statistics according to layout parameters such as print-tip or plate is useful for the purpose of identifying printing, hybridization, and scanning artifacts, as demonstrated in Subsection 3.4.3.

2D spatial images. In a *2D spatial image*, shades of gray or colors are used to represent the values of a statistic for each spot on the array. Each rectangle in the grid corresponds to a particular spot, and its coordinates reflect the location of the spot on the array. The statistic can be the intensity log ratio M, a spot quality measure (e.g., spot size or shape), or a test statistic. These pseudo-images may be used to explore spatial biases in the values of a particular statistic due to, for example, print-tip or cover-slip effects.

Boxplots. *Boxplots*, also called *box-and-whisker plots*, were first proposed by John Tukey in 1977 as simple graphical summaries of the distribution of a variable. The summary consists of the median, the upper and lower quartiles, the range, and, possibly, individual extreme values. The central box in the plot represents the *inter-quartile range* (IQR), which is defined as the difference between the *upper quartile* and *lower quartile* (i.e., the difference between the 75th and 25th percentiles). The line in the middle of the box represents the *median* or 50th percentile, a measure of central location of the data. Extreme values greater than 1.5 IQR above the 75th percentile and less than 1.5 IQR below the 25th percentile are usually plotted individually.

Scatterplots. Single-slide expression data are typically displayed by plotting the log intensity $\log_2 R$ in the red channel versus the log intensity $\log_2 G$ in the green channel. Such plots tend to give an unrealistic sense of concordance between the red and green intensities and can mask interesting features of the data. We thus prefer to plot the intensity log ratio $M = \log_2 R/G = \log_2 R - \log_2 G$ vs. the mean log intensity $A = \log_2 \sqrt{RG} = (\log_2 R + \log_2 G)/2$. An *MA-plot* amounts to a 45° counterclockwise rotation of the $(\log_2 G, \log_2 R)$-coordinate system followed by scaling of the coordinates. It is thus another representation of the (R, G) data in terms of the log ratios M, which directly measure differences between the red and green channels and are the quantities of interest to most investigators. We have found MA-plots to be more revealing than their $\log_2 R$ vs. $\log_2 G$ counterparts in terms of identifying spot artifacts and for normalization purposes (Yang et al., 2001; Dudoit et al., 2002; Yang et al., 2002b). Applications of MA-plots are also discussed in Irizarry et al. (Chapter 4, this volume), Lee and O'Connell (Chapter 7, this volume), Colantuoni et al. (Chapter 9, this volume), and Wu et al. (Chapter 14, this volume).

3.2.4   Normalization Using Robust Local Regression

The purpose of *normalization* is to identify and remove the effects of systematic variation, other than differential expression, in the measured fluorescence intensities (e.g., different labeling efficiencies and scanning properties of the Cy3 and Cy5 dyes; different scanning parameters, such as PMT (photomultiplier tube) settings; print-tip, spatial, or plate effects). It is necessary to normalize the fluorescence intensities before any analysis that involves comparing expression measures within or between slides (e.g., classification, multiple testing) in order to ensure that differences in intensities are indeed due to differential expression and not experimental artifacts. The need for normalization can be seen most clearly in *self-self experiments*, in which two identical mRNA samples are labeled with different dyes and hybridized to the same slide (Dudoit et al., 2002). Although there is no differential expression and one expects the red and green intensities to be equal, the red intensities often tend to be lower than the green intensities. Furthermore, the imbalance in the red and green intensities is usually not constant across the spots within and between arrays and can vary according to overall spot intensity A, location on the array, plate origin, and possibly other variables.

Location normalization. We have developed *location normalization* methods that correct for intensity, spatial, and other dye biases using *robust locally-weighted regression* (Cleveland, 1979; Cleveland and Devlin, 1988; Yang et al., 2001; Yang et al., 2002b). Local regression is a *smoothing* method for summarizing multivariate data using general curves and surfaces. The smoothing is achieved by fitting a polynomial function of the predictor variables *locally* to the data in a fashion that is analogous to computing a moving average. *Robust* fitting guards against deviant points distorting the smoothed points. For the *lowess* and *loess* procedures, polynomials are fitted locally using iterated weighted least squares. In the context of microarray experiments, robust local regression allows us to capture the nonlinear dependence of the intensity log ratio $M = \log_2 R/G$ on the overall intensity $A = \log_2 \sqrt{RG}$ while ensuring that the computed normalization values are not driven by a small number of differentially expressed genes with extreme log ratios. The `marrayNorm` and `marrayPlots` packages rely on the R `loess` (`modreg` package) and `lowess` functions; greater details can be found in the help files for these functions.

Scale normalization. For *scale normalization*, a robust estimate of scale, such as the *median absolute deviation* (MAD), may be used (Yang et al., 2001; Yang et al., 2002b). For a collection of numbers, $x_1, \ldots, x_n$, the MAD is the median of their absolute deviations from the median $m = \text{median}\{x_1, \ldots, x_n\}$; that is, $\text{MAD} = \text{median}\{|x_1 - m|, \ldots, |x_n - m|\}$. The R function for MAD is `mad`.

Location and scale normalized intensity log ratios M_{norm} are given by

$$M_{norm} = \frac{M - l}{s},$$

where l and s denote the location and scale normalization values, respectively. The location value l can be obtained, for example, by robust local regression of M on A within print-tip-groups. The scale value s could be the within-print-tip-group MAD of location normalized log ratios. Note that related approaches to normalization are implemented in the software package `SNOMAD`, described in Colantuoni et al. (Chapter 9, this volume) and available at `pevsnerlab.kennedykrieger.org/snomadinput.html`.

3.3   Application: Swirl Microarray Experiment

We demonstrate the functionality of the four `marray` R packages using gene expression data from the Swirl experiment. These data were provided by Katrin Wuennenberg-Stapleton from the Ngai Lab at the University of California, Berkeley. (The swirl embryos for this experiment were provided by David Kimelman and David Raible at the University of Washington.) This experiment was carried out using zebrafish (*Brachydanio rerio*) as a model organism to study early development in vertebrates. Swirl is a point mutant in the BMP2 gene that affects the dorsal/ventral body axis. Ventral fates such as blood are reduced, whereas dorsal structures such as somites and the notochord are expanded. A goal of the Swirl experiment is to identify genes with altered expression in the swirl mutant compared to wild-type zebrafish. A total of four hybridizations were performed with dye-swap. For each of these hybridizations, target cDNA from the swirl mutant was labeled using one of the Cy3 or Cy5 dyes, and wild-type target cDNA was labeled using the other dye. Target cDNA was hybridized to microarrays containing 8,448 probes, including 768 control spots (negative, positive, and normalization controls). Microarrays were printed using 4×4 print-tips and are thus partitioned into a 4×4 grid matrix. Each grid consists of a 22×24 spot matrix that was printed with a single print-tip. Here, spot row and plate coordinates coincide, as each row of spots corresponds to probe sequences from the same 384-well plate ($384 = 16 \times 24$).

Each of the four hybridizations produced a pair of 16-bit images, which were processed using the image analysis software package Spot (Buckley, 2000; Yang et al., 2002a). Raw images of the Cy3 and Cy5 fluorescence intensities for all four hybridizations are available at fgl.lsa.berkeley.edu/ Swirl/index.html. The swirl dataset in the marryinput package includes four output files, swirl.1.spot, swirl.2.spot, swirl.3.spot, and swirl.4.spot, from the Spot package. Each of these files contains 8,448 rows and 30 columns; rows correspond to spots and columns to different statistics from the Spot image analysis output. The file fish.gal, a Gene Array List file (or .gal file) for the GenePix package, was generated by the program GalFileMaker, version 1.2 (www.microarrays.org/ software.html). It contains information on individual probe sequences, such as gene names, spot IDs, and spot coordinates. Hybridization information for the mutant and wild-type target samples is stored in SwirlSample.txt. The function data is used to load the swirl dataset into R. To view a description of the experiments and dataset, type ?swirl.

```
> data(swirl)
```

3.4 Software

3.4.1 Package *marrayClasses*—Classes and Methods for cDNA Microarray Data

The following microarray *classes* were defined to represent pre-and post-normalization fluorescence intensity data and data on probes and targets for batches of arrays. Here, a batch of arrays refers to a set of arrays with the same layout, as described in Subsection 3.1.2.

Microarray Classes

Class marrayLayout. Keeping track of array layout information is essential for quality assessment of fluorescent intensity data and for normalization purposes. Important layout parameters are the dimensions of the spot and grid matrices and, for each probe on the array, its grid matrix and spot matrix coordinates. In addition, it is useful to keep track of gene identifiers, the plate origin of the probes, and information on the spotted control sequences (e.g., probe sequences that should have equal abundance in the two target samples, such as housekeeping genes). The class marrayLayout was designed to keep track of these various layout parameters and contains the following slots (the classes of the slots are listed below the slot names).

```
> getClassDef("marrayLayout")
```

Slots:

Name:	maNgr	maNgc	maNsr	maNsc
Class:	numeric	numeric	numeric	numeric

Name:	maNspots	maSub	maPlate	maControls
Class:	numeric	logical	factor	factor

Name:	maNotes
Class:	character

Here, maNgr and maNgc store the dimensions of the grid matrix, maNsr and maNsc store the dimensions of the spot matrices, naNspots refers to the total number of spots on the array, maSub keeps track of the subset of spots currently being considered, maPlate is a vector of plate labels, maControls is a vector of spot control labels, and maNotes can be used to store any character string describing the array. In addition, a number of methods were defined to compute other important layout parameters, such as print-tip, grid matrix, and spot matrix coordinates: maPrintTip, maGridRow, maGridCol, maSpotRow, and maSpotCol (see discussion below). No slots were defined for these quantities for memory management reasons. For details on slots and methods associated with the marrayLayout class, type ?marrayLayout.

Class marrayInfo. Information on the target mRNA samples co-hybridized to the arrays is stored in objects of class marrayInfo. Such objects may include the names of the arrays, the names of the Cy3- and Cy5-labeled samples, notes on the hybridization and scanning conditions, and other textual information. Descriptions of the spotted probe sequences (e.g., gene identifiers, annotation, notes on printing conditions) are also stored in objects of class marrayInfo. The marrayInfo class is not specific to the microarray context and has the following definition:

```
> getClassDef("marrayInfo")
```

Slots:

Name:	maLabels	maInfo	maNotes
Class:	character	data.frame	character

Class marrayRaw. Prenormalization intensity data for a batch of arrays are stored in objects of class marrayRaw, which contain slots for the matrices of Cy3 and Cy5 background and foreground intensities (maGb, maRb, maGf, maRf), spot quality weights (maW), layout parameters of the arrays (maLayout), descriptions of the probes spotted onto the arrays (maGnames), and mRNA target samples hybridized to the arrays (maTargets).

```
> getClassDef("marrayRaw")

Slots:
```

Name:	maRf	maGf	maRb
Class:	matrix	matrix	matrix

Name:	maGb	maW	maLayout
Class:	matrix	matrix	marrayLayout

Name:	maGnames	maTargets	maNotes
Class:	marrayInfo	marrayInfo	character

Class marrayNorm. Postnormalization intensity data are stored in similar objects of class marrayNorm. These objects store the normalized intensity log ratios maM, the location and scale normalization values (maMloc and maMscale), and the average log intensities (maA). In addition, the marrayNorm class has a slot for the function call used to normalize the data, maNormCall. For more details on the creation of normalized microarray objects, the reader is referred to Subsection 3.4.4 below.

```
> getClassDef("marrayNorm")

Slots:
```

Name:	maA	maM	maMloc
Class:	matrix	matrix	matrix

Name:	maMscale	maW	maLayout
Class:	matrix	matrix	marrayLayout

Name:	maGnames	maTargets	maNotes
Class:	marrayInfo	marrayInfo	character

Name:	maNormCall
Class:	call

Class marraySpots. This class is used to store information on the spotted probe sequences for a batch of arrays. The class contains slots for the layout of the arrays, maLayout, and a description of the probe sequences spotted onto the arrays, maGnames.

```
> getClassDef("marraySpots")

Slots:
```

Name:	maGnames	maLayout
Class:	marrayInfo	marrayLayout

Class marrayTwo. The `marrayTwo` class can be viewed as a leaner version of the `marrayRaw` and `marrayNorm` classes. It contains slots for only two types of spot statistics (`maX` and `maY`), the layout of the arrays (`maLayout`), and a description of the target samples hybridized to the arrays (`maTargets`). The two spot statistics can be, for example, the unnormalized green and red foreground intensities or the normalized log ratios M and average log intensities $A = \log_2 \sqrt{RG}$.

```
> getClassDef("marrayTwo")
```

```
Slots:
```

```
Name:           maX     maY     maLayout
Class:          matrix  matrix  marrayLayout
```

```
Name:    maTargets
Class:   marrayInfo
```

Most microarray objects contain an `maNotes` slot, which can be used to store any string of characters describing the experiments, such as notes on the printing, hybridization, or scanning conditions.

Creating and Accessing Slots of Microarray Objects

Creating new objects. The function `new` from the `methods` package may be used to create new objects from a given class. For example, to create an instance of the class `marrayInfo` describing the target samples in the Swirl experiment, one could use the following code:

```
> zebra.RG <- as.data.frame(cbind(c("swirl",
+       "WT", "swirl", "WT"), c("WT", "swirl",
+       "WT", "swirl")))
> dimnames(zebra.RG)[[2]] <- c("Cy3", "Cy5")

> zebra.samples <- new("marrayInfo", maLabels = paste("Swirl
+       array ", 1:4, sep = ""), maInfo = zebra.RG, maNotes =
            "Description of targets for Swirl experiment")
> zebra.samples
```

```
Object of class marrayInfo.

        maLabels   Cy3   Cy5
1 Swirl array 1 swirl    WT
2 Swirl array 2    WT swirl
3 Swirl array 3 swirl    WT
4 Swirl array 4    WT swirl
```

```
Number of labels: 4
Dimensions of maInfo matrix:    4   rows by   2   columns
```

```
Notes:
Description of targets for Swirl experiment
```

Slots that are not specified in **new** are initialized to the *prototype* for the corresponding class. This is usually an "empty" object (e.g., `matrix(0,0,0)`). In most cases, microarray objects can be created automatically using the input functions and their corresponding widgets in the `marrayInput` package (Subsection 3.4.2). These were used to create the object `swirl` of class `marrayRaw`.

Accessing slots. Different components or slots of the microarray objects may be accessed using the operator @ or, alternatively, the function `slot`, which evaluates the slot name. For example, to access the `maLayout` slot in the object `swirl` and the `maNgr` slot in the layout object L,

```
> L <- slot(swirl, "maLayout")
> L@maNgr
```

The function `slotNames` can be used to get information on the slots of a formally defined class or an instance of the class. For example, to get information on the slots for the `marrayLayout` class and on the slots for the object `swirl`, use

```
> slotNames("marrayLayout")
> slotNames(swirl)
```

Basic Microarray Methods

The following basic *methods* were defined to facilitate manipulation of microarray data objects. To see all methods available for a particular class (e.g., the class `marrayLayout`) or just the print methods, use

```
> showMethods(classes = "marrayLayout")
> showMethods("print", classes = "marrayLayout")
```

Printing methods for microarray objects. Since there is usually no need to print out fluorescence intensities for thousands of genes, **print** methods were defined to provide brief summaries of microarray data objects. For an overview of the available microarray *printing methods*, type `methods?print`. For example, summary statistics for an object of class `marrayRaw`, such as `swirl`, can be obtained by `print(swirl)` or simply `swirl`.

```
> swirl
```

```
Prenormalization intensity data:    Object of class marrayRaw.
```

```
Number of arrays:          4 arrays.

A) Layout of spots on the array:
Array layout:         Object of class marrayLayout.

Total number of spots:                         8448
Dimensions of grid matrix:              4 rows by 4 cols
Dimensions of spot matrices:           22 rows by 24 cols

Currently working with a subset of 8448 spots.

Control spots:
There are    2 types of controls :
Control        N
    768      7680

Notes on layout:
No Input File

B) Samples hybridized to the array:
Object of class marrayInfo.

        maLabels     # of slide           Names      experiment Cy3
1           81              81      swirl.1.spot              swirl
2           82              82      swirl.2.spot          wild type
3           93              93      swirl.3.spot              swirl
4           94              94      swirl.4.spot          wild type


experiment Cy5                   date    comments
1           wild type      2001/9/20        NA
2           swirl          2001/9/20        NA
3           wild type      2001/11/8        NA
4           swirl          2001/11/8        NA

Number of labels: 4
Dimensions of maInfo matrix: 4    rows by   6    columns

Notes:
C:/GNU/R/rw1041/library/marrayInput/data/SwirlSample.txt

C) Summary statistics for log-ratio distribution:
                Min. 1st Qu. Median  Mean 3rd Qu. Max.
swirl.1.spot -2.73   -0.79  -0.58  -0.48  -0.29  4.42
swirl.2.spot -2.72   -0.15   0.03   0.03   0.21  2.35
```

```
swirl.3.spot -2.29    -0.75  -0.46  -0.42  -0.12  2.65
swirl.4.spot -3.21    -0.46  -0.26  -0.27  -0.06  2.90
```

D) Notes on intensity data:

Subsetting methods for microarray objects. In many instances, one is interested in accessing only a subset of arrays in a batch and/or spots in an array. *Subsetting methods* "[" were defined for this purpose. For an overview of the available microarray subsetting methods, type methods?"[" or, to see all subsetting methods for the session, showMethods("["). When using the "[" operator, the first index refers to spots and the second to arrays in a batch. Thus, to access data on the first 100 probe sequences in the second and third arrays in the batch swirl, use

```
> swirl[1:100, 2:3]
```

Methods for accessing slots of microarray objects. A number of simple *accessor methods* were defined to access slots of the microarray classes. Using such methods is more general than using the slot function or @ operator. In particular, if the class definitions are changed, any function that uses the @ operator will need to be modified. When using a method to access the data in the slot, only that particular method needs to be modified. Accessor methods are named after the slot; thus, to access the layout information for the array batch swirl, one may also use maLayout(swirl).

In addition, various methods were defined to compute basic statistics from microarray object slots. For instance, for memory management reasons, objects of class marrayLayout do not store the spot coordinates of each probe. Rather, these can be obtained from the dimensions of the grid and spot matrices by applying methods maGridRow, maGridCol, maSpotRow, and maSpotCol to objects of class marrayLayout. Print-tip-group coordinates are given by maPrintTip. Similar methods were also defined to operate directly on objects of classes marrayRaw, marrayNorm, marraySpots, and marrayTwo. The commands below may be used to display the number of spots on the array, the dimensions of the grid matrix, and the print-tip-group coordinates.

```
> swirl.layout <- maLayout(swirl)
> maNspots(swirl)

[1] 8448

> maNspots(swirl.layout)

[1] 8448

> maNgr(swirl)

[1] 4
```

```
> maNgc(swirl.layout)
```

```
[1] 4
```

```
> maPrintTip(swirl[525:534, 3])
```

```
 [1] 1 1 1 1 2 2 2 2 2 2
```

Methods for assigning slots of microarray objects. A number of methods were defined to replace slots of microarray objects without explicitly using the @ operator or slot function. These *assignment methods* make use of the setReplaceMethod function from the R methods package. Like the accessor methods just described, the assignment methods are named after the slots. For example, to replace the maNotes slot of swirl.layout, use

```
> maNotes(swirl.layout)
```

```
[1] "No Input File"
```

```
> maNotes(swirl.layout) <- "New value"
> maNotes(swirl.layout)
```

```
[1] "New value"
```

To initialize slots of an empty marrayLayout object, use

```
> L <- new("marrayLayout")
> maNgr(L) <- 4
```

Similar methods were defined to operate on objects of classes marrayInfo, marrayRaw, marrayNorm, marraySpots, and marrayTwo.

Methods for coercing microarray objects. To facilitate navigation between different classes of microarray objects, we have defined methods for converting microarray objects from one class into another. These *coercing methods* make use of the setAs function from the R methods package. A list of such methods can be obtained by methods?coerce. For example, to coerce an object of class marrayRaw into an object of class marrayNorm, use

```
> swirl.norm <- as(swirl, "marrayNorm")
```

It is also possible to convert objects of classes marrayRaw or marrayNorm into objects of class exprSet (see definition in the Biobase package) using

```
> as(swirl, "exprSet")
```

```
Expression Set (exprSet) with
        8448 genes
        4 samples
```

```
phenoData object with 6 variables and 4 cases
varLabels
            : # of slide
            : Names
            : experiment Cy3
            : experiment Cy5
            : date
            : comments
```

Functions for computing layout parameters. In some cases, plate information is not stored in `marrayLayout` objects when the data are first read into R. We have defined a function, `maCompPlate`, which computes plate indices from the dimensions of the grid matrix and number of wells in a plate. For example, the zebrafish arrays used in the Swirl experiment were printed from 384-well plates, but the plate IDs were not stored in the `fish.gal` file. To generate plate IDs (arbitrarily labeled by integers starting with 1) and store them in the `maPlate` slot of the `maLayout` slot for `swirl`, use

```
> maPlate(swirl) <- maCompPlate(swirl, n = 384)
```

Similar functions were defined to generate and manipulate spot coordinates: `maCompCoord`, `maCompInd`, `maCoord2Ind`, `maInd2Coord`. The function `maGeneTable` produces a table of spot coordinates and gene names for objects of classes `marrayRaw`, `marrayNorm`, and `marraySpots`.

3.4.2 Package *marrayInput*—Data Input for cDNA Microarrays

We begin our analysis of microarray data with the fluorescence intensities produced by image processing of the microarray scanned images. Microarray image quantification data are typically stored in tables whose rows correspond to the spotted probe sequences and columns to different spot statistics (e.g., grid row and column coordinates, spot row and column coordinates, red and green background and foreground intensities for different segmentation and background adjustment methods, spot quality measures). For the `GenePix` image-processing software, these data are stored in the `.gpr` files, and for `Spot`, they are stored in the `.spot` files. We also consider probe and target textual information stored, for example, in `.gal` and `.gdl` (god list) files. The `marrayInput` package provides functionality for reading such microarray data into R. The main functions are `read.marrayLayout`, `read.marrayInfo`, and `read.marrayRaw`, which create objects of classes `marrayLayout`, `marrayInfo`, and `marrayRaw`, respectively. In addition, widgets are provided for each of these functions to facilitate data entry.

Textual information and fluorescence intensity data from processed images for the Swirl experiment (Section 3.3) were included as part of the `marrayInput` package and can be accessed as follows (here, `datadir` is the name of the R package subdirectory containing the data files):

```
> datadir <- system.file("data", package = "marrayInput")
> dir(datadir)
```

```
[1] "00Index"         "fish.gal"
[3] "swirl.1.spot"    "swirl.2.spot"
[5] "swirl.3.spot"    "swirl.4.spot"
[7] "swirl.RData"     "SwirlSample.txt"
```

Main Input Functions

read.marrayLayout. This function may be used to read in and store layout information for a batch of arrays. The following commands store layout parameters for the Swirl experiment in the object `swirl.layout` of class `marrayLayout`. The location of the control spots is extracted from the fourth (`ctl.col=4`) column of the file `fish.gal`.

```
> swirl.layout <- read.marrayLayout(fname = file.path(datadir,
+      "fish.gal"), ngr = 4, ngc = 4, nsr = 22,
+      nsc = 24, skip = 21, ctl.col = 4)
> ctl <- rep("Control", maNspots(swirl.layout))
> ctl[maControls(swirl.layout) != "control"] <- "N"
> maControls(swirl.layout) <- factor(ctl)
```

read.marrayInfo. This function creates objects of class `marrayInfo`, to store, for example, textual information on probe sequences and target samples for a batch of arrays. The following commands create such objects for the Swirl experiment by reading in the text files `SwirlSample.txt` and `fish.gal` supplied by the experimenter.

```
> swirl.targets <- read.marrayInfo(file.path(datadir,
+      "SwirlSample.txt"))
> swirl.gnames <- read.marrayInfo(file.path(datadir,
+      "fish.gal"), info.id = 4:5, labels = 5,
+      skip = 21)
```

read.marrayRaw. This function creates objects of class `marrayRaw` for a batch of arrays. It takes as its main arguments a list of file names for the intensity data (e.g., GenePix output files, `.gpr`) and the names of already created layout, probe, and target description objects (e.g., `swirl.layout`, `swirl.gnames`, and `swirl.targets` for the Swirl experiment). The following commands read in all the `.spot` files residing in the `datadir` directory. The arguments further specify that the red and green foreground intensities are stored under the headings `Rmean` and `Gmean`, and that the red and green background intensities are stored under the headings `morphR` and `morphG`, respectively.

```
> fnames <- dir(path = datadir, pattern = paste("*",
+      "spot", sep = "."))
```

```
> swirl.raw <- read.marrayRaw(fnames, path = datadir,
+      name.Gf = "Gmean", name.Gb = "morphG",
+      name.Rf = "Rmean", name.Rb = "morphR",
+      layout = swirl.layout, gnames = swirl.gnames,
+      targets = swirl.targets)
```

Widgets for Data Input

To facilitate data input and automate the creation of microarray-specific objects, each of the three input functions above has a corresponding `tcltk` *widget*, providing a point-and-click graphical interface: `widget.marrayRaw`, `widget.marrayLayout`, and `widget.marrayInfo`. A screen shot of the `marrayRaw` widget is shown in Figure 3.1 (see color insert); the command to launch the widget and read in data from Spot image output files is

```
> widget.marrayRaw(path = datadir, ext = "spot")
```

Wrapper Input Functions

The functions `read.Spot`, `read.GenePix`, and `read.SMD` automate the creation of `marrayRaw` objects from Spot (`.spot`) and GenePix (`.gpr`) image analysis files and from the Stanford Microarray Database (SMD) raw data files (`.xls`). The main arguments to these functions are a list of file names and the directory path for these files. The following commands read in two specific files from the `datadir` directory

```
> fnames <- dir(path = datadir, pattern = paste("*",
+      "spot", sep = "."))[1:2]
> swirl <- read.Spot(fnames, path = datadir,
+      layout = swirl.layout, gnames = swirl.gnames,
+      targets = swirl.targets)
```

Alternatively, without specifying any arguments, the functions `read.spot` and `read.GenePix` will read in by default all Spot and GenePix files, respectively, within a current working directory. One has the option of setting the layout, probe, and target information manually at a later stage.

```
> swirl <- read.Spot()
> test.raw <- read.GenePix()
> slot(swirl, "maLayout") <- swirl.layout
> slot(swirl, "maGnames") <- swirl.gnames
> slot(swirl, "maTargets") <- swirl.targets
```

3.4.3 Package *marrayPlots*—Diagnostic Plots for cDNA Microarray Data

Three main functions were defined to produce boxplots, scatterplots, and 2D spatial images of spot statistics for pre- and postnormalization intensity data. The main arguments to these functions are microarray objects of classes `marrayRaw`, `marrayNorm`, or `marrayTwo` and arguments specifying which spot statistics to display (e.g., Cy3 and Cy5 background intensities, intensity log ratios M) and which subset of spots to include in the plots. Default graphical parameters are chosen for convenience using the function `maDefaultPar` (e.g., color palette, axis labels, plot title), but the user has the option to overwrite these parameters at any point. Note that by default the plots are done for the first array in a batch; that is, the first array in a microarray object of class `marrayRaw`, `marrayNorm`, or `marrayTwo`. To produce plots for other arrays in a batch, subsetting methods may be used. For example, to produce diagnostic plots for the second array in the batch of zebrafish arrays `swirl`, the argument `swirl[,2]` should be passed to the plot functions.

Spatial Plots of Spot Statistics—maImage

The function `maImage` uses the R base function `image` to create *2D spatial images* of shades of gray or colors that correspond to the values of a statistic for each spot on an array. Details on the arguments of the function are given in the help file `?maImage`. In addition to existing color palette functions, such as `rainbow` and `heat.colors`, a new function, `maPalette`, was defined to generate color palettes from user-supplied low-, middle-, and high-color values. To create white-to-green, white-to-red, and green-to-white-to-red palettes for microarray images, use

```
> Gcol <- maPalette(low = "white", high = "green",
+      k = 50)
> Rcol <- maPalette(low = "white", high = "red",
+      k = 50)
> RGcol <- maPalette(low = "green", high = "red",
+      mid = "white", k = 50)
```

Useful diagnostic plots are spatial images of the Cy3 and Cy5 background intensities; these images may reveal hybridization artifacts such as scratches on the slides and cover-slip effects. The following commands produce spatial images of the Cy3 and Cy5 background intensities for the Swirl 93 array using white-to-green and white-to-red color palettes, respectively (Swirl 93 refers to the third array in the batch `swirl`, with label `"93"`, as given by `maLabels(maTargets(swirl))`).

```
> tmp <- maImage(swirl[, 3], x = "maGb", subset = TRUE,
+      col = Gcol, contours = FALSE, bar = FALSE)

> tmp <- maImage(swirl[, 3], x = "maRb", subset = TRUE,
+      col = Rcol, contours = FALSE, bar = FALSE)
```

Note that the same images can be obtained using the default arguments of the `maImage` function by the shorter commands

```
> maImage(swirl[, 3], x = "maGb")
> maImage(swirl[, 3], x = "maRb")
```

If bar=TRUE, a calibration color bar is displayed to the right of the images. Other options include displaying contours and altering graphical parameters such as axis labels and plot title. The maImage function returns the values and corresponding colors used to produce the color bar as well as a six-number summary of the spot statistics (see function summary).

The 2D spatial images of background intensities for the Swirl 93 array are shown in Figure 3.2 (see color insert). It can be noted that the Cy3 and Cy5 background intensities are not uniform across the slide and are higher in the top right-hand corner, perhaps due to cover-slip effects or tilting of the slide during scanning. Such patterns were not as clearly visible in the individual Cy3 and Cy5 TIFF images. Similar displays of the Cy3 and Cy5 foreground intensities do not exhibit such strong spatial patterns. For the Swirl 81 array (first array in the batch), background images uncovered the existence of a scratch with very high background in print-tip-groups (3,2) and (3,3). Note that spatial images of background intensities revealed a bug (since fixed) in the image-processing package Spot, whereby the morphological opening background estimate was sometimes set to a constant and artificially large value for the last row of spots on the array.

The maImage function may also be used to generate images of the pre- and postnormalization log ratios M using a green-to-red color palette. Panel (b) in Figure 3.3 (see color insert) displays such an image for the Swirl 93 array, highlighting only those spots with the highest and lowest 10% prenormalization log ratios M. The image suggests the existence of spatial dye biases in the intensity log ratio, with higher values in grid (3,3) and lower values in grid column 1 of the array.

```
> tmp <- maImage(swirl[, 3], x = "maM", bar = FALSE,
+      main = "Swirl 93 array: image of pre-normalization M")

> tmp <- maImage(swirl[, 3], x = "maM", subset = maTop
       (maM(swirl[,
+      3]), h = 0.1, l = 0.1), col = RGcol, contours = FALSE,
+      bar = FALSE, main = "Swirl 93 array: image of
       pre-normalization
       M for % 10 tails")
```

Note that the maImage function (and the functions maBoxplot and maPlot to be described next) can be used to display statistics other than fluorescence intensities, such as spot quality measures and layout parameters (e.g., plate IDs in maPlate slot).

Boxplots of Spot Statistics—maBoxplot

Boxplots of spot statistics by plate, print-tip-group, or slide can also be useful to identify spot or hybridization artifacts. The function maBoxplot, based on the R base function `boxplot`, produces boxplots of microarray spot statistics for the classes `marrayRaw`, `marrayNorm`, and `marrayTwo` (see details in `?maBoxplot`). The function `maBoxplot` has three main arguments:

m: Microarray object of class `marrayRaw`, `marrayNorm`, or `marrayTwo`.

x: Name of accessor method for the spot statistic used to stratify the data, typically a slot name for the microarray layout object such as `maPlate` or a method such as `maPrintTip`. If x is NULL, the data are not stratified.

y: Name of accessor method for the spot statistic of interest, typically a slot name for the microarray object m, such as `maM`.

Panel (a) in Figure 3.4 (see color insert) displays boxplots of prenormalization log ratios M for each of the 16 print-tip-groups for the Swirl 93 array. This plot was generated by the following commands:

```
> maBoxplot(swirl[, 3], x = "maPrintTip", y = "maM",
+     main = "Swirl 93 array: pre-normalization")
```

The boxplots clearly reveal the need for normalization since most log ratios M are negative in spite of the fact that only a small proportion of genes are expected to be differentially expressed in the mutant and wild-type zebrafish. As is often the case, this corresponds to a higher signal in the Cy3 channel than in the Cy5 channel, even in the absence of differential expression. In addition, the boxplots uncover spatial dye biases in the log ratios. In particular, print-tip-group (3,3) clearly stands out from the remaining ones, as suggested also in the images of Figure 3.3. The function `maBoxplot` may also be used to produce boxplots of spot statistics for all arrays in a batch. Such plots are useful when assessing the need for between-array normalization—for example, to deal with scale differences among different arrays. The following command produces a boxplot of the prenormalization intensity log ratios M for each array in the batch `swirl`. Panel (a) in Figure 3.5 (see color insert) suggests that different normalizations may be required for different arrays, including possibly scale normalization.

```
> maBoxplot(swirl, y = "maM", main = "Swirl arrays:
      pre-normalization")
```

Scatterplots of Spot Statistics—`maPlot`

The function `maPlot` produces *scatterplots* of microarray spot statistics for the classes `marrayRaw`, `marrayNorm`, and `marrayTwo`. It also allows the user to highlight and annotate subsets of points on the plot and display fitted curves from robust local regression or other smoothing procedures (see details in `?maPlot`). It relies on the R base functions `plot`, `text`, and `legend`. The function `maPlot` has seven main arguments:

m: Microarray object of class `marrayRaw`, `marrayNorm`, or `marrayTwo`.

x: Name of accessor function for the abscissa spot statistic, typically a slot name for the microarray object m, such as `maA`.

y: Name of accessor function for the ordinate spot statistic, typically a slot name for the microarray object m, such as `maM`.

z: Name of accessor method for the spot statistic used to stratify the data, typically a slot name for the microarray layout object such as `maPlate` or a method such as `maPrintTip`. If z is NULL, the data are not stratified.

lines.func: Function for computing and plotting smoothed fits of y as a function of x separately within values of z (e.g., `maLoessLines`). If `lines.func` is NULL, no fitting is performed.

text.func: Function for highlighting a subset of points (e.g., `maText`). If `text.func` is NULL, no points are highlighted.

legend.func: Function for adding a legend to the plot (e.g., `maLegendLines`). If `legend.func` is NULL, there is no legend.

As usual, optional graphical parameters may be supplied, and these will overwrite the default parameters set in the plot functions. A number of functions for computing and plotting the fits can be used, such as `maLowessLines` and `maLoessLines` for robust local regression using the R functions `lowess` and `loess`, respectively. Functions are also provided for highlighting points (e.g., `maText`) and adding a legend to the plot (e.g., `maLegendLines`).

Panel (a) in Figure 3.6 (see color insert) displays the prenormalization MA-plot for the Swirl 93 array, with the sixteen lowess fits for each of the print-tip-groups (using a smoother span $f = 0.3$ for the `lowess` function). The figure was generated with the following commands:

```
> defs <- maDefaultPar(swirl[, 3], x = "maA",
+     y = "maM", z = "maPrintTip")
> legend.func <- do.call("maLegendLines", defs$def.legend)
> lines.func <- do.call("maLowessLines", c(list(TRUE,
+     f = 0.3), defs$def.lines))
> maPlot(swirl[, 3], x = "maA", y = "maM", z = "maPrintTip",
+     lines.func, text.func = maText(), legend.func,
+     main = "Swirl 93 array: pre-normalization MA-plot")
```

The same plot can be obtained using the default arguments of the `maPlot` function by the shorter command

```
> maPlot(swirl[, 3])
```

Figure 3.6 illustrates the nonlinear dependence of the log ratio M on the overall spot intensity A and thus suggests that an intensity- or A-dependent normalization method is preferable to a global one (e.g., median normalization). Also, the lowess fits vary among print-tip-groups, again revealing the existence of spatial dye biases. To highlight, say, the spots with the highest and lowest 5% log ratios using purple "O" symbols, set

```
text.func=maText(subset=maTop(maM(swirl[,3]),
h=0.05,l=0.05),labels="O",col="purple").
```

Wrapper Functions for Basic Sets of Diagnostic Plots

Three wrapper functions are provided to automatically generate a standard set of diagnostic plots: maDiagnPlots1, maRawPlots, and maNormPlots. For example, maDiagnPlots1 produces eight plots of pre- and postnormalization cDNA microarray data: 2D spatial images of Cy3 and Cy5 background intensities and of pre- and postnormalization log ratios M; boxplots of pre- and postnormalization log ratios M by print-tip-group; and MA-plots of pre- and postnormalization log ratios M by print-tip-group. All three functions provide options for saving the figures to a file in PostScript or JPEG format.

```
R> maDiagnPlots1(swirl[,2], title="Swirl 93 array: Diagnostic
   plots", save=TRUE, fname="swirl93.jpeg", dev="jpeg")
```

3.4.4 Package *marrayNorm*—Location and Scale Normalization for cDNA Microarray Data

General Normalization Function—maNormMain

The main function for location and scale normalization of cDNA microarray data is maNormMain. It has eight arguments described in detail in the help file ?maNormMain. The main arguments are: mbatch, an object of class marrayRaw or marrayNorm containing intensity data for the batch of arrays to be normalized; f.loc and f.scale, lists of location and scale normalization functions; and a.loc and a.scale, functions for computing the weights used in composite normalization (Yang et al., 2002b). Other arguments mainly deal with controlling output. Normalization is performed simultaneously for each array in the batch using the location and scale normalization procedures specified by the lists of functions f.loc and f.scale. Typically, only one function is given in each list; otherwise, composite normalization is performed using the weights given by a.loc and a.scale. The maNormMain function returns objects of class marrayNorm.

The marrayNorm package contains functions for median (maNormMed), intensity- or A-dependent (maNormLoess), and 2D spatial (maNorm2D) location normalization. The R robust local regression function loess is used for intensity-dependent and 2D spatial normalization. The package also contains a function for scale normalization using the median absolute deviation, or MAD (`maNormMAD`). The functions allow normalization to be done separately within values of a layout parameter, such as plate or print-tip-group, and using different subsets of probe sequences (e.g., dilution series of control probe sequences). Arguments are available for controlling the local regression when applicable.

Simple Normalization Function—`maNorm`

A simple wrapper function, `maNorm`, is provided for users interested in applying a standard set of normalization procedures using default parameters. This function returns an object of class `marrayNorm` and has seven arguments described in detail in the help file `?maNorm`. The main arguments are: `mbatch`, an object of class `marrayRaw` or `marrayNorm` containing intensity data for the batch of arrays to be normalized; `norm`, a character string specifying the normalization procedure (e.g., `"p"` or `"printTipLoess"` for within-print-tip-group intensity-dependent location normalization using the `loess` function); and `subset`, a logical or numeric vector indicating the subset of points used to compute the normalization values.

Simple Scale Normalization Function—`maNormScale`

A simple wrapper function, `maNormScale`, is provided for users interested in applying a standard set of scale normalization procedures using default parameters. This function returns an object of class `marrayNorm` and has six arguments described in detail in the help file `?maNormScale`. The main arguments are: `mbatch`, an object of class `marrayRaw` or `marrayNorm` containing intensity data for the batch of arrays to be normalized; `norm`, a character string specifying the normalization procedure; and `subset`, a logical or numeric vector indicating the subset of points used to compute the normalization values. This function performs in particular between-slide scale normalization using the MAD (`norm = "g"` or `"globalMAD"`).

Normalization for the Swirl Experiment

The prenormalization MA-plot for the Swirl 93 array in panel (a) of Figure 3.6 reveals the nonlinear dependence of the log ratio M on the overall spot intensity A and the existence of spatial dye biases. Only a small proportion of the spots are expected to vary in intensity between the two channels. We thus perform within-print-tip-group loess location normalization using all 8, 448 probes on the array.

Using main function `maNormMain`. The following command normalizes all four arrays in the Swirl experiment simultaneously and stores the results in the object `swirl.norm` of class `marrayNorm`. A summary of the normalized data can be viewed using the print method; that is, by typing `print(swirl.norm)` or simply `swirl.norm`.

```
> swirl.norm <- maNormMain(swirl, f.loc = list(maNormLoess(x
    = "maA", y = "maM", z = "maPrintTip", w = NULL,
+    subset = TRUE, span = 0.4)), f.scale = NULL,
+    a.loc = maCompNormEq(), a.scale = maCompNormEq())
```

This is the default normalization procedure in `maNormMain`; thus, the same results could be obtained by calling

```
> swirl.norm <- maNormMain(swirl)
```

Using simple function maNorm. Alternatively, the simple wrapper function could be used to perform the same normalization:

```
> swirl.norm <- maNorm(swirl, norm = "p")
```

Using simple function maNormScale. The `maNormScale` function may be used to perform scale normalization separately from location normalization. The following examples do not necessarily represent a recommended analysis but are simply used for demonstrating the software functionality. Within-print-tip-group intensity-dependent *location* normalization followed by within-print-tip-group MAD *scale* normalization could be performed in one step by

```
> swirl.norms <- maNorm(swirl, norm = "s")
```

or sequentially by

```
> swirl.norm1 <- maNorm(swirl, norm = "p")
> swirl.norm2 <- maNormScale(swirl.norm1, norm = "p")
```

For between-slide scale normalization using MAD scaled by the geometric mean of MAD across slides (Yang et al., 2001; Yang et al., 2002b), use

```
> swirl.normg <- maNormScale(swirl.norm, norm = "g")
```

Plots for normalized intensity data. The diagnostic plot functions of Subsection 3.4.3 may also be applied to objects of class `marrayNorm` using the same commands as with objects of class `marrayRaw`. For example, the postnormalization boxplots and MA-plots in panel (b) of Figures 3.4, 3.5, and 3.6 were produced by

```
> maBoxplot(swirl.norm[, 3], x = "maPrintTip",
+  y = "maM", main = "Swirl 93 array: post-normalization")

> maBoxplot(swirl.norm, y = "maM", main = "Swirl arrays:
    post-normalization")

> maPlot(swirl.norm[, 3], main = "Swirl 93 array:
    post-normalization MA-plot")
```

Note that normalization was performed using the `loess` function, but the fitted lines in the MA-plots were produced using `lowess`. To see the effect of within-print-tip-group location normalization, compare panels (a) and (b) in each of these figures. Normalized log ratios M are now evenly distributed about zero across the range of intensities A for each print-tip-group. Furthermore, the nonlinear location normalization seems to have eliminated, to some extent, the scale differences among print-tip-groups and arrays.

3.5   Discussion

Access to efficient, extensible, and interoperable statistical software is an essential aspect of the analysis of microarray data. We have described four packages for exploratory analysis and normalization of cDNA microarray data and illustrated their functionality using gene expression data from the Swirl zebrafish experiment. This case study highlighted the importance of exploratory data analysis using diagnostic plots of various spot statistics. Spatial color images (function `maImage`) of foreground and background intensities and log ratios of intensities may be used to reveal hybridization (e.g., scratches and cover-slip effects) and printing artifacts (e.g., small spots with little probe material). Boxplots may assist in identifying spatial or other types of dye biases in the fluorescence intensities in addition to scale differences within and between arrays (function `maBoxplot`). Scatterplots such as MA-plots (function `maPlot`) are useful in revealing intensity-dependent biases and experimental artifacts. For instance, examination of the range of intensities A may reveal saturation due to improper scanner settings.

Normalization is a key step in the preprocessing of cDNA microarray data and one that can have a large impact on the results of downstream analyses, such as classification or the identification of differentially expressed genes. Normalization is required to ensure that observed differences in fluorescence intensities are indeed reflecting differential gene expression and not some printing, hybridization, or scanning artifact. The simplest approach to within-slide location normalization is to subtract a constant from all intensity log ratios, typically their mean or median. Such global normalization methods are still widely used in spite of the evidence of spatial and intensity-dependent dye biases in numerous experiments, including the Swirl experiment. We thus recommend more flexible normalization procedures, based on robust locally-weighted regression, which take into account the effects of predictor variables such as spot intensity A, location, and plate origin (e.g., using the lowess or loess procedures). We are currently investigating the application of single-channel normalization methods initially developed for the analysis of Affymetrix chip data (see Irizarry et al. (Chapter 4, this volume) for a discussion of the Bioconductor package `affy`). The design of microarray classes and methods in Bioconductor packages should facilitate the sharing of code between packages.

Normalized microarray data consist primarily of pairs (M, A) of intensity log ratios and average log intensities for each spot in each of several slides, in addition to probe and target textual information, and possibly spot quality weights. We are now in a position to address the main question for which the Swirl microarray experiment was designed: the identification of genes that are differentially expressed between swirl mutant and wild-type zebrafish. The Bioconductor packages Biobase, edd, genefilter, multtest, ROC, and sma may be used to address this question. In general, appropriate methods for the main statistical analysis will depend largely on the question of interest to the investigator and can span the entire discipline of statistics (e.g., linear and nonlinear modeling, time series analysis, multiple testing, classification). The implementation of suitable statistical methodology will require the development of question-specific R packages; a number of such packages are already available on the Bioconductor Web site.

We envisage two main classes of users for these and other Bioconductor packages. The first class, primarily biologists, will be interested in applying a standard set of procedures from the packages. Researchers in the second group will likely be interested in writing their own functions and packages in addition to using existing functions. To accommodate these two types of users, the packages were designed at two levels. Regarding the first group, particular emphasis is placed on facilitating access to the statistical methodology implemented in R. The definition of microarray-specific classes and methods for storing and manipulating different types of data greatly reduces the complexity associated with handling microarray data (see also Biobase and annotate for classes and methods for microarray and annotation data). Other steps have been taken to reduce the barrier of entry into R and allow biologists to readily exploit its power and versatility for the analysis of genomic data. New developments include the design of widgets, or small-scale graphical interfaces, for basic input and analysis procedures. The packages AnnBuilder and annotate provide tools for associating microarray data in real time to metadata from biological information resources such as the National Center for Biotechnology Information (NCBI) Entrez system (www.ncbi.nlm.nih.gov/Entrez/, e.g., GenBank, LocusLink, PubMed) or the Gene Ontology (GO) Consortium (www.geneontology.org). Regarding the second class of users, the R language allows the design of functions and packages that are extensible and interoperable.

Acknowledgments. We are grateful to Katrin Wuennenberg-Stapleton from the Ngai Lab at the University of California, Berkeley, for providing data from the Swirl microarray experiment. We would also like to thank Robert Gentleman and Vincent Carey of Harvard University for many helpful discussions and feedback on the design of the packages.

References

Brazma A, Hingamp P, Quackenbush J, Sherlock G, Spellman P, Stoeckert C, Aach J, Ansorge W, Ball CA, Causton HC, Gaasterland T, Glenisson P, Holstege FCP, Kim IF, Markowitz V, Matese JC, Parkinson H, Robinson A, Sarkans U, Schulze-Kremer S, Stewart J, Taylor R, Vilo J, Vingron M (2001). Minimum information about a microarray experiment (MIAME)–toward standards for microarray data. *Nature Genetics* 29:365–371.

Brown PO, Botstein D (1999). Exploring the new world of the genome with DNA microarrays. In: *The Chipping Forecast*, volume 21, 33–37. Supplement to Nature Genetics.

Buckley MJ (2000). *The Spot user's guide.* CSIRO Mathematical and Information Sciences, Sydney, Australia. //www.cmis.csiro.au/IAP/Spot/spotmanual.htm.

Cleveland WS (1979). Robust locally weighted regression and smoothing scatterplots. *Journal of the American Statistical Association* 74(368):829–836.

Cleveland WS, Devlin SJ (1988). Locally-weighted regression: An approach to regression analysis by local fitting. *Journal of the American Statistical Association* 83:596–610.

Dudoit S, Yang YH, Callow MJ, Speed TP (2002). Statistical methods for identifying differentially expressed genes in replicated cDNA microarray experiments. *Statistica Sinica* 12(1):111–139.

Ihaka R, Gentleman R (1996). R: A language for data analysis and graphics. *Journal of Computational and Graphical Statistics* 5:299–314.

Leisch F (2002). Dynamic generation of statistical reports using literate data analysis. Technical Report 69, SFB Adaptive Information Systems and Modelling in Economics and Management Science, Vienna University of Economics and Business Administration: Vienna.

Schena M (ed.) (2000). *Microarray Biochip Technology.* Eaton.

Yang YH, Buckley MJ, Dudoit S, Speed TP (2002a). Comparison of methods for image analysis on cDNA microarray data. *Journal of Computational and Graphical Statistics* 11(1):108–136.

Yang YH, Dudoit S, Luu P, Lin DM, Peng V, Ngai J, Speed TP (2002b). Normalization for cDNA microarray data: A robust composite method addressing single and multiple slide systematic variation. *Nucleic Acids Research* 30(4):e15.

Yang YH, Dudoit S, Luu P, Speed TP (2001). Normalization for cDNA microarray data. In: ML Bittner, Y Chen, AN Dorsel, ER Dougherty (eds.), *Microarrays: Optical Technologies and Informatics*, volume 4266 of *Proceedings of SPIE*, 141–152. SPIE: Bellingham, WA.

4

An R Package for Analyses of Affymetrix Oligonucleotide Arrays

RAFAEL A. IRIZARRY
LAURENT GAUTIER
LESLIE M. COPE

Abstract

We describe an extensible, interactive environment for data analysis and exploration of Affymetrix oligonucleotide array probe–level data. The software utilities provided with the Affymetrix analysis suite summarize the probe set intensities and makes available only one *expression measure* for each gene. We have developed this package because much can be learned from studying the individual probe intensities or, as we call them, the *probe–level data*. We provide some examples demonstrating that having access to and methods for probe–level data results in improvements to quality control assessments, normalization, and expression measures. The software is implemented as an add-on package, conveniently named `affy`, to the freely available and widely used statistical language/software R (Ihaka and Gentleman, 1996). The development of this software as an add-on to R allows us to take advantage of the basic mathematical and statistical functions and powerful graphics capabilities that are provided with R. Our package is distributed as open source code for Linux, Unix, and Microsoft Windows. It is is released under the GNU General Public License. It is part of the Bioconductor project and can be obtained from `http://www.bioconductor.org`.

4.1   Introduction

High-density oligonucleotide expression array technology is widely used in many areas of biomedical research. Affymetrix GeneChip arrays are the most popular of such arrays. As described, for example, in Affymetrix (1999) and Lockhart et al. (1996), oligonucleotides with a length of 25 base pairs are used to probe genes. These oligonucleotides are referred to as probes. Typically, each gene, or more generally genomic sequence of interest, is represented by a *probe set* composed of 11 to 20 pairs of oligonucleotides. The first type of probe is referred to as a perfect match (PM). Each PM probe is paired with a mismatch (MM) probe that is created by changing the middle (13th) base with the intention of measuring nonspecific binding. These two probes (PM, MM) are referred to as a *probe pair*. RNA samples are prepared, labeled, and hybridized with arrays. Arrays are scanned and images are produced and analyzed to obtain an intensity value for each probe. These intensities represent how much hybridization occurred for each oligonucleotide. The software utilities provided with the Affymetrix suite summarize the probe set intensities to form one *expression measure* for each gene. This package was developed for the analysis of the probe intensities, or as we call them, the *probe–level data*. Improving the way in which Affymetrix summarizes the probe–level data to obtain an expression measures was one of the motivations for developing this package. In Section 3 we demonstrate how the robust multiarray average (RMA) measure (Irizarry et al., in press), developed using this package outperforms Affymetrix's default. In Sections 4.2.5 and 4.2.6 we give some examples of the exploratory data analyses that motivated this expression measure. Throughout the chapter we give other examples of advantages having access to probe–level data provides.

4.2   Methods

The package provides two approaches to working with probe–level data. The first is via probe–level objects (`Plob`). One can automatically read in all the relevant information with one function and store it in a `Plob`. Various built–in functions permit the user to easily view graphical displays of the data and compute expression values using different techniques. The second approach uses objects of the classes `Cdf` and `Cel` described in Section 2.2. The `Plob` convention is easier to use than the `Cel/Cdf` convention; however, the `Cel/Cdf` convention is more flexible. We describe and give examples for both. More details are given in the help files for these classes.

4.2.1   Notation

In this chapter we use notation different from other chapters for consistency with the papers on Affymetrix probe–level data. We denote the intensities obtained for each probe as

$$PM_{ijg} \text{ and } MM_{ijg}, i = 1, \ldots, I, j = 1, \ldots, J, \text{ and } g = 1, \ldots, G,$$

with g representing the different genes, i representing different arrays, and j representing the probe pair number (this number is related to the physical position of the oligonucleotide in the gene). For simplicity, we assume $J =$

20 is the same for all i. Note that G is usually on the order of 10^4 so the number of probes on each array is on the order of $2 \times 20 \times 10^4$. Throughout the text, indexes are suppressed when there is no ambiguity.

We will denote examples of R input and output with typewriter font as below:

```
R> cat("This is R output.\n")
This is R output.
```

Here, R> denotes the R prompt.

4.2.2 The CEL/CDF Convention

Affymetrix provides various chips, for example, the human HGU95 chip and mouse MGU74 chip. We will refer to these as *chip types*. One can hybridize various arrays of a specific chip type. For each of them, millions of molecules of a particular probe are attached to a 400 μm^2 area on the chip. After processing the raw image produced by the Affymetrix scanner, each probe is represented by about 100 pixels at a specific location of the image. At the final stage, the image processing–software stores the location and two summary statistics, a mean and standard deviation (SD), for each probe in a file denoted with the extension CEL.

Our package defines a class for storing what we consider to be relevant information in a CEL file. The mean and SD of intensities are stored in matrices. The x,y entry in these matrices contains the probe intensity and SD in position x,y on the array. These objects also contain information on the position of *masked* and *outlier* probes (defined by the Affymetrix image processing system), the name of the array, and relevant information added by the user. Since it is not clear how the pixel–level SD information is useful, the default behavior was set not to load it.

The function read.celfile() is used to read CEL file information into R. An object of class Cel is returned. In this example, we read in a CEL file to create a Cel object:

```
R> CEL1 <- read.celfile("filename.cel") #CEL1 now class Cel
```

The method image() is available for Cel objects. By default it creates an image of the intensities. The images shown in Figure 4.1 showing both the raw (left panels) and log-transformed values (right panels), were made using image in this way:

```
R> image(CEL1)
R> image(CEL1,transfo=log())
```

These images are quite useful for quality control. We recommend examining these images as a first step in data exploration. The top panels in Figure 4.1 show the effect of an "air bubble" trapped in the hybridization

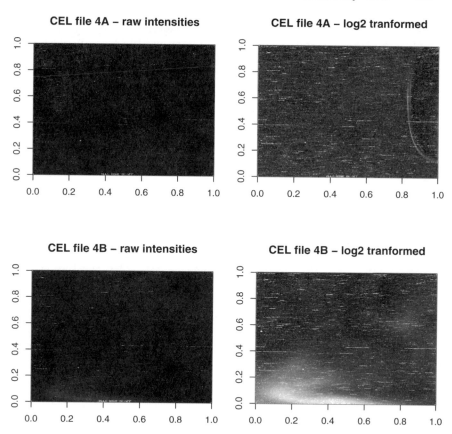

FIGURE 4.1. Image of raw (left) and log-transformed (right) intensities for two CEL files.

chamber of the array. In the bottom panels, we see "cloudy" areas that were caused by a leak in an assembly joint around the devices for injecting solutions onto the chip. In our experience, the log scale is more useful for detecting biases. This seems to be the case in Figure 4.1. In this example, displaying transformations of probe–level data intensities, which is straightforward using our package, leads to the discovery of problems with the data. Once expression measures are computed location, information is lost; thus having access to probe–level data is useful for finding artifacts such as these. The data shown in Figure 4.1 are described in Knudsen et al. (2002).

The information relating genes and probe numbers to locations on the chip are stored in a file with extension .cdf. This implies that a CDF file is needed to decode each CEL file. Our package provides the class Cdf for storing the information in these files. The main slots of this class are name and name.levels. The different gene identifiers of the chip represented by

a `Cdf` object are in the `name.levels` slot, which contains a character vector
of length G. To determine what gene is represented in a particular location,
one looks at the `name` slot, containing a matrix of the same dimensions as
those storing the `intensity` information. If the `x,y` entry is `g`, then that
location is representing `name.levels[g]`. The other slots of `Cdf` objects
contain further information about the probes and the nature of the nucleic
acid bases involved in the PM/MM discrimination.

The function `read.cdffile` is available for reading CDF files:

```
R> CDF <- read.cdffile("filename.CDF") ##CDF now class Cdf
```

Each chip type has a unique CDF file. Each hybridization has its unique
CEL file. A typical experiment will have various hybridized arrays all from
the same chip type. Therefore, the complete probe–level data will be stored
in various CEL files, and one common CDF file needed to decode them.

Objects of class `Cel` can be put together in a *container*. Although it
can be considered as a `container` object as defined in the Biobase package-
age (Gentleman and Carey, chapter 2), the implementation of the class is
specifically adapted for numerical computations. More details can be found
in the help page for the class `Cel.container`.

In principle, all the `Cel` objects in such a container have a common `Cdf`.
They represent hybridizations of the same chip.

4.2.3   Probe Pair Sets

The probe pair set (`PPSet`) class holds the information of all the probes
related to a gene (or a fraction of a gene).[1] The two slots are currently
`probes` and `name`. The slot `probes` is a data frame in which rows represent
probe pairs and the two columns represent PM and MM. The `name` is a
character string, usually the Affymetrix name for the gene.

The function `get.PPSet()` can be used to obtain a `PPSet` object

```
R> CDF <- read.cdffile("filename.CDF")
R> CEL1 <- read.celfile("filename1.cel")

R> num.of.genes <- length(name.levels(CDF))
R> mylevel <- "100000_at" #lets look at gene with this ID
R> myname <- name.levels(CDF)[mylevel] ##get name of gene

R> mypps <- get.PPSet(myname, CDF, CEL1)
```

Notice that `name.levels()` is an accessor method for the `name.levels`
slot.

---

[1]Affymetrix internal controls for the reverse transcription step are fractions of
genes. For such a control gene, one `PPSet` corresponds to the 5-prime end of the
gene, while another `PPSet` corresponds to the 3-prime end.

The methods `barplot()` and `plot()` can be used to graphically explore the intensities of a `PPSet` object.

```
R> par(mfrow=c(2,1))
R> barplot(mypps)
R> plot(mypps)
```

*PM*s are in red and *MM* in blue. See Figure 4.2 (see color insert) for the result.

4.2.4   Probe–Level Objects

The `Plob` class combines the information of various objects of class `Cel` with a common `Cdf`. This class is designed to keep information of one experiment. It is especially useful for users who want to obtain expression values without much tinkering with `Cel` and `Cdf` objects. However, the probe–level data is available.

The main slots of this class are `pm`, `mm`, and `name`. The slots `pm` and `mm` are matrices with rows representing probe pairs and columns representing arrays. The gene name associated with the probe pair in row i can be found in the ith entry of `name`. Other information about the experiment is available in this class. See the `Plob` help file for details.

If called with no arguments, the function `ReadAffy()` reads a CDF file and all the CEL files in the working directory and returns an object of class `Plob`. For example

```
R> Data <- ReadAffy()
```

creates an object of class `Plob` that contains all the relevant information for an experiment using the CEL files and CDF file in the working directory.

The methods `summary()`, `boxplot()`, `hist()`, `plot()`, and `image()`, among others, are available for objects of this class.

For example one can type

```
R> par(mfrow=c(2,2))
R> boxplot(Data)
R> hist(Data)      ##see figure 2b
R> legend(13,11500,c("PM","MM"),col=c("red","blue"),
          pch=c(15,15))
```

Figure 4.3 (see color insert) shows the results of the above commands. The fact that the last bin in the histogram is large shows a saturation problem. This is another example of the importance of having access to probe–level data.

Alternatively, if one has already read in CEL files and CDF files into `Cdf` and `Cel` objects we can use the function `convert()` to create a `Plob` object. For example,

```
R> CELS <- read.container.celfile("filename1.cel",
                                  "filename2.cel")
R> CDF <- read.cdffile("filename.CDF")
R> Data <- convert(CDF,CELS)
```

produces the same object `Data` as in the previous example.

4.2.5 Normalization

In many of the applications of high-density oligonucleotide arrays, the goal is to learn how RNA populations differ in expression in response to genetic and environmental differences. Observed expression levels also include variation introduced during the sample preparation, during manufacture of the arrays, and during the processing of the arrays (labeling, hybridization, and scanning). These sources of variation can have many different effects on data. See Hartemink et al. (2001) for a more detailed discussion. Normalization at the probe–level has been show to work well (Bolstad et al., in press). In this section we give some examples of normalization procedures for the `Plob` approach.

In this section we will use a dataset included in the package for the users convenience. The user can load it by typing

```
R> data(Dilution)
```

This will create a `Plob` named `Dilution` containing a subset of dilution study data described in detail in Irizarry et al. (in press). `Dilution` represents human arrays (HGU95A) hybridized to source human liver tissue cRNA in a range of dilutions starting with 2.5 μg cRNA and rising through 5.0, to 10.0 μg. We include three replicate arrays for each generated cRNA for a total of 9 arrays. Three scanners have been used for the arrays in this example. Each array replicate was processed in a different scanner. The subset contains 100 genes.

The two plots in Figure 4.4 were created using

```
R> boxplot(Dilution[,1:3],range=0)   #range=0 so no outliers
```

and shows the three arrays hybridize with 2.5μg. For the same data, Figure 4.5 shows log ratios $M = \log_2(PM'/PM)$ versus average log intensity, $A = \log_2 \sqrt{PM' \times PM}$, (MVA) plots for the 2 pairwise array comparisons. PM and PM' represent the intensities obtained for the same probe on two different arrays. These plots have been used by, for example, Dudoit et al. (2001) to explore intensity related biases. We can use the function `mva.pairs` to create Figure 4.5

```
R> mva.pairs(Dilution[,1:3])
```

Notice that we are using the subset operator [ ]. The indexes given inside this operator will represent which genes (first dimension) and which arrays

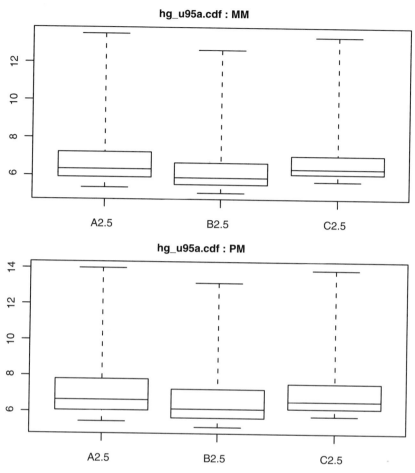

FIGURE 4.4. Boxplot of the first three arrays in dilution data.

(second dimension) are returned. In the above example all genes will be returned for arrays 1 through 3. These three arrays are replicates of the 2.5 μg concentration.

These three arrays are replicates, so we expect the boxplot to show them having similar quantiles and the mva pairs plots to scatter around the horizontal line at 0. It is clear from Figures 4.4 and 4.5 that these data need normalization.

For these oligonucleotide arrays, the approach presented in Dudoit et al. (2001) is not appropriate because, unlike for cDNA arrays, there are no pairs. A procedure that normalizes each array against all others is needed. Åstrand (2001) presents an approach that extends the ideas presented in Dudoit et al. (2001). Another approach to normalization is to make the PM, MM quantiles of all arrays agree. We refer to this as quan-

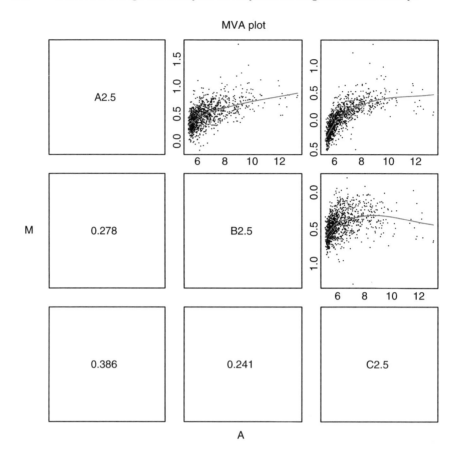

FIGURE 4.5. MVA pairs for the first three arrays in dilution data.

tile normalization. See Bolstad et al. (in press) for more details. Various procedures are available for normalization via the method `normalize` (see the help file). The default is quantile normalization.

```
R> normalized.Dilution <- normalize(Dilution[,1:3])
R> boxplot(normalized.Dilution)
R> mva.pairs(normalized.Dilution)
```

Figures 4.6 and 4.7 show the boxplot and mva pairs plot after normalization. The normalization routine seems to correct the boxplots and mva plots. Notice how the normalization has removed the bias seen in Figure 4.5. The benefits of this normalization at the probe–level carry over to the expression level (Irizarry et al., in press).

Although we include some predefined normalization routines, we have developed the package to make it relatively easy for the user to add nor-

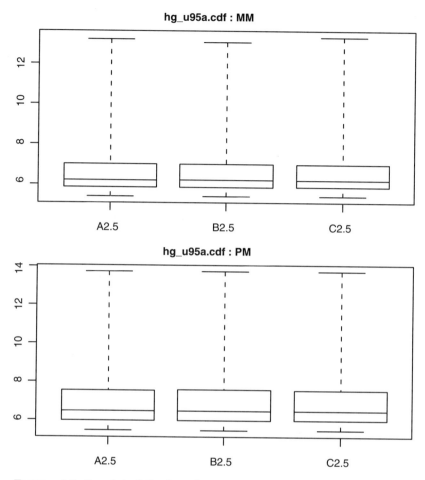

FIGURE 4.6. Boxplot of the first three arrays in normalized dilution data.

malization methods of their own. This is described in more detail in Section 4.4.

4.2.6 Exploratory Data Analysis of Probe–Level Data

Many expression measures, for example Affymetrix's AvDiff (Affymetrix, 1999) and dChip's reduced model (Li and Wong, 2001), consider the difference $PM - MM$ with the intention of correcting for nonspecific binding. Some researchers, for example Naef et al. (in press), propose expression measures based only on the PM. Here we give an example of how one can use the CDF/CEL approach of our package to assess the behavior the MM.

For the users convenience, we include an example of a PPSet container containing intensities read from 12 arrays where one of the control cRNAs,

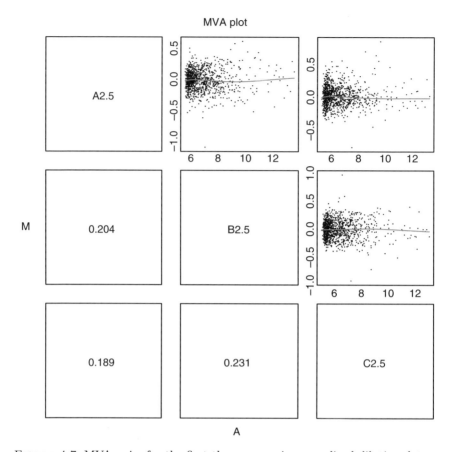

FIGURE 4.7. MVA pairs for the first three arrays in normalized dilution data.

AFFX-BioB-5_at, was added at known concentrations. The concentrations were varied from chip to chip. The different concentration values are also available through the example dataset:

```
R> data(SpikeIn) #loads objects: concentrations and SpikeIn
R> print(concentrations)
 [1]   0.50  0.75  1.00   1.50   2.00   3.00   5.00    12.50
 [9]  25.00 50.00 75.00 150.00 ##units are picoMolar
```

Notice that the concentrations are growing exponentially.

In Figure 4.8 (see color insert) we show the PM, MM, PM/MM, and $PM - MM$ values for each of the 20 probes. The 20 different probe pairs are represented with different symbols and colors. To create this plot, we used the following code (we show it only for PM, the others are similar):

```
R> plot(concentrations,concentrations,ylim=c(20,20000),
        main="PM",ylab="Intensity",log="xy",type="n")
R> for(i in 1:12)
R>     points(rep(concentrations[i],20),pm(SpikeIn[[i]]),
            col=rainbow(20),pch=1:20)
```

Notice that pm is a method for extracting PM values for the class PPSet.

As expected, the PM values are growing in proportion to the concentration. Notice also that the lines representing the 20 probes are close to being parallel, showing that there is a strong additive (in the log scale) probe–specific effect. The fact, seen in Figure 4.8 (in color insert), that the additive probe–specific effect is also detected by the MM is motivation for subtracting them from the PM. However, parallel lines are still seen in the $PM - MM$ plot, demonstrating that subtracting does not remove the probe effect. The lack of parallel lines in the PM/MM plot shows that dividing by MM removes, to some degree, the probe effect. However, the fact that the MM also grow with concentrations, and therefore detect signal as well as nonspecific binding, results in an attenuated signal. Notice in particular that PM/MM is unable to distinguish between concentrations of 25 and 150 picoMolar. Since subtracting probe-specific MM adds noise with no obvious gain in bias and because PM/MM results in a biased signal, one may conclude that alternative measure to those based on $PM - MM$ will provide an improvement. Notice that this discovery was possible only through exploration of probe–level data.

4.3 Application

4.3.1 Expression Measures

To perform gene expression analyses, we need to summarize the probe-set data available for each gene into one expression measure. Various approaches on how to do this have been proposed; see for example Affymetrix (1999), Li and Wong (2001), and Irizarry et al. (in press). Our package permits the user to construct a procedure for computing expression measures. Routines for the three expression measures mentioned above are included in the package.

The function express() automatically computes expression measures for any Plob. For example,

```
R> E <- express(Data)
```

returns an object of class exprSet (defined in the Biobase package). The exprs slot of the class exprSet is a matrix with rows representing genes and columns representing arrays. This is the format that many other packages provide to analyze gene expression data. Many of the Bioconductor packages have methods available for objects of class `exprSet`. However, if the user wants to use another package, the method `write.exprs()` creates an ASCII flat file that can easily be read by most packages for the analysis of microarray data. The objects created by `express()` also provides standard error estimates of the expression measures through the `se.exprs` slot.

The function `express` uses the RMA (robust multichip average) expression measure presented in Irizarry et al. (in press) as a default. However, the function is quite flexible. An expression measure for a gene can be considered a summary of the background–corrected and normalized PMs of the probe set representing that gene.

One can choose from various normalization routines through the argument `normalize.method`. The default is quantile normalization, but one can also use constant, loess, invariantset, or contrasts normalization (see help file for normalize for more information on these routines). The option not to normalize is also available via the logic argument `normalize`.

The way in which PM values are "background" corrected can be changed through an arbitrary function sent via the argument `bg.correct`. The current default is based on a background correction method described in Irizarry et al. (in press) and available through the function `bg.correct.rma`. Functions to perform background correction can be easily added. Such functions must take two matrices as first arguments. The first matrix will contain the PM values for a *probes set* (one row per probe, one column per array), while the second matrix will contain the corresponding MM values. They must return a matrix of background–corrected probe intensities. The writing of the function that corrects the PM signal by subtracting the MM signal will help to clarify this:

```
R> bg.correct.subtractmm <- function(pm, mm){
+       return(pm-mm)
+ }
```

Once background corrected PMs are available, we compute expression measures via some summary statistic. The functions computing expression values summarize probe intensities by one *expression value*. They take as a first parameter a matrix of probe intensities for a *probe set* (one row per probe, one column per array), and return a vector of *expression values* (one per array). Notice that the summary statistic can be the parameter estimate obtained from fitting a model to the probe–level data. For example, Li and Wong (2001) fit a multiplicative model

$$PM_{ij} - MM_{ij} = \theta_i \phi_j + \epsilon_{ij}$$

with ϵ_{ij} independent identically–distributed normal errors. The MLEs $\hat{\theta}_i, i = 1, \ldots, I$ is considered the "summary statistic" that defines the expression measure. R's extensive functionality for statistical methods gives the user endless possibilities of models to fit and procedures to fit them. For example, to compute RMA (Irizarry et al., in press), we fit a linear additive model to background corrected and normalized $\log(PM_{ij})$ using median polish. However, the user can easily change the fitting procedure from median polish (medpolish from the package eda) to robust regression (rlm from the package MASS).

Expression values can also be obtained for some other classes. A way to include new approaches is to go through the PPSet objects, or more precisely PPSset.container objects. As described above, a container is used to bundle several PPSet objects and is used here to collect the PPSet for a given gene in different hybridizations (i.e., on different arrays). A method that computes expression values is defined for PPSet.container objects. Computing expression values is done like shown below:

```
## pps.cont is a PPSet.container
ev <- express.summary.stat(pps.cont,
                   bg.correct="bg.correct.pmonly",
                   method="liwong")
```

The parameters method and bg.correct correspond respectively to the parameters summary.stat and bg.correct described for the function express. As mentioned earlier, user-written functions to correct for background or to compute expression values can be added easily.

Examples are given in the documentation that comes with the package.

4.4   Software

4.4.1   A Case Study

The dilution dataset provided with the package is a great resource for assessing the technology because we know that measures of expression for genes that are expressed should grow in direct proportion to the amount of RNA hybridized (which is known) and that replicates should be like each other.

As seen in Section 4.2, these data need normalization. However, in this case we can only normalize within replicates. Normalizing across replicates is not recommended in this cases because, owing to the different concentrations, the assumption needed for most normalization procedures (most genes, or at least a large group of genes, don't change much) is not satisfied. We therefore normalize the replicates and then use the union method to put them together.

```
R> data(Dilution) #load dilution data
R> tmp1 <- normalize(Dilution[, 1:3])
R> tmp2 <- normalize(Dilution[, 4:6])
R> tmp3 <- normalize(Dilution[, 7:9])
```

```
R> normalized.Dilution <- union(tmp1, tmp2)
R> normalized.Dilution <- union(normalized.Dilution, tmp3)
```

We are now ready to compute measures of expression. Keep in mind that the `express` function normalizes by default, so we need to avoid this by using the `normalize` argument.

```
R> e <- express(normalized.Dilution, normalize = F)
R> f <- express(Dilution, normalize = F)
```

To see if normalizing makes a difference, let's look at expressions for two randomly selected genes for the 9 arrays plotted against known concentration. If the gene is expressed, these values should be growing with concentration but the replicates should be close if not equal. Figure 4.9 shows the advantages of normalization: we still capture the signal and reduce the variance.

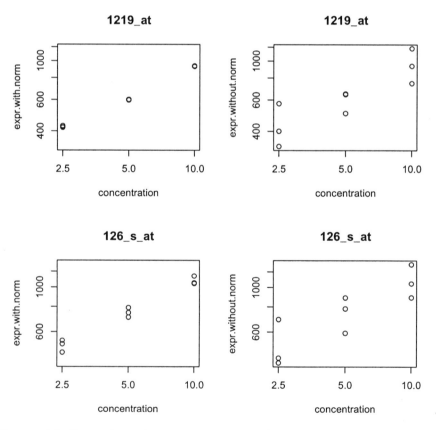

FIGURE 4.9. Comparison of expression measures obtained with and without normalization.

Now let's assess the variance of three different measures of expression: a version of Affymetrix's average difference, Li and Wong's model–based expression index, and the default (RMA). In the code below we use `express` to obtain these expression measures for three replicate arrays.

```
R> Dilution <- Dilution[, 1:3]
R> rma <- express(Dilution)
R> LiWong <- express(Dilution, bg = subtractmm,
                     summary.stat = li.wong)
R> Avdiff <- express(Dilution, normalize = F, bg = subtractmm,
                     summary.stat = avdiff)
```

Then we create a function (not part of the package), which we call *assess*, to compare the IQR of the log fold changes obtained between these.

```
R> assess <- function(w) {
+       x <- exprs(w)[, 1]
+       y <- exprs(w)[, 2]
+       z <- exprs(w)[, 3]
+       Index <- x > 0 & y > 0
+       signif(c(IQR(log2(x[Index]/y[Index]), na.rm = T),
+                IQR(log2(x[Index]/z[Index]), na.rm = T),
+                IQR(log2(y[Index]/z[Index]), na.rm = T)),
+            2)
+ }
```

We want these IQR to be small because we are comparing replicates:

```
R> rma@exprs <- 2^exprs(rma) ##RMA is a log scale measure
R> assess(rma)
```

```
[1] 0.027 0.018 0.026
```

```
R> assess(LiWong)
```

```
[1] 0.26 0.27 0.39
```

Affymetrix software normalizes by forcing expression from all arrays to have the same median as a reference array. In R this is easy to do:

```
R> refmedian <- median(exprs(Avdiff)[,1])
R> Avdiff@exprs <- sweep(exprs(Avdiff),2,
          esApply(Avdiff,2,median),"/")*refmedian
   ##esApply is apply for class exprSet.
R> assess(Avdiff)
```

```
[1] 0.49 0.35 0.58
```

Notice RMA has the smallest IQR between replicates. See Irizarry et al. (in press) for assessments such as the one presented in this section.

4.4.2   Extending the Package

The package was also designed to offer simple and convenient ways of extending it with new methods and algorithms. This section covers technical aspects of the package that will aid those who wish to do this through an example related to normalization. We must warn the reader that minor difference between this documentation and the package may occur in the near future. The differences will be reported on the Web site `http://www.bioconductor.org`.

Different approaches to normalize array data have already been suggested, but new methods are very likely be developed too. We tried to make the addition of new normalization methods straightforward. It should be the case if the following indications are carefully observed.

- The normalization method must obey a naming convention. The name of the function should be something like `normalize.XYZ.abc`, where `abc` is the *nickname* of the method and `XYZ` is the class of the object the normalization methods works on. Currently XYZ can only be `Cel.container` or `Plob`.[2] You are free to implement a normalization method working on one object without having to implement it for the others.

- The first argument passed to the function must be of class `XYZ`. Optional parameters can be appended as long as they have default values.

- The *nickname* of the new normalization method must be registered in the variable `normalize.XYZ.methods` (a vector of mode *character*).

The documentation that comes with the package has some examples.

4.5   Conclusion

We have described an add-on package for R that permits the user to manipulate probe–level data obtained from Affymetrix GeneChip arrays. We present two approaches: the easy–to–use `Plob` approach and the flexible `Cel/Cdf` approach. For both approaches, we include various useful functions in the package but expect the user to change these to their liking as well as create their own. As mentioned, the add-on package, and R, are open–source. The user can easily view the code of our functions. In this version we describe version 1.0 of the package. We expect future versions to include many other features developed by us and other users.

---

[2] A general scheme for normalization of arrays, whether Affymetrix or cDNA is currently being evaluated.

Acknowledgments. We would like to thank Magnus Åstrand, Ben Bolstad, Emmanuel Lazardris, William J. Lemon, Cheng Li, and Fred Wright for contributing or sharing ideas and code; Robert Gentleman and Karl Broman for helping better understand R; Terry Speed, Sandrine Dudoit, and Jean Yang for suggestions regarding methodology; the scientists who inspired us to make the code somewhat user friendly, in particular Tom Coppola and Skip Garcia; Francois Collin, Yasmin D. Beazer-Barclay, Kristen J. Antonellis, and Uwe Scherf from Gene Logic for providing some of the data; and last but least to the R core for creating such a wonderful programing environment. The work for this book was in part supported by the PGA U01 HL66583, a travel grant from the first INCOB conference and the Danish Biotechnology Instrument Center.

References

Affymetrix (1999). *Affymetrix Microarray Suite User Guide.* Affymetrix, Santa Clara, CA, version 4 edition.

Åstrand M (2001). Normalizing oligonucleotide arrays. *Unpublished manuscript.*

Bolstad BM, Irizarry RA, Åstrand M, Speed TP (in press). A comparison of normalization methods for high density oligonucleotide array data based on variance and bias. *Bioinformatics.*

Dudoit S, Yang YH, Callow MJ, Speed TP (2001). Statistical methods for identifying genes with differential expression in replicated cDNA microarray experiments. *Statistica Sinica*, 12(1):111–139.

Hartemink AJ, Gifford DK, Jaakola TS, Young RA (2001). Maximum likelihood estimation of optimal scaling factors for expression array normalization. In *SPIE BiOS.*

Ihaka R, Gentleman R (1996). R: A language for data analysis and graphics. *Journal of Computational and Graphical Statistics*, 5(3):299–314.

Irizarry RA, Hobbs B, Collin F, Beazer-Barclay YD, Antonellis KJ, Scherf U, Speed TP (in press). Exploration, normalization, and summaries of high density oligonucleotide array probe level data. *Bioinformatics.*

Knudsen S, Gautier L, Nielsen H, Nielsen C, Fomsgaard A, Tirstrup K, Blom N, Ponten TS, Workman C, Brunak S (2002). Hiv-1 induces transcription of cytoskeletal genes in t cells. *Submitted.*

Lockhart DJ, Dong H, Byrne MC, Follettie MT,Gallo MV, Chee MS, Mittmann M, Wang C, Kobayashi M, Horton H, Brown EL (1996). Expression monitoring by hybridization to high-density oligonucleotide arrays. *Nature Biotechnology*, 14:1675–1680.

Li C, Wong WH (2001). Model-based analysis of oligonucleotide arrays: Expression index computation and outlier detection. *Proceedings of the National Academy of Science USA*, 98:31–36.

Naef F, Lim DA, Patil N, Magnasco MO (in press). From features to expression: High density oligonucleotide array analysis revisited. *Proceedings of DIMACS Workshop on Analysis of Gene Expression Data.*

5

DNA-Chip Analyzer (dChip)

CHENG LI
WING HUNG WONG

Abstract

DNA-Chip Analyzer (dChip) is a software package implementing model-based expression analysis of oligonucleotide arrays and several high-level analysis procedures. The model-based approach allows probe-level analysis on multiple arrays. By pooling information across multiple arrays, it is possible to assess standard errors for the expression indexes. This approach also allows automatic probe selection in the analysis stage to reduce errors due to cross-hybridizing probes and image contamination. High-level analysis in dChip includes comparative analysis and hierarchical clustering. The software is freely available to academic users at www.dchip.org.

5.1 Introduction

Affymetrix oligonucleotide arrays (Lockhart et al., 1996; Lipshutz et al., 1999; Parmigiani et al., Chapter 1, this volume) have been applied to many gene expression studies in the past few years (Tavazoie et al., 1999; Cho et al., 2001; Hakak et al., 2001; Gentleman and Carey, Chapter 2, this volume; Irizarry et al., Chapter 4, this volume). Many sets of 11–20 pairs of perfect match and mismatch oligonucleotides are used to measure the underlying mRNA concentrations of genes in a sample. Besides linear normalization and average difference and signal methods provided by MAS software (Affymetrix, 2001), researchers have proposed alternative low-level analysis methods such as feature extraction (Schadt et al., 2002), normalization (Hill et al., 2001), and expression index computation (Li and Wong, 2001a; Holder et al., 2001; Irizarry et al., in press; Lazaridis et al., 2002; Zhou and Abagyan, 2002) in the attempt to improve on these aspects. Hoffmann et al. (2002) have shown that such low-level analysis methods can have a large impact on high-level results such as sample comparisons. In the rest of this chapter, we will describe the DNA-Chip Analyzer (dChip) software, which provides several low-level and high-level analysis methods.

5.2   Methods

Irizarray et al. (Chapter 4, this volume) introduce the Affymetrix array design, CEL and CDF files, and probe-level data. Interested readers can refer to Chapter 4 or Lipshutz et al. (1999) for details on oligonucleotide array data.

5.2.1   Normalization of Arrays Based on an "Invariant Set"

Since array images usually have different overall image brightnesses (Figure 5.1A, see color insert), especially when they are generated at different times and places, proper normalization is required before comparing the expression levels of genes between arrays. Model-based expression computation (Li and Wong, 2001a) requires normalized probe-level data (from Affymetrix's DAT or CEL files). For a group of arrays, we normalize all arrays (except the baseline array) to a common baseline array having the median overall brightness (as measured by the median CEL intensity in an array).

A normalization relation can be understood as a curve in the scatterplot of two arrays with the baseline array drawn on the y-axis and the array to be normalized on the x-axis. A line running through the origin is a multiplicative normalization method (Affymetrix, 2001; the scaling method in Affymetrix Microarray Suite software (MAS)), and a smoothing spline through the scatterplot can also be used (Figure 5.1A; Schadt et al., 2001).

We should base the normalization only on probe values that belong to nondifferentially expressed genes, but generally we do not know which genes are nondifferentially expressed (control or housekeeping genes may also be variable across arrays). Nevertheless, we expect that probes of a nondifferentially expressed gene in two arrays to have similar intensity ranks (ranks are calculated in the two arrays separately).

We use an iterative procedure to identify a set of probes (called an "invariant set"), which presumably consists of points from nondifferentially expressed genes (Figure 5.1B). Specifically, we start with points of all *PM* probes (about 140,000 for the HU6800 array). If a point's proportional rank difference (absolute rank difference in the two arrays divided by $n = 140,000$) is small enough, it is kept for the new set. By small enough we mean < 0.003 when its average intensity rank in the two arrays is small and < 0.007 when it is large, accounting for fewer points at the high-intensity range, and this threshold is interpolated in between them. These parameters were chosen empirically to make the selected points in the "invariant set" thin enough to naturally determine a normalization relation. In this way, we may obtain a new set of 10,000 points, and the same procedure is applied to the new set iteratively until the number of points in the new set no longer decreases. A piecewise-linear running median curve is then calculated and used as the normalization curve. After normalization, the two arrays have similar overall brightnesses (Figure 5.1C, D).

5.2.2   Model-Based Analysis of Oligonucleotide Arrays

After normalization we can then compute the expression level of each gene in all samples. Affymetrix MAS software uses the direct average ("average difference") or robust average ("signal") of $PM-MM$ differences of all probes in a probe set as the expression level (Affymetrix, 2001); we have proposed a statistical model for one probe set in multiple arrays to account for the probe variability in computing expression levels (Li and Wong, 2001a):

$$p_{ij} = PM_{ij} - MM_{ij} = \theta_i \phi_j + \varepsilon_{ij}, \quad \sum \phi_j^2 = J, \quad \varepsilon_{ij} \sim N\left(0, \sigma^2\right). \quad (5.1)$$

It states that the difference $PM-MM$ in array i, probe j (equivalent to l used in the rest of the book) of this probe set is the product of the model-based expression index (MBEI) in array i (θ_i; equivalent to x_i used in the rest of the book) and the probe-sensitivity index of probe j (ϕ_j), plus random error. Here, J is the number of probe pairs in the probe set. Fitting the model, we can identify cross-hybridizing probes (ϕ_j's with large standard error, which are excluded during iterative fitting) and arrays with image contamination at this probe set (θ_i's with large standard error) as well as single outliers (image spikes), which are replaced by the fitted values (Figure 5.2, see color insert). In effect, the estimated expression index $\hat{\theta}_i$ is a weighted average of $PM-MM$ differences, $\hat{\theta}_i = (\sum_j p_{ij}\phi_j)/J$, with larger weights given to probes with larger ϕ. The outlier image (Figure 5.3, see color insert) can be used to assess the quality of an experiment and to identify unexpected problems such as a misaligned corner of a DAT file (Li and Wong, 2001a).

5.2.3   Confidence Interval for Fold Change

After obtaining expression indexes using MAS's signal method or MBEI, fold changes can be calculated between two arrays for every gene and used to identify differentially expressed genes. Usually, low or negative expressions are truncated to a small number before calculating fold changes, and MAS also cautions against using fold changes when the baseline expression is absent.

The availability of standard errors for MBEI allows us to obtain confidence intervals for fold changes. Suppose $\hat{\theta}_1 \sim N\left(\theta_1, \sigma_1^2\right)$ and $\hat{\theta}_2 \sim N(\theta_2, \sigma_2^2)$, where θ_1 and θ_2 are the real expression levels in two samples, and $\hat{\theta}_1$ and $\hat{\theta}_2$ are the model-based estimates of expression levels. We substitute the model-based standard errors for σ_1 and σ_2. Letting $r = \theta_1/\theta_2$ be the real fold change, the inference on r can be based on the quantity

$$Q = \frac{(\hat{\theta}_1 - r\hat{\theta}_2)^2}{\sigma_1^2 + \sigma_2^2 r^2}.$$

It can be shown that Q has a χ^2 distribution with one degree of freedom, irrespective of the values of θ_1 and θ_2 (Wallace, 1988). Thus Q is a pivotal quantity involving r. We can use Q to construct fixed-level tests and to invert them to obtain confidence intervals (CI) for fold changes (Cox and Hinkley, 1974).

Table 5.1 presents the estimated expression indexes (with standard errors) in two arrays and the 90% confidence intervals of the fold changes for 14 genes. Although all the genes have similar estimated fold changes, the confidence intervals are very different. For example, gene 1 has fold change 2.47 and a tight confidence interval (2.07, 3.03). In contrast, gene 11 has a similar fold change of 2.49 but a much wider confidence interval (0.96, 18.18). Thus, the fold change around 2.5 for gene 11 is not as trustworthy as that for gene 1. Further examination reveals that this is due to the large standard errors relative to the expression indexes for gene 11. This agrees with the intuition that when a gene has one or both expression levels close to 0 in two samples, the fold change cannot be estimated with much accuracy. In addition, when image contamination results in unreliable expression values with large standard errors, the fold changes calculated using these expression values are attached with wide CIs. In this manner, the

TABLE 5.1. Using expression levels and associated standard errors to determine confidence intervals of fold changes.

Gene	MBEI 1	SE 1	MBEI 2	SE 2	Fold change	Lower CB	Upper CB
1	859.64	41.78	347.57	36.09	2.47	2.07	3.03
2	405.72	31.23	164.01	44.25	2.47	1.67	4.49
3	283.93	28.53	114.70	18.47	2.48	1.84	3.48
4	45.98	64.24	18.57	84.53	2.48	0	∞
5	225.18	57.49	90.90	36.18	2.48	1.18	7.49
6	247.00	50.65	99.66	19.54	2.48	1.51	4.02
7	49.97	21.53	20.15	23.57	2.48	0.49	∞
8	276.49	18.69	111.37	36.10	2.48	1.59	5.35
9	436.07	32.98	175.38	21.07	2.49	1.99	3.19
10	75.69	17.72	30.44	17.97	2.49	1.07	86.17
11	80.67	25.31	32.43	16.96	2.49	0.96	18.18
12	181.52	42.48	72.88	28.17	2.49	1.25	7.12
13	1122.28	99.28	449.89	63.28	2.49	1.92	3.35
14	168.23	40.63	67.44	30.30	2.49	1.18	9.82

measurement accuracy of expression values propagates to the estimation of fold changes. In practice, we find it useful to sort genes by the lower confidence bound ("Lower CB" in Table 5.1), which is a conservative estimate of the fold change.

5.2.4 Pooling Replicate Arrays Considering Measurement Accuracy

For a gene, the signal or model-based expression index (MBEI) is estimating the "expression level" specific to a scanned array. However, variation is introduced in the experimental steps, such as sample amplification, hybridization and staining, and the analysis steps, such as normalization and expression index computation. For example, the relative concentration of a particular mRNA in the total mRNA sample and the amplified sample may not be the same. We are ultimately interested in the comparison between two biological samples rather than that between two scanned images.

Performing replicate experiments helps us to estimate the experimental and analysis variations. For example, one may extract mRNA from a biological sample, split it into two portions, and then do the subsequent experimental steps and analysis separately. The variation between the resultant two arrays reflects the variation introduced after the sample splitting. Since expression levels calculated from the two arrays are estimating the same expression level in the biological sample, averaging the replicates can provide less biased expression estimates.

The standard errors for MBEI are the measurement accuracy of the array-specific expression. When averaging replicates, we can down-weight the expression values with large standard errors instead of simple averaging. For example, one array may have a local image contamination, and the affected probe sets have unreliable expression values with large standard errors; the weighted average will be closer to the expression level in the other replicate array, where the expression values for these probe sets are more trustworthy.

Statistical formulation of this weighted averaging scheme is as follows. Assuming that array-specific expression levels of a gene in replicate arrays are independently observed, $\hat{\theta}_i \sim N(\theta_i, \hat{\sigma}_i^2)$, where θ_i is the array-specific expression in replicate array i, and $\hat{\theta}_i, \hat{\sigma}_i$ are the observed model-based expression value and its standard error. In addition, we assume that the real array-specific expression is also a random variable, $\theta_i \sim N(\mu, \tau^2)$, where μ is the expression level of this gene before replication (for example, if the total mRNA is split, μ is the expression level in the total mRNA), and τ is the variation introduced after the replication. Taken together, $\hat{\theta}_i \sim N(\mu, \tau^2 + \hat{\sigma}_i^2)$.

We are mainly interested in estimating μ. If τ^2 was known, a linear unbiased minimum-variance estimate of μ would be

TABLE 5.2. Example of pooling replicate arrays considering measurement accuracy.

Gene	$\hat{\theta}_1$	$\hat{\sigma}_1$	$\hat{\theta}_2$	$\hat{\sigma}_2$	$\hat{\mu}$	$std(\hat{\mu})$	Comment
1	100	20	200	20	150	52.8	$\hat{\mu}$ is the same as simple average
2	100	20	200	300	124.3	178.9	$\hat{\mu}$ is close to $\hat{\theta}_1$
3	100	20	101	20	100.5	19.9	$std(\hat{\mu})$ is closer to $\hat{\sigma}_1$ or $\hat{\sigma}_2$

$$\hat{\mu} = \frac{\sum_i a_i \hat{\theta}_i}{\sum_i a_i} \quad \text{where} \quad a_i = \frac{1}{\tau^2 + \hat{\sigma}_i^2}, i = 1, \ldots n \tag{5.2}$$

with variance $\sigma_{\hat{\mu}}^2 = 1/\sum_i a_i$ smaller than any $\tau^2 + \hat{\sigma}_i^2$.

One may use the sample variance of $\hat{\theta}_i$ to estimate τ^2 when $\hat{\sigma}_i^2$ is considerably smaller than τ^2 and the number of replications is large, but often the number of replicate arrays is small (for example, in duplicates), and using $\hat{\theta}_i$ alone may underestimate τ^2 if $\hat{\theta}_i$ happen to be close to each other. dChip uses a resampling approach to perturb $\hat{\theta}_i$ to estimate τ^2 multiple times. Specifically, the observed expression values are resampled from $\tilde{\theta}_i \sim N(\hat{\theta}_i, \hat{\sigma}_i^2)$, and the sample variance of the resampled $\tilde{\theta}_i$ is a "resampled estimate of τ^2," denoted $\tilde{\tau}^2$. After resampling 20 times, 20 such $\tilde{\tau}^2$ are averaged to obtain $\hat{\tau}^2$, an estimate of τ^2. Finally, μ is estimated using formula (5.2) by substituting τ^2 by $\hat{\tau}^2$. Table 5.2 shows examples where the resampling method helps for genes 2 and 3.

5.3   Software and Applications

dChip is a single executable program running on Windows NT/2000. The computer memory in megabytes should be as large as the number of arrays (for example, if there are 100 arrays to analyze together, we need > 100MB memory). In the following, we describe the main functions and outputs of the software.

5.3.1   Reading in Array Data Files

On startup, dChip enters the "Analysis View," which can also be accessed at any time by clicking the "Analysis" icon on the left panel or selecting the menu "View/Analysis." The "Analysis View" displays information such as the status of analysis processes and the error messages (colored in red).

dChip analysis is based on a group of array data files that a researcher generates, either in DAT or CEL format. All the arrays to be used in a single

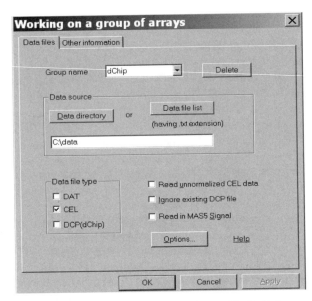

FIGURE 5.4. The "Open group/Data files" dialog.

analysis should be of the same chip type. The current limit on the number of arrays is 400. To read in the data, select the menu "Analysis/Open Group" (Figure 5.4). Type in a group name in the "Group name" drop-down list, or click the down-arrow button to select a previous group. Click the "Data directory" button to choose the directory containing the data (DAT, CEL) files to be analyzed. Alternatively, a "Data file list" can be used when the data files are stored in several directories or we want to specify individual data files. Click the "Working directory" button or directly input to specify a "Working directory" for dChip to output analysis files. Next, specify whether you want to read in DAT or CEL files. dChip will automatically look for MAS's analysis result for absolute/detection calls (text output of the CHP file, with the same file name as the DAT or CEL file except with the extension .txt).

After completing the "Analysis/Open group/Data files" dialog, click the "Other information" tab on the top (Figure 5.5). The CDF file for the current chip type can be obtained from the MAS or Affymetrix library CD. An optional but desirable "Gene information file" can be specified in this dialog. We have organized gene annotation information in a set of "Gene information files" downloadable from the dChip Web site. They combine Affymetrix's annotation and some gene functional classifications from NCBI's LocusLink database (annotation is also discussed by Gentleman and Carey in Chapter 2 of this volume), which are used to automatically identify functionally significant clusters in the clustering analysis (Subsec-

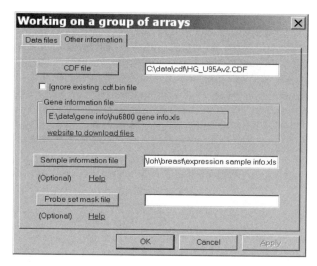

FIGURE 5.5. The "Open group/Other information" dialog.

tion 5.3.7). Finally, if array data file names are not informative, we can specify alternative names for arrays in a "Sample information file."

Click "OK" after filling in file names. dChip reads in the CDF file, CEL files, and other information files. An "array summary file" is saved after all arrays are read in. The file is a tab-delimited text file but has a .xls extension for easy opening by Microsoft Excel. The binary version of the CDF file and the CEL files are saved in .cdf.bin or .dcp files; thus, the next time "Open group" is selected, the data can be read in faster.

5.3.2   Viewing an Array Image

Next, click the "CEL Image" icon in the left panel or select the menu "View/CEL Image" to expand the blue image icons for arrays. Click an array name on the left panel to display its image (Figure 5.6). Since the displaying dynamic range is from the 1% quantile (below this intensity, the color is black) to the 95% qauntile (beyond this intensity, the color is the brightest yellow) of all the *PM* and *MM* probe intensities, images for different arrays have similar overall brightness visually, regardless of the actual brightness of the array images.

Click inside the array image to select a probe set and activate the "Image View." The blinking bar (or the scattered blinking probe cells for arrays with "distributed probe set format") indicates the currently selected probe set. Click other image areas or use the "Home" and "End" keys to highlight other probe sets. In the bottom status bar, some information is displayed: the array name, the current probe-set name (with number of probe pairs and the presence call), and the intensity value at the current cursor position.

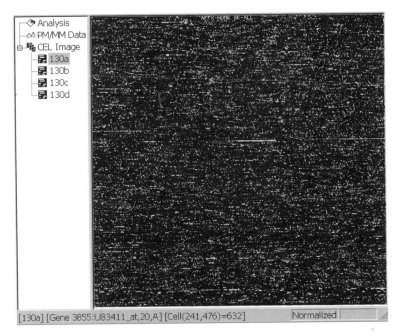

FIGURE 5.6. The "CEL image view."

Use the arrow keys to zoom the array image: "right" or "down" keys for a larger image and "left" or "up" keys for a smaller image. Select the menu "View/Export image" to export the array image into .jpg or .bmp file.

The latest arrays have the "distributed probe set format," where the probe pairs of the same probe set are scattered into the various places on the chip. This is to prevent local image contamination from completely destroying the *PM* and *MM* information of a probe set. Toggle the menu "Image/Unscrambled" to reorganize the array layout so the probe pairs for the same probe set appear adjacent in the array image.

Some images may have obvious local contamination. If not handled properly, such contamination can affect downstream normalization and expression level comparison. Assuming that the contamination is "additive" on the true signals and behaves like a semitransparent layer (Schadt et al., 2001), we implemented an image gradient correction algorithm in dChip. In the "CEL Image" view (check "Use unnormalized data" at the "Open Group" step), one can right-click to outline a contaminated image region (Figure 5.7, color insert, left picture; left-click to cancel, double-right-click to finish). Then select "Image/Gradient Correction" to adjust the background brightness of this region to a similar level as the background of the surrounding region. The background of a CEL is defined as the median of the CEL values in the 7×7 square centering around this CEL and is calculated for CELs in the outlined region as well as the surrounding region extending out by 7 CELs (Figure 5.7, middle picture). Then, a CEL value in the outlined contaminated region is adjusted by the difference between its background and the median background of the extended surrounding region (Figure 5.7, right picture).

5.3.3 Normalizing Arrays

Since scanned images may have different overall brightnesses, it is important to normalize arrays to make them comparable either before or after computing expression values. The MAS software analyzes one array at a time so that we can scale or normalize the expression values after calculating them. However, the model (5.1) assumes that arrays are already at comparable brightnesses so we need to normalize arrays at the PM/MM intensity level before computing the MBEI. In addition, the "PM/MM Data View" (Subsection 5.3.4) can best be viewed after normalization. Otherwise, during animation it is not difficult to find probe sets where the MM curves jump up and down for the same probe set, indicating the different overall background brightnesses of different arrays.

To perform normalization, select the menu "Analysis/Normalize." By default, the array with median overall intensity is chosen as the baseline array against which other arrays are normalized. In the dialog, we can specify a different array as the baseline. This is useful if the default baseline array has problems such as image contamination. Check "Ignore the normalized data" to force dChip to renormalize. Click "OK" to perform the "invariant set" normalization (subsection 5.2.1).

After normalization, the normalized CEL values are saved in the DCP files and a "Normalized" mark is shown at the lower-right-hand corner (Figure 5.6). The next time that dChip is started, the normalized CEL values will be read in by default.

5.3.4 Viewing PM/MM Data

Click the "PM/MM Data" icon in the left-hand panel (or select the menu "View/PM/MM Data" or simply press Enter at the "Image View") to view the probe-level data for the current probe set in the current array as well as the model-based expression indexes (θ), the probe sensitivity indexes (ϕ), and the fitted values (red curve) (Figure 5.8; see color insert).

There are six grids in the "PM/MM Data View." Let us use (x, y) to denote different grids, with $(1, 1)$ the upper-left grid and $(2, 3)$ the lower-right grid. Grid $(1, 1)$ displays the PM and MM data in blue and green curves, with the x-axis ordering probe sets from 1 to 20 and y-axis for probe intensities with range $(0, 2686)$. Grid $(1, 2)$ is the $PM-MM$ difference curve with the horizontal blue line $y = 0$. Grid $(1, 3)$ overlays the red fitted curve of model (5.1) to the blue $PM-MM$ difference curve and also shows the residual curve in light gray color. From this grid, we can also read that the explained variance is 97.21% after fitting the model to the $PM - MM$ difference data of this probe set (a data table of 4 arrays by 20 probes), and it takes three rounds of iterations for model fitting and outlier identification. Here we found 0 array outliers, 0 probe outliers, and 0 single outliers. Grid (2, 1) is the intensity image of the current probe set in the current array, and the intensity brightnesses are determined in such a way that the images for the same probe set across all arrays are comparable. Grid (2, 2) displays the scatterplot of the standard error of θ versus θ. The black dot represents the current array, and its value (709, 10) is displayed. Grid (2, 3) displays the scatterplot of the standard error of ϕ (has range (0, 0.19) in this case) versus ϕ. The value of the standard error of ϕ is also shown as vertical blue lines in grids (1, 1), (1, 2), and (1, 3). Probes with larger standard errors behave inconsistently with the remaining probes across arrays, as seen in the animated view ("Data/Animate"). The points representing array or probe outliers are colored blue in grids (2, 2) and (2, 3) (not shown in Figure 5.8).

Here are several visual clues in Grid (1, 3) indicating different outliers identified by the model. Single outliers do not have small circles (Figure 5.2, left picture, pointed by two black arrows), and their values are replaced by imputed values (red curve), thus leading to large residuals (these "imputed" residuals are treated as 0 when calculating standard errors for θ's and ϕ's). Probe outliers do not have fitted red curves at the corresponding probe locations for all arrays (Figure 5.2, middle picture, pointed by arrow). The data of this probe are not used to estimate the expression index θ, although the ϕ value for this probe and its standard error are still calculated using the fitted θ's (Li and Wong, 2001a). Array outliers can only be identified through blue points in Grid (2, 2), and generally the fitted values in the red curve have an obvious deviation from the original data in the blue curve (Figure 5.2, right picture). This means that the probe response pattern of this probe set in the current array is inconsistent with the patterns seen in other arrays (maybe due to image contamination, such as in Figure 5.7). Although expression values θ are calculated for such array outliers, they are attached with large standard errors, indicating that they are not reliable. The standard errors can be used to propagate the accuracy measurement into the downstream analysis, such as by downweighting unreliable expression values when pooling replicates (Subsection 5.2.4) or producing wider confidence intervals for unreliable fold changes (Subsection 5.2.3).

Click anywhere in the right panel to activate the "PM/MM Data View." Press the "PageDown" and "PageUp" keys to go to the data view of this probe set in other arrays. Select "Data/Animate" to sequentially display data curves for the current probe set in different arrays by the order of fitted θ values. In this way, we can visualize how PM and MM responses increase as mRNA level in the sample increases. Viewing such animation is instrumental in the development of model (5.1). Select "Data/Pause" to stop animation and "Data/Faster or Slower" for different speeds.

Press the "Home" and "End" keys to go to the previous or the next probe set, respectively. Toggle the menu "Data/Jump to Present" to make the "Home" and "End" keys go through each probe set or jump to the probe sets called "Present" in more than half of the arrays (it is more interesting to look at such probe sets). To view a particular probe set, select "View/Find Gene" and then input the probe-set name or probe-set number.

5.3.5   Calculating Model-Based Expression Indexes

The model-based expression indexes (MBEI) in Grid (2, 2) of the "PM/MM Data View" are calculated on the fly. To fit the model for all the probe sets and store the results, select the menu "Analysis/Model-based Expression." By default, no option is checked. Check "Ignore existing calculated expressions" to force recalculation of MBEIs instead of reading the existing values. Click OK to proceed. Some statistics from model fitting are reported in the "Array summary file": the percentage of probe sets called "array outlier" in one array, and the percentage of "probe pairs" called "single outlier" in one array. Calculated expression values and standard errors are stored in DCP files and can be exported for use with other software. Select "Tools/Export data/Expression value" to export the MBEI values.

A user is encouraged to click image icons to visually check the image after model-based expression calculation. These images will have "array outlier" (white bars) and "single outlier" (pink dots) superimposed (use "Image/Array Outlier" to hide or show outliers on the image) (Figure 5.3). Probe sets called as "array outlier" are tagged, and their expression values can be treated as missing data when clustering or exporting expression data by checking "Tools/Options/Analysis/Treating outlier expression as missing values." Single outliers (such as image spikes) have been replaced by imputed values in the model fitting, and their adverse effects have been eliminated.

The Affymetrix MAS software uses the percentage of probe sets called present (usually 30%–50%) in an array, 5′ to 3′ ratio (to validate the IVT procedure), and the signal-to-background ratio as the quality-control assessment. In contrast, dChip cross-references one array with other arrays through a modelling approach to identify problematic arrays. Having fewer outliers is a sign of better array quality and more consistency with other arrays. Arrays with large numbers of array/single outliers ($>5\%$) deserve special attention. If the number of array outliers exceeds 5% for an array, a "*" will be shown in the "Warning" column of the "Array summary file," and the corresponding array icon in the left panel will become dark blue for such "outlier arrays." The problem may be due to image contamination or that the sample was contaminated at some experimental stage. A visually clean chip may have a lot of array outliers. It is recommended to discard arrays with a large number of array outliers ($>15\%$). To do this, we can either physically remove the CEL or DCP files from the directory and then redo the normalization and model-based expression using only the good arrays, or we can deselect these arrays in the "Tools/Array list file" to exclude them from the high-level analysis. Although the array outliers do not affect the expression index calculation in other good arrays since they are excluded during iterative model fitting, the first approach is safer since an "outlier array" may be used as the baseline array for normalization.

If the image contamination is severe in one array but using the array data is still desirable, the image contamination correction (Subsection 5.3.2) may be used to alleviate the situation.

5.3.6   Filter Genes

After obtaining model-based expression indexes, we can perform some high-level analysis such as filtering genes and hierarchical clustering (Subsection 5.3.7). Generally, it is desirable to exclude genes that show little variation across the samples or are "absent" in the majority of the samples from the clustering analysis. Select the menu "Analysis/Filter genes" to filter genes by several criteria (Figure 5.9). Criterion (1) requires that the ratio between the standard deviation and the mean of a gene's expression values across all samples be greater than a certain threshold (1 in this example; the upper limit 10 is a reasonably large number that typically is satisfied). Criterion (2) requires a gene to be called "Present" in more than a percentage of arrays. Criterion (4) selects genes whose expression values are larger than a threshold in more than a percentage of arrays. See the dChip online manual for other criteria.

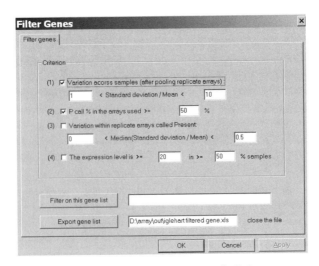

FIGURE 5.9. The "Filter genes" dialog.

The filtering can be restricted to an existing gene list if a "Gene list file" (a tab-delimited text file with the first column of each row being the probe-set name) is specified by the "Filter on" button. The genes satisfying the filtering criterion will be exported to a "Gene list file" specified by the "Export gene list" button. This file can be used for hierarchical clustering.

5.3.7 Hierarchical Clustering

Select the menu "Analysis/Hierarchical clustering" to perform hierarchical clustering (Eisen et al, 1998). Specify a "Gene list file" (which can be generated by "Analysis/Filter genes," "Analysis/Compare samples" (see Subsection 5.3.8) or "Tools/Gene list file") or a "Tree file" (saved by the "Clustering/Save tree" function so that an existing tree structure saved before can be used). dChip will use genes in the file for clustering. We can choose whether to cluster samples or genes.

Before clustering, the expression values for a gene across all samples are standardized (linearly scaled) to have mean 0 and standard deviation 1, and these standardized values are used to calculate correlations between genes and samples and serve as the basis for merging nodes. If desired, we can check "Standardize columns" to standardize the raw expression data columnwise for sample clustering instead of using rowwise standardized values for sample clustering. A user is advised to try clustering samples with or without "Standardize columns" checked to judge which option yields more reasonable sample clustering.

If the number of genes is large (e.g., 10,000), dChip may report "out of memory" or perform slowly since storing all the pairwise distances requires too much memory and may cause virtual memory swapping. The solution is to uncheck the "Tools/Options/Clustering/Pre-calculate distances" option to calculate the pairwise distances between genes on the fly. Click "OK" to start clustering. Select "Analysis/Stop Analysis" or press "ESC" to stop the ongoing analysis.

Following the clustering information output, the clustering picture will be displayed immediately in the right panel (Figure 5.10, see color insert). It can also be accessed at any time by clicking the "Clustering" icon in the left panel or selecting the menu "View/Clustering." In the clustering picture, each row represents a gene and each column represents a sample. The clustering algorithm of genes is as follows: the distance between two genes is defined as $1 - r$, where r is the correlation coefficient between the standardized values of the two genes across samples. Two genes with the closest distance are first merged into a supergene and connected by branches with length representing their distance and are then deleted for future merging. The expression level of the newly formed supergene is the average of standardized expression levels of the two genes across samples. Then, the next pair of genes (supergenes) with the smallest distance is chosen to merge, and the process is repeated $n - 1$ times to merge all the

n genes. The dendrogram on the left illustrates the final clustering tree, where genes close to each other have a high similarity in their standardized expression values across all the samples. A similar procedure is used to cluster samples.

The colored region on the right-hand side of the clustering picture represents the "functional category classification," with different colors representing distinct functional descriptions (use "Control + Click" to change the color of the functional blocks). Such information comes from the NCBI LocusLink (http://www.ncbi.nlm.nih.gov/LocusLink) database, which classifies a gene according to molecular function, biological process, and cellular component using GeneOntology (http://www.geneontology.org) terms (annotation is also discussed by Gentleman and Carey in Chapter 2 of this volume). We trace to the top levels of category classification for each gene and store them in the "Gene information file." After the hierarchical clustering, dChip searches all branches with at least four functionally annotated genes to assess whether a local cluster is enriched by genes having a particular function. Such assessment has been used in Tavazoie et al. (1999) and Cho et al. (2001) for K-mean or supervised clustering. Here, dChip systematically assesses the significance of all functional categories in all branches of the hierarchical clustering tree.

Click a "Functional cluster" icon below the "Clustering" icon in the left panel to highlight a functionally significant cluster in blue (Figure 5.10, see color insert). We use the hypergeometric distribution to measure the significance: "there are n annotated genes on the array, of which m genes have a certain function; if we randomly select k genes, what is the probability that x or more genes (of the k genes) have this certain function?"; this probability is used as the p-value to indicate the significance of seeing x genes of a certain function occurring in a cluster of k genes. In Figure 5.10, the blue

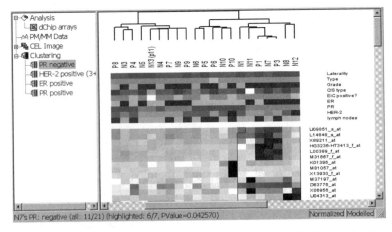

FIGURE 5.11. Hierarchical clustering of samples and enriched sample clusters.

cluster has 49 functionally annotated genes (genes without annotation are not counted), of which 6 are chaperone genes; considering that there are 61 chaperone genes in the 5009 functionally annotated genes on the array, this cluster is significantly enriched by chaperone genes. A p-value of $2.36e-05$ is calculated. P-values smaller than 0.005 are considered significant by default, and the corresponding functional clusters will be represented by the "Functional cluster" icons below the "Clustering" icon. This p-value threshold can be set to other values in the "Tools/Options/Clustering" dialog. Sometimes, different "Functional cluster" icons may represent the same cluster of genes since this cluster is enriched by several different functional terms.

Similarly, during sample clustering, the sample information specified in the "Sample information file" is used to find the sample clusters enriched by samples of a certain description (Figure 5.11).

5.3.8 Comparing Samples

Another high-level analysis performed by dChip is the comparison analysis. Given two samples or two groups of samples, we want to identify genes that are reliably differentially expressed between the two groups. dChip provides several filtering criteria to filter interesting genes. In the "Analysis/Compare samples" dialog (Figure 5.12), select one or more arrays in

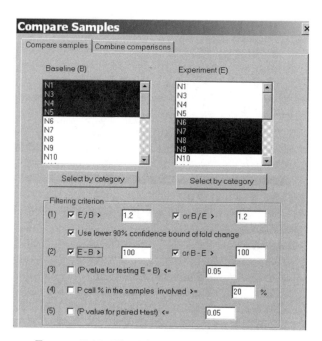

FIGURE 5.12. The "Compare samples" dialog.

the listbox of the Baseline group and the Experiment group. "E" and "B" in the dialog stand for the group means, which are the mean expression levels of arrays in the group computed by considering measurement accuracy (Subsection 5.2.4).

Next, we can select the filtering criteria to be applied when comparing the two groups. Filtering criterion (1) requires that the fold change between the group means exceed a specified threshold. We can also check the "Use lower 90% confidence bound" to specify use of the lower confidence bound of fold changes (see Subsection 5.2.3). Filtering criterion (2) concerns the absolute difference between group means. Since the down-regulated genes and the up-regulated ones have different change magnitudes, we may specify different fold-change criteria for E/B and B/E and different mean difference criteria for E−B and B−E. Note that if the expression values are log-transformed at the "Analysis/Model-based expression" step, we need to use E−B, B−E instead of E/B, B/E for fold-change threshold. Filtering criterion (3) tests the hypothesis that the two groups have the same mean using the unpaired t-test. Filtering criterion (4) requires the percentage of samples called "Present" in arrays of both groups to be larger than a threshold. Filtering criterion (5) specifies the p-value threshold for the paired t-test. The "paired t-test" assumes the first baseline sample to be compared with the first experiment sample, and so on.

After specifying these criteria, click the "Combine comparisons" tab on the top (Figure 5.13). The comparison parameters in the "Compare sam-

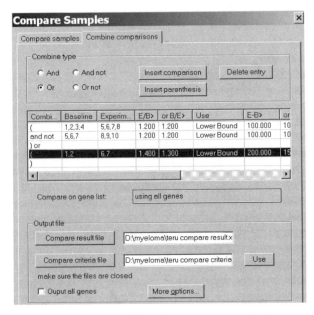

FIGURE 5.13. The "Combine comparisons" dialog.

ples" tab are automatically added in the grid sheet. We can add multiple comparisons and combine them using logical relations. To do this, click the last line of the grid sheet (the line with a single right parenthesis ")") to highlight this line, then click the "Compare samples" tab on the top to specify other comparing samples and filtering criteria, then click the "Combine comparisons" tab on the top to come back to this dialog, select the "Combine Type" to be "And," "Or," "And not," or "Or not," and click the "Insert comparison" button to insert the new comparison. The "Not" relation is useful when one wants a list of genes that are up-regulated by condition A when comparing to the control but not up-regulated by condition B. Parentheses can also be inserted after the highlighted line in the grid sheet using the "Insert parenthesis" button to specify two-level logic combination of comparisons.

Specify an output "Compare result file" and then click "OK" to proceed with the comparisons. The "Compare result file" contains genes that pass the filtering criterion (marked by "*" in the "Filtered" columns) and the comparison statistics for these genes, and it can be used as the "Gene list file" in the "Analysis/Hierarchical clustering" dialog to view the expression patterns of the filtered genes and identify functionally enriched gene clusters.

Sometimes we may wonder why some known differentially expressed genes are not selected by the comparison. In such cases, it is helpful to check the button "Output all genes" to output the statistics of all genes and then check the statistics for known genes to suggest refined filtering criteria. Typically, one may need to estimate the number of differentially expressed genes beforehand, look at the comparison statistics for known differentially expressed genes, and experiment with different parameters to finally filter a set of interesting genes.

5.3.9 Mapping Genes to Chromosomes

Mapping a list of genes generated by "Analysis/Compare samples" or "Analysis/Filter genes" may help to identify chromosomal translocations or duplications. Choose the menu "Analysis/Map chromosome" and specify a "Genome information file" (downloadable from the dChip Web site, containing chromosome number, transcription starting and ending sites, and strand information) and a "Gene list file." When a hierarchical clustering already exists, the genes used for clustering will be used for mapping. Click "OK" to enter the "Chromosome View" (Figure 5.14; see color insert). In the "Chromosome View," genes in the gene list are colored in black (or the same color as the highlighted gene branches in the "Clustering View"), and the other genes on the array are colored light gray. Genes on each chromosome are placed proportionally from chromosomal position 0 to the gene with the maximal chromosomal position in the "Genome information file." The transcription starting site is used for gene position. The p-value is calculated for every stretch of the selected genes to assess the significance of "proximity," and significant p-values are reported in the "Analysis View" (for method see the dChip online manual), such as "Chromosome 6, the stretch of gene 1 to 9 has p-value 0.021799." These significant gene stretches are also outlined in blue boxes. If a significant longer stretch contains a significant shorter stretch, only the longer one is reported and outlined.

5.3.10 Sample Classification by Linear Discriminant Analysis

Linear discriminant analysis (LDA) is a classical statistical approach for classifying samples of unknown classes based on training samples with known classes. LDA has been previously applied to the sample classification of microarray data in Hakak et al. (2001) and Dudoit et al. (2002).

The dChip "Analysis/LDA Classification" function requires the installation of R software (Ihaka and Gentleman, 1996; see the dChip online manual for details). A dialog opens for specifying the sample classes and a list of genes used as features (Figure 5.15). The gene list can be obtained by "Analysis/Filter genes" or "Analysis/Compare samples." For example, if we want to use the three sample groups in Figure 5.15 to classify the unknown samples, it is reasonable to use "Analysis/Compare samples" to compare every two of the three sample groups to obtain genes distinguishing every two groups, or more simply use "Analysis/Filter genes" to obtain genes with large variation across all samples. The "LDA result file" will

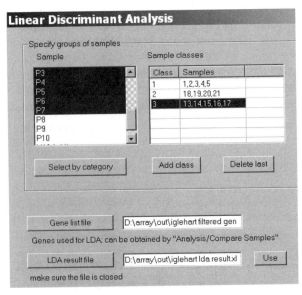

FIGURE 5.15. The "LDA classification" dialog.

hold the LDA sample class specification as well as the classification result. Previous sample class specifications stored in an "LDA result file" can be loaded by clicking the "Use" button. Select samples belonging to the same class in the left "Sample" listbox and then click the "Add class" button to add a known class. Use the "Delete last" button to delete the last sample class. Samples not added in any of the classes will be regarded as "unknown" samples, and their class labels will be predicted after LDA is performed.

Clicking the "OK" button will start R software and call its `fig5-15` and `predict.lda` functions to perform the LDA training on the known classes and predict the class labels for the unknown samples. In the output "LDA result file," LD1 and LD2 are the first two linear discriminants that map the samples with known class from the n-dimensional (n is the number of genes) space to the plane in such a way that the ratio of the between-group variance and the within-group variance is maximized. If there are only two known classes, then only LD1 is meaningful, and LD2 is arbitrarily set to the order of samples for visualization purposes. Also reported is the percentage of the correction prediction for the samples with known class. Using different gene lists may produce different prediction powers. Note here that the cross-validation is not used and that the prior is the class proportion for the training samples.

An "LDA" icon is added to the left panel after the analysis; clicking the icon or selecting "View/Classification" displays the scatterplot of LD1 versus LD2 (Figure 5.16, see color insert). Each square represents a sample, and the colors of the first four sample classes are blue, red, green, and cyan, with further colors generated randomly. The gray squares represent unknown samples, and their predicted class labels are indicated by a smaller colored square inside. Use "Arrow" or "Control + Arrow" keys to zoom and the "Enter" key to switch to other views. Select the menu "View/Export Image" to export the picture.

5.4 Discussion

There are many challenging low-level and high-level issues in analyzing oligonucleotide arrays: obtaining more accurate expression values from intensity-level data, replication schemes to best estimate the expression changes in the original biological samples, and correlating the microarray data with biological databases and human-curated knowledge bases to automate biological hypothesis formation. We will continue efforts on the methodology and software development to provide useful tools for biologists and statisticians.

Acknowledgments We thank Eric Schadt, Dan Tang, and Andrea Richardson for providing data used in this chapter and many dChip users for feedback and suggestions. The subsection 5.2.1, 5.2.2, and 5.2.3 are adapted from Li and Wong (2001b) and we thank the BioMed Central Ltd. for granting the permission. This research is supported in part by NIH Grant 1 RO1 HG02341 and NSF Grant DBI-9904701.

References

Affymetrix, Inc. (2001). Microarray Suite 5.0, Affymetrix, Inc.: Santa Clara, CA.

Cho RJ, Huang M, Campbell MJ, Dong H, Steinmetz L, Sapinoso L, Hampton G, Elledge SJ, Davis RW, Lockhart DJ (2001). Transcriptional regulation and function during the human cell cycle. *Nature Genetics*, 27:48–54.

Cox, DR, Hinkley DV (1974). *Theoretical Statistics*. Chapman and Hall: London.

Dudoit S, Fridlyand J, Speed TP (2002). Comparison of discrimination methods for the classification of tumors using gene expression data. *Journal of the American Statistical Association*, 97(457):77–87.

Eisen MB, Spellman PT, Brown PO, Botstein D (1998). Cluster analysis and display of genome-wide expression patterns. *Proceedings of the National Academy of Sciences USA*, 95:14863–14868.

Hakak Y, Walker JR, Li C, Wong WH, Davis KL, Buxbaum JD, Haroutunian V, Fienberg AA (2001). Genome-wide expression analysis reveals dysregulation of myelination-related genes in chronic schizophrenia. *Proceedings of the National Academy of Sciences USA*, 98:4746–4751.

Hill AA, Brown EL, Whitley MZ, Tucker-Kellogg G, Hunter GP, Slonim DK (2001). Evaluation of normalization procedures for oligonucleotide array data based on spiked cRNA controls. *Genome Biology*, 2(12):research0055.1–0055.13.

Hoffmann R, Seidl T, Dugas M (2002). Profound effect of normalization on detection of differentially expressed genes in oligonucleotide microarray data analysis. *Genome Biology*, 3(7):research0033.1–0033.11.

Holder D, Raubertas RF, Pikounis VB, Svetnkik V, Soper K (2001). Statistical analysis of high density oligonucleotide arrays: A SAFER approach. *Proceedings of the American Statistical Association*.

Ihaka R, Gentleman R (1996). R: A language for data analysis and graphics. *Journal of Computational and Graphical Statistics*, 5(3):299–314.

Irizarry RA, Hobbs B, Collin F, Beazer-Barclay YD, Antonellis KJ, Scherf U, Speed T (in press). Exploration, normalization, and summaries of high density oligonucleotide array probe level data. *Biostatistics*.

Lazaridis EN, Sinibaldi D, Bloom G, Mane S, Jove R (2002). A simple method to improve probe set estimates from oligonucleotide arrays. *Mathematical Biosciences*, 176(1):53–58.

Li C, Wong WH (2001a). Model-based analysis of oligonucleotide arrays: Expression index computation and outlier detection. *Proceedings of the National Academy of Sciences USA*, 98:31–36.

Li C, Wong WH (2001b). Model-based analysis of oligonucleotide arrays: Model validation, design issues and standard error application. *Genome Biology*, 2(8): research0032.1–0032.11.

Lipshutz RJ, Fodor S, Gingeras T, Lockhart D (1999). High density synthetic ologonucleotide arrays. *Nature Genetics Supplement*, 21:20–24.

Lockhart D, Dong H, Byrne M, Follettie M, Gallo M, Chee M, Mittmann M, Wang C, Kobayashi M, Horton H, Brown E (1996). Expression monitoring by hybridization to high-density oligonucleotide arrays. *Nature Biotechnology*, 14:1675–1680.

Schadt EE, Li C, Su C, Wong WH (2001). Analyzing high-density oligonucleotide gene expression array data. *Journal Cellular Biochemistry*, 80:192–202.

Schadt EE, Li C, Ellis B, Wong WH (2002). Feature extraction and normalization algorithms for high-density oligonucleotide gene expression array data. *Journal of Cellular Biochemistry*, 84(S37):120–125.

Tavazoie S, Hughes JD, Campbell MJ, Cho RJ, Church GM (1999). Systematic determination of genetic network architecture. *Nature Genetics*, 22:281–285.

Wallace D (1988). The Behrens–Fisher and Fieller–Creasy Problems. In: Fienberg SE, Hinkley DV (eds) *R.A. Fisher: An Appreciation*. pp.119–117. Lecture Notes in Statistics, Volume 1, Springer-Verlag: New York.

Zhou Y, Abagyan R (2002). Match-only integral distribution (MOID) algorithm for high-density oligonucleotide array analysis. *BMC Bioinformatics*, 3:3.

6

Expression Profiler

JAAK VILO
MISHA KAPUSHESKY
PATRICK KEMMEREN
UGIS SARKANS
ALVIS BRAZMA

Abstract

Expression Profiler (EP, http://ep.ebi.ac.uk/) is a set of tools for the analysis and interpretation of gene expression and other functional genomics data. These tools perform expression data clustering, visualization, and analysis, integration of expression data with protein interaction data and functional annotations, such as GeneOntology, and the analysis of promoter sequences for predicting transcription factor binding sites. Several clustering analysis method implementations and tools for sequence pattern discovery provide a rich data mining environment for various types of biological data. All the tools are Web-based, with minimal browser requirements. Analysis results are cross-linked to other databases and tools are available on the Internet. This enables further integration of the tools and databases; for instance, such public microarray gene expression databases as ArrayExpress.

6.1   Introduction

Gene expression data measuring the absolute or relative transcript abundance of potentially every gene in the cell provide valuable insights into the global functioning of organisms. There is a need for various tools and methods for analyzing these data, as well as for correlating them with other functional genomics data such as the whole genome sequences, functional annotations, protein–protein interaction data, and others.

Expression Profiler (EP) is a set of tools for answering some of the aspects of these various analysis needs. Expression Profiler consists of many tools, and each provides different ways of obtaining insights into gene expression and other genome-wide data. Here, we shall focus on those tools and methods that are directly useful for analyzing the gene expression data.

These are EPCLUST, URLMAP, EP:GO, and EP:PPI. Other tools give extra information about gene function or are used in promoter analysis: GENOMES, SPEXS, PATMATCH, and SEQLOGO.

We started the development of EP originally for identifying sets of co-expressed genes and predicting their potential coregulation mechanisms (Vilo et al., 2000). The main design principle of EP has been rapid prototyping of new analysis methods and integration of different data types and tools for analyzing these data. More recently, we have focused on providing a reliable service by offering Web-based access to different analysis methods. Access to the tools over the Web allows users to share their data and analysis results with other people and to create their own collaborative research environments.

6.2   EPCLUST

EPCLUST (Expression Profile data CLUSTering and analysis) is the module that provides the tools for gene expression data matrix analysis, including data selection and filtering, clustering, visualization, and similarity searches. Additionally, it supports missing-data imputation, data randomization, and data rescaling.

EPCLUST provides a folder-based analysis environment where the datasets are kept on the Web server in directories that are presented as folders to the user (i.e., universal containers of the data and the data analysis results). These files can be downloaded from the server-side directories for further use.

We describe the main functionality of EPCLUST by following analysis steps that may be needed in a real-life situation, starting from getting the data into the analysis tools, filtering them, performing the actual analysis, and then interpreting the results.

6.2.1   EPCLUST: Data Import

Raw data may be uploaded to EPCLUST in a number of different formats, either from the users' own computers or from gene expression databases available on the Web. These raw data then should be filtered to create an analysis dataset.

There are several ways to import raw data into the EPCLUST system, ranging from simple copy-paste or type-in and the standard file upload to the more advanced URL submission option that enables downloading the data from online gene expression databases.

copy-pasting the data is a quick way to import small datasets, perhaps to get acquainted with the tools and methods available in EPCLUST. Using artificial data randomization and generation utilities, a small dataset can be used to construct larger ones in order to study the effects and performance of various analysis techniques.

file-based upload is the customary way of uploading the data. A file is uploaded to the EPCLUST server using the form-based browser file upload capability.

URL-based submission allows the user to import into EPCLUST the data from other Web servers simply by providing a URL to these data. This will be most useful when more standard reference datasets and databases become available on the Web.

Data Formats

For quick data import, EPCLUST operates with basic text-based "spreadsheet" formats, where data items are separated by delimiters (e.g., spaces, tabs, commas, or others). Microsoft Excel is frequently used for primary data analysis; it can export the data in CSV (comma separated values) format, or as tab-delimited files, for example. First rows and columns of these spreadsheet files may be designated as the main row and column names (identifiers) for the dataset.

If users have very special, custom-formatted files, we encourage them to write perl data converters for these data types and convert the data prior to upload. For future use, these perl converters can be incorporated into EPCLUST.

With the development and expanding utilization of gene expression data XML-based exchange formats (e.g., MAGE-ML), EPCLUST will in future releases support them as one of the standard data upload options, allowing also more detailed data annotation.

6.2.2   EPCLUST: Data Filtering

Upon the transfer of a large dataset to an EPCLUST server, the user will often select only part of the data for a particular purpose. For example, after uploading the dataset of the expression of an entire genome in a time-course experiment with numerous measurements, the user may desire to focus on specific time points or experiments and only on those genes that have shown significant differential expression under certain specified conditions.

Prior to filtering the data, some basic descriptive statistics are computed automatically. These serve to help the user make more intelligent, educated data selections.

Experiment Selection

The initial list of experiments is currently obtained from the column identifiers. For each column (typically an individual hybridization representing all the genes under a specific condition), the descriptive statistics (e.g., mean, median, standard deviation) are presented, and users can choose which of the columns to use from the data matrix.

Additionally, regular expression queries may be issued against this list of experiment titles to create easier preselections.

To assist selecting the columns, EPCLUST will allow incorporation of more metadata (i.e., annotations of experimental conditions and value types). When the data are downloaded directly from a gene expression database (ArrayExpress), the database will support the retrieval of only a selection of columns (e.g., with some desired value quantitation type). For other users, XML column and row annotations will be supported, with MAGE-ML as the preferred format.

Gene Selection

The standard way to select genes from the list of all available genes (rows) is to specify suitable value ranges for different experiments, to eliminate genes that do not show enough differential expression (various numerical criteria can be used), to eliminate genes for which there are too many missing data points, or simply to allow direct regular expression searches over the row identifiers. The user can also specify a list of desired row identifiers directly.

In recent versions of EPCLUST, we have developed methods for rigorous systematic and precise ways of combining different selection criteria. This includes ways of selecting the columns of interest for each criterion, the actual condition of the criteria, and logical combination operations. These steps may be repeated as many times as necessary to create the complete query.

This mechanism allows the user to create complex queries such as:

- Select the genes that have not changed in the first six time points and then are either up-regulated or down-regulated.

- Select the genes that are more than two standard deviations away from the mean in at least 10% of the hybridizations.

Because the expressive power and hence the representation language for these queries can be quite complex, we developed a user interface where most queries can be performed by selecting the query type from a pull-down menu and specifying the criteria values (e.g., thresholds, missing value counts, and so forth) in the text boxes provided.

The query (i.e., the selected criteria) is represented as a tree in the interface (see Figure 6.1), showing its Boolean AND/OR/NOT structure. To enter a new criterion, the user provides the criterion type, required values, and a place in the query tree where it should be placed. The query is updated by clicking on the respective Boolean operator button.

The query can be previewed at any time during the process, and interim row counts are calculated, giving the user an overview of the effect of ap-

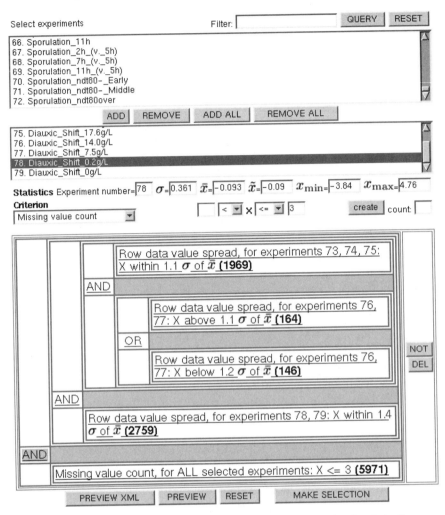

FIGURE 6.1. An example of the experiment and gene selection interface.

plying the query to a particular dataset. Finally, the query can be executed and a new dataset can be created in a set folder, where it would be analyzed further.

6.2.3   EPCLUST: Data Annotation

The rows and columns of data have tags by which they can be identified. The row annotations are read from a file that provides a description for every row ID. These descriptions may contain HTML markup to link to corresponding entries in a database or to highlight the key biological roles known for the genes. The annotations will be shown together with the clustering results, facilitating data interpretation, if the annotations were created carefully and with detail.

Users can also change the annotation information, which lets them concentrate on different aspects of data by substituting alternative annotation files.

The gene IDs themselves are used to identify gene clusters. A cluster is represented by a list of IDs of the objects in it. These IDs can also be carried to URLMAP, which makes cross-links from EPCLUST to any tool or database that has a Web interface (see Section 6.3).

6.2.4   EPCLUST: Data Environment

The prepared data files are stored in data folders, where various analyses can be performed. The intermediate (e.g., the all against all distance matrices) as well as the final analysis results (clustering results and visualizations) are also stored in the same folders, where they can be viewed later without having to redo the same analysis steps.

Users can generate new datasets within the same folders either by going through the full filtering and data selection procedure of selecting different subsets of data, by storing cluster contents as new data files, by generating randomized versions of the data, by transposing data matrices, or by other means.

Due to disk space limitations and machine resources, the data in the server-side folders have a limited life span, preset by the system, after which the data and the analysis files are deleted from the server. This period may be extended by the user manually for as long as the data remain needed.

Data security is organized by protecting access to folders by hiding them from public viewing and by selecting hard-to-guess names. Password or IP-address security will be implemented in future releases of Expression Profiler, with the possibility of employing additional, more advanced security mechanisms. The existing protection should be sufficient for academic use as well as for the analysis of published data. For users with higher security demands, however, we offer the intranet installation option.

EPCLUST offers a simple publishing environment for commonly used datasets. Most important, the data analysis can be presented to other users, and the datasets can be used as reference data for simple queries and visualization. For example, the similarity search allows linking from any gene to any particular dataset for the retrieval of other genes that are similarly expressed. This, combined with other tools for gene cluster annotation, sequence pattern discovery, and protein–protein interaction data analysis, provides a powerful mechanism for combining and analyzing different data types within a single environment.

6.2.5 EPCLUST: Data Analysis

Distance Measures

Cluster analysis is a type of classification algorithm that is intended to organize a collection of objects into meaningful structures. This meaning is given by the relative proximity of objects within one cluster in comparison to objects in other clusters. There exist many different distance measures for the comparison of expression profiles. These measures are grouped in EPCLUST on the basis of their general properties.

The first group consists of the standard metric distances: the Euclidean distance, the Manhattan distance (city block), and the average distance (Euclidean distance that has been normalized by the vector length).

The second group consists of distances that essentially measure the correlation (or angle) between vectors. Note that these distances capture the idea that the direction and relative intensity of the values are what is important and not the absolute values and magnitudes of the change. This corresponds well with the tasks of expression analysis, where the absolute change can be meaningless while the direction and relative scale of the change are crucial.

The third group is based on measures that work on ranked expression vectors instead of on the original values (rank correlation) or measures that are applicable only to discrete (e.g., vectors of 0's and 1's) data.

Clustering Methods

We implemented fast versions of standard agglomerative (bottom-up) hierarchical clustering methods and partitioning-based K-means methods and incorporated them into the Web interface with extensive visualization options.

The hierarchical clustering method requires that all pairwise distances be calculated first. For longer vectors (e.g., for many hybridizations analyzed simultaneously) this can be the most time-consuming step of the analysis. The distance matrices can be reused by different clustering methods after being calculated only once. The hierarchical clustering methods include:

complete linkage using maximum distance between members of two clusters for defining the distance between clusters.

average linkage using weighted and unweighted groupwise average methods WPGMA and UPGMA.

single linkage using minimum distance methods.

The K-means clustering method requires users to provide ab initio the number of clusters as well as a method for choosing the initial cluster centers. Starting from potential cluster centers, the K-means procedure assigns all genes to their respective clusters (the cluster to whose center they are closest) and refines cluster centers to be the geometric centers of gravity for each defined cluster. This process is repeated until the clustering stabilizes or the number of allowed cycles is reached.

Other clustering methods are currently being added to EPCLUST. These include the linear-time top-down divisive hierarchical clustering method SOTA (self-organizing trees (Herrero et al., 2001)), and some variants of the standard self-organizing map (SOM) algorithms (Kohonen, 1997). For extensive coverage of clustering methods, see, for example, Legendre and Legendre (1998) and Everitt et al. (2001).

Similarity Searches

Similarity searches provide users with alternative straightforward ways to study gene expression without requiring clustering of the whole dataset first.

The user can select genes of interest by typing in their IDs or by querying from dataset annotations and then use these genes to query the entire dataset for the similar ones.

Genes whose expression profiles are similar under a chosen distance measure (a variety are available) are reported. To reduce the size of the meaningful answer, users can set cutoff thresholds for the number of genes to be displayed or for the maximal distance that is relevant to their query, such as the estimate of the significance threshold for a particular dataset (Kruglyak and Tang, 2001).

There exists also an "opposite" search in the sense that only the genes of farthest distance are displayed in the result set. In this way, the user can study the genes that are anticorrelated with the gene of interest.

The similarity search can be performed either on one gene at a time or on a set of genes simultaneously. If many genes are used in the search together, the results can be displayed either for each one individually or as a combined result merging all the individual results. In this way, for example, starting from a tight cluster of coexpressed genes, one can search for other genes that have profiles similar to any of the original ones. This can be seen as generating a "supercluster."

Visualization Methods

The results of the cluster analysis and similarity searches are presented visually in the heat-map format, made popular in the microarray community by Mike Eisen (Eisen et al., 1998). The results are presented in PNG (Portable Network Graphics) or GIF formats, with clickable regions for drilling into the expression data (e.g., exploring subtrees in a hierarchical clustering).

The main objective of using the heat-map presentation is to represent the numerical values by colors, allowing the investigators to distinguish the general features and trends in the data. The continuous spectrum of gene expression values is discretized into a relatively small number of value slots, and then each slot is assigned a color on a single unifying image, thus enabling the viewer to quickly spot areas of important variation in the data (see Figure 6.2 in color insert).

The assignment of specific colors to these discrete value slots is a problem that is complicated by the fact that the visual perception of the RGB (red-green-blue) color model does not map linearly to the data variation that needs to be visualized. We call the complete assignment of colors to discrete slots a color schema. It is the mapping of what we know as a continuous color palette to the discrete set of values that represent the distribution of data in the underlying gene expression matrix.

Because the gene expression data distribution can vary wildly from one microarray chip to another, it is impossible to design one color schema that would fit all chips and would visualize the data successfully. What does it mean to visualize the data well? There are a few benchmarks by which we can judge a heat map:

• Outliers must be clearly visible.
• Differing values should have distinguishably different colors.
• Color representations of opposite values must be seen as opposites by people with color vision deficiencies.

Sometimes it also is important that points that are very close be dispersed into different colors instead of all blurring into an indistinct blotch.

The main difficulty in the design of a successful color schema lies in applying the appropriate discretization of the palette. In Figure 6.2 we demonstrate alternative visualizations for the same dataset. The sizes of the discrete slots determine the relative resolution of the color representation of the underlying values. The narrower and the more numerous the slots, the more different colors will be assigned to closely neighboring values. The wider and the fewer the slots, the fewer colors will be used to group the values together.

The heat-map color schemas for EPCLUST can be chosen by the user from a selection of standard choices or designed according to specific needs or preferences. The premade palettes include the green-black-red, the blue-white-pink, and black-and-white palettes and so should be visually accessible to most users.

Palettes can be used by choosing different numbers of colors, thus essentially providing different resolution visualizations. For selecting the requested number of different colors from the "continuous" palettes we provide different methods. The linear schema chooses the colors at equal intervals from the palette, whereas harmonic and exponential schemas allow varying the distance on the palette. This is important because the "equal distances" along the palette may have different separation capabilities for humans in different regions of the palette. Typically, we want to avoid the situation where most of the values in the center of the distribution would be represented with almost indistinguishable shades of color. The histogram-based discretization is an experimental feature that attempts to create the discretization automatically, using a histogram of the distribution of the data; that is, assigning an equal number of data values to each of the colors.

To create a customized color schema, the user first will need to define a color palette and then create the discretization.

To define the color palette, users are presented with a color cube, where each click identifies the next color to be placed in the palette. The chosen colors are placed at equal intervals on the linear palette representation, and all the intermediate values are linearly extrapolated between the neighboring color values. For example, to create the familiar green-black-red palette, first one clicks on a green square, then on a black square and, finally on a red square.

To create a discretization, or the assignment of different numerical ranges to different color values on the palette, users can choose one of the discretization types (linear/exponential/harmonic/histogram) and enter the number of slots into which the palette will need to be discretized.

The newly designed color schema becomes immediately available in the EPCLUST clustering visualization options page, where it can be used for visualizing the data.

6.3   URLMAP: Cross-Linking of the Analysis Results Between the Tools and Databases

From the very early days of development of EPCLUST, it was obvious that expression data analysis, however important, cannot be viewed in isolation from other available data and information about the genes and experiments. The question is how to integrate expression data analysis results (e.g., the list of genes in a particular cluster) to genomic databases, metabolic pathways, available annotations, putative promoter sequences, and their analysis (i.e., to all the resources available on the Internet, scattered between many different sites and locations).

Rather than trying to build into the EPCLUST clustering tool all the possible cross-links, we wanted to concentrate the cross-linking functionality into one central tool and to reduce the complexity of other tools by linking to that tool alone.

The main problem is that all these resources have different Web interfaces, and users often need to copy-paste the gene IDs into these tools. Moreover, each of these interfaces have different parameters and choices that need to be filled in for the queries to be meaningful. Finally, different databases often use different naming conventions for the same genes, so semiautomatic translations between these IDs are necessary. URLMAP

handles query generation automatically; users are presented with preconfig-
ured hyperlinks and/or buttons, making it easy to try out different database
queries and tools.

URLMAP is configurable on the server side with simple text file formats. It is up to the server administrator to provide most meaningful and
reasonable links in a systematic order. Links can be grouped into files by
knowledge domain (e.g., an individual file can be kept for yeast links), and
these links are available at separate URLs. When these link lists become
excessive, the tools lose their usefulness. End users, however, can assist in
developing and maintaining these lists. For example, one can design and
test custom-made mappings, which can later be incorporated into the server
set of links.

A major problem that most databases suffer from is that they do not
allow queries with multiple IDs simultaneously, and this is the typical case
when studying gene clusters. Rather, these databases require the queries to
be performed one-by-one for each of the genes separately. We hope this will
change in the future when more whole-genome data become available and
the analysis of many genes simultaneously becomes a standard practice.
For these cases, URLMAP creates multiple individual queries, and users
can click on each one separately.

Up to now we have maintained mostly the links for the yeast *Saccharomyces cerevisiae*, as there exist many useful complementary databases,
and the gene expression data are most systematic. The yeast community
has done a good job in standardizing gene identifiers: the yeast ORF names
provide the most commonly used references for each gene. However, some
databases use the gene names as the primary identifier or even provide their
own identifiers. In SwissProt, for example, the proteins corresponding to
yeast genes also have SwissProt accession numbers. The Stanford Saccharomyces Genome Database (SGD) uses its own unique IDs, which are also
used as primary identifiers in the *S. cerevisiae* GeneOntology association
files.

In most cases, there exists a one-to-one mapping between these different
naming conventions. URLMAP provides a mapping between yeast ORF
names, gene names, SwissProt IDs, and SGD identifiers, for example. When
creating links to tools and databases, if the site administrator knows which
of these naming conventions is most appropriate for the given database,
he or she may provide an automatic mapping to that naming convention
before creating the query.

6.4   EP:GO GeneOntology Browser

EP:GO is a tool for browsing and using the ontologies developed by
the Gene Ontology consortium (The Gene Ontology Consortium, 2000).
EP:GO allows study of the ontology by searching for specific entries and by

traversing the ontology directly. It provides cross-links to other databases, mapping the IDs when necessary. For example, EP:GO maps GO entries to enzyme databases by the EC numbers.

EP:GO incorporates information about gene associations (i.e., the lists of genes corresponding to each ontology entry for each annotated organism). Thus, it becomes easy to extract expression information for all genes in any particular GO category. To get all the genes in any particular category, all the subcategories have to be queried and combined. These extracted gene lists can then be analyzed by other methods; for example, their expression profiles can be looked up with the EPCLUST tool, or their respective upstream sequences can be analyzed for identifying GO category-specific transcription factor binding-site motifs. Links between these analysis tools are provided by URLMAP.

Gene associations are also useful in another respect—namely, in trying to understand the biological relevance of a cluster of coexpressed genes. Given a list of genes, it is natural to ask in which processes these genes are involved, what are the possible common functions of the genes, and whether they might be expressed in the same subcellular locations. EP:GO takes as input a list of genes and ranks the GO categories based on how well they correlate with the genes from the cluster.

6.5   EP:PPI: Comparison of Protein Pairs and Expression

High-throughput functional genomics are generating a wealth of information. Inherent is the artificial nature of many of these assays and the heterogeneity in data quality. Therefore, additional verification is necessary to reduce the number of false positives and to provide a more accurate functional annotation.

Using EP:PPI, high-throughput protein–protein interactions (PPI) can be compared with mRNA expression data. This comparison enables the prioritization of protein–protein interactions. A more reliable functional annotation can then be obtained for uncharacterized proteins, using a significance threshold, based on previously known protein interactions (Kemmeren et al., 2002).

EP:PPI uses existing PPI datasets and compares them with mRNA expression datasets selected from EPCLUST. The protein pairs are ranked according to their expression distance and plotted on a graph. Other genes that show a similar pattern in mRNA expression levels can be obtained using the similarity-based search (a direct hyperlink to EPCLUST similarity search). This also shows whether genes closely related according to mRNA expression also belong to the same functional class.

Another advantage of using mRNA expression data to prioritize the PPI is that it gives insights into particular conditions under which these proteins are coexpressed. In the future, EP:PPI will be expanded to include more different types of functional genomics data so that they can be used to accurately pinpoint false positives as well as speed up the functional annotation efforts.

6.6   Pattern Discovery, Pattern Matching, and Visualization Tools

The numerical data analysis methods for gene expression data are complemented within Expression Profiler with the sequence analysis tools. In the context of gene expression data analysis, these tools facilitate primarily the discovery, analysis, and visualization of putative transcription factor binding site motifs. This type of analysis is illustrated in the example in the next section. Here, we provide a brief description of the available tools.

SPEXS (Sequence Pattern EXhaustive Search) is a tool for discovery of novel patterns from sets of unaligned sequences. The pattern classes that can be discovered are substring patterns of any length and patterns that can contain wildcard positions, character groups (e.g., positions such as [AT], denoting either A or T), or variable-width wildcards.

PATMATCH is a tool for pattern matching and visualization used for matching regular expression patterns in sequences and for visualizing the results. PATMATCH also facilitates approximate matching of patterns for certain pattern classes. Visualization of PATMATCH search results shows the sequences and where along these sequences the matches occur. Sequence and pattern visualization can be combined with gene expression clustering and heat-map visualizations, thus allowing one to see the correlations between sequences, sequence motifs, and respective gene expression profiles.

SEQLOGO is a pattern visualization tool for DNA and protein motifs (position weight matrices) that can be used in combination with PATMATCH and SPEXS. It takes as input a set of sequences or a position count matrix and outputs a visual representation of the respective position weight matrix in the form of a sequence logo.

6.7   An Example of the Data Analysis and Visualizations Performed by the Tools in Expression Profiler

Here we present an example that illustrates how the Web-based tools in Expression Profiler can be used for analyzing the various aspects of gene expression.

Consider a yeast gene, YGR128C; it recently was given a reserved name UTP8. This gene has been classified as being of unknown function; deletion of this gene, however, results in a lethal phenotype (Winzeler et al., 1999), suggesting that this gene plays an important role.

Searching for Profiles by Similarity

As has been suggested before (Eisen et al., 1998), proteins with related functions often show coexpression on the mRNA level. In this case, we want to identify genes that are possibly functionally related (i.e., have expression profiles similar to YGR128C). For that we use the dataset from Eisen et al. (1998), available in EPCLUST in the folder All_PB. In EPCLUST, we first go to the folder named All_PB. Then, after choosing the dataset called All_genes, we select the action "Search profiles by their similarity."

Upon entering the gene name YGR128C and searching for the 100 most similar genes using correlation distance (noncentered), we receive a list of 101 genes: YGR128C and 100 other genes whose expression profiles are most similar to it. These can be viewed in the results under the expression heat map and the profile graph, where there will be a listing of 101 genes, with YGR128C at the top. It should be noted that the majority of these genes have been annotated as being of unknown function. The heat map of the expression of these 101 genes is presented in Figure 6.3 (see color insert).

Analyzing Annotations

To analyze this set further, we click the "Submit to URLMAP" button below the brief annotations. We choose the category "Bioinformatics for yeast ORFnames" from URLMAP and press "Redirect" to get the links to different databases for these 101 genes.

We select "Annotate a cluster of yeast ORFnames by GO." This analysis shows that the genes in the cluster of 101 that have an annotated function are mostly related to ribosomal RNA, transcription, the Pol I promoter, rRNA processing, ribosome biogenesis, RNA binding, RNA processing, RNA metabolism, and cytoplasm organization.

Promoter Analysis

It has been shown that by analyzing promoter regions for sets of coexpressed genes it is possible to identify putative transcription factor binding site motifs that cannot be explained by chance (see, for example, Brazma et al., 1998; van Helden et al., 1998; Vilo et al., 2000; Vilo and Kivinen, 2001). We demonstrate how this approach can be used to analyze the promoter region of the gene YGR128C and provide bioinformatical evidence for two independent binding sites that possibly regulate the expression of this particular gene.

Sequence Extraction

We choose "Genome tools: Yeast, full table" to get to the GENOMES tool, where we can study the MIPS annotations for these genes and extract upstream sequences of suitable length (start and end positions of the sequences can be determined relative to the ORF start position). These extracted sequences can be analyzed with SPEXS to identify the patterns that are overrepresented in the sequence set.

Pattern Discovery

We choose "SPEXS pattern discovery" to start the pattern discovery process. SPEXS can see the sequences extracted by the GENOMES tool; we recommend using the 600bp-long sequences, upstream relative to the ORF start (i.e., the dataset `Yeast_-600_+2_W_all.fa`). It may be that not all gene IDs from the array data will be matched with genes in the genomic sequence. In our example, only 98 genes were matched instead of 101. These sequences will be stored in a new folder.

SPEXS is able to study two datasets simultaneously, and we want to discover motifs that occur more frequently in the set of 98 upstream sequences of coexpressed genes than in the set of *all* upstream sequences in the yeast genome. In fact, a random sample of, say, 1000 upstream sequences would be sufficient for this analysis, so for the background dataset users may want to use 600bp upstream sequences of 1000 randomly chosen genes, all mapped to the same sense strand (dataset `Yeast_-600_+ 2_W_random_1000_all.fa`).

When SPEXS searches for patterns, it follows a strictly defined pattern language specified by a collection of parameters describing the acceptable patterns. Some of these parameters are the maximum motif length, inclusion of stretches of wild-card characters, length, and number of these stretches in a pattern.

For the current dataset, we require SPEXS to search for unrestricted length patterns that occur in at least 20 sequences within the cluster, allowing up to two wild-card characters within a motif, and to report the patterns that are significantly overrepresented. For overrepresentation, we require it to be at least twice as frequent within the cluster as in the random set of sequences, with the respective binomial probability less than 1e−08.

Here is the list of the top ten most significant patterns, as provided by the SPEXS pattern report:

	Pattern	Cluster	Background	Ratio	Binomial Prob.
1.	G.GATGAG.T	1:39/49	2:23/26	R:17.3026	BP:1.12008e-37
2.	G.GATGAG	1:45/60	2:44/50	R:10.436	BP:1.61764e-34
3.	GATGAG.T	1:52/70	2:72/78	R:7.36961	BP:2.79148e-33
4.	TG.AAA.TTT	1:53/61	2:79/84	R:6.84578	BP:1.83509e-32
5.	AAAATTTT	1:63/77	2:137/154	R:4.69239	BP:1.19109e-30

6.	TGAAAA.TTT	1:45/53	2:59/61	R:7.78277	BP:3.86086e-29
7.	AAA.TTTT	1:79/145	2:264/392	R:3.05349	BP:5.66833e-29
8.	G.AAA.TTTT	1:51/62	2:84/94	R:6.19534	BP:5.69933e-29
9.	TG.GATGAG	1:30/35	2:19/22	R:16.1117	BP:9.35765e-28
10.	TG.AAA.TTTT	1:40/43	2:46/48	R:8.87311	BP:1.11240e-27

This output means that, for instance, the top-ranking pattern G.GAT-GAG.T occurs in 39 sequences in the first dataset of 98 sequences, with 49 matches in total (i.e., some sequences have multiple occurrences of the motif). It also occurs in 23 out of 1000 sequences in the second, background dataset; hence, the relative ratio is $39/98 \cdot 1000/23 = 17.3$. The probability $1.12008e-37$ is the binomial probability of observing 39 sequences out of 98 that contain the pattern given the background probability that on average only $23/1000$ (i.e., 2.3%) of the randomly chosen sequences would contain that very motif.

From this list, the first and the fourth patterns represent the most significant distinct motifs (the second and third are variations of the first). These two motifs have been identified previously as the PAC and RRPE motifs, respectively, and thought to be involved in rRNA transcription and processing (Pilpel et al., 2001).

Matching Discovered Patterns to Sequences

To analyze further the discovered motifs, we use the pattern-matching and visualization tool PATMATCH. The significance of the PAC and RRPE motifs is well-supported, as PATMATCH shows that both of them are in fact well-conserved and occur on average between 50 and 250bp upstream from the ORF start (see Figure 6.4 in color insert).

In PATMATCH, users can search the sequence data for specific regular expression type patterns and see the graphic visualizations of the occurrences of these patterns on the sequences. Several patterns can be matched simultaneously and the occurrences of each pattern visualized by a different color. For simple patterns approximate matching is implemented: executing a match for −1:TGAAAA.TTT is equivalent to matching the pattern TGAAAA.TTT and allowing one mismatch.

The sequence visualizations by PATMATCH can be combined with gene expression data visualization from EPCLUST. This is illustrated in the Figures 6.4 and 6.5 (see color insert).

Sequence Logos

The motif-matching regions extracted from PATMATCH can be submitted to EP:SEQLOGO, a tool that creates position-weight matrices and respective sequence logos from the aligned or unaligned sequences. Alignments are made by looking for conserved motifs, if necessary. The position matrix is then formed by taking counts of occurrence of each nucleotide in each of the motif's positions within the matching sequence regions.

Pattern	In cluster	Total nr	Ratio	Probability
G.GATGAG.T	39	193	13.24	$2.490e-33$
TG.AAA.TTT	53	538	6.46	$3.248e-31$
TGAAAA.TTT	45	333	8.86	$1.699e-31$
−1:G.GATGAG.T	61	1295	3.09	$1.441e-19$
−1:TG.AAA.TTT	89	3836	1.52	$6.126e-12$
−1:TGAAAA.TTT	76	2190	2.27	$1.654e-18$
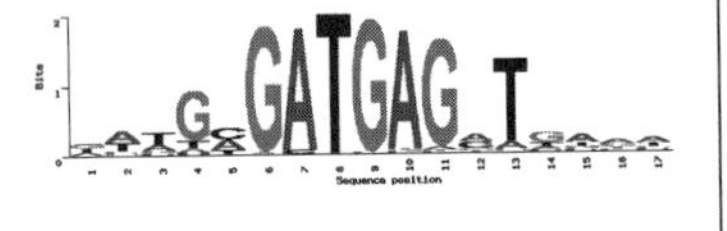	62	395	10.29	$6.909e-50$
	83	1227	4.43	$1.703e-44$
	69	593	7.63	$1.585e-48$

FIGURE 6.6. Putative yeast transcription factor binding site motifs and statistics of their occurrences in the 600bp ORF upstream regions. The second column shows the number of sequences in the cluster that match the pattern. The third column shows the total number of upstream sequences matched by the motif (including the cluster). The last two columns show the relative frequency of matches in the cluster versus the genome and the probability of such events, respectively. The results are grouped by matching patterns exactly as found by SPEXS (the top three), matching the same patterns with one mismatch (the middle three), and finally by matching of position-weight matrices derived using approximate matching within the cluster (represented by the sequence logos). Note that, with one mismatch, the number of motif occurrences increases dramatically, creating also many false matches. When approximate matching within the sequences of one cluster is used to create position-weight matrices, false-positive matches are reduced and the probability scores are actually improved over those of the exact patterns discovered by SPEXS.

To test the goodness of weight matrices derived in such a way, we used the tool ScanACE (Hughes et al., 2000) with default options to match the generated position-weight matrix against the sequences in the cluster, as well as against all the upstream sequences, to obtain comparable statistics for the top three motifs above. The results are summarized in Figure 6.6.

Assessing the Quality of the Motifs

By visual inspection, it seems that most sequences have occurrences of both of the patterns. The query G.GATGAG.T W/40 TG.AAA.TTT (pattern G.GATGAG.T within at most 40bp from TG.AAA.TTT) performed against the upstream sequences of all yeast genes shows that there are only 52 yeast genes that have both of these motifs within 40bp from each other in their upstream sequences. See Figure 6.4 for the query results and the gene expression profiles for these 52 genes. We can hypothesize that the presence of both of the motifs together determines the majority of the gene expression responses for this set of genes, including the YGR128C (Figure 6.4). This has also been suggested previously by Pilpel et al. (2001), where they report strong correlation between these two motifs during the cell cycle, sporulation, heat shock, and DNA-damage experiments.

6.8   Integration of Expression Profiler with Public Microarray Databases

ArrayExpress is a public repository for microarray gene expression data housed in the EBI (see http://www.ebi.ac.uk/arrayexpress/). It can accommodate microarray design descriptions, experiment annotations, and experiment results, satisfying the requirements posed by MIAME (Minimal Information About Microarray Experiments; (Brazma et al., 2001)). An interface has been implemented that allows Expression Profiler to import experiment results from ArrayExpress for analysis.

ArrayExpress is based on the MAGE object model, which is a standard model for microarray gene expression experiment domains. Please see http://www.mged.org/ for the details. In MAGE, the experimental data are represented as three-dimensional matrices, with every measurement having corresponding

- microarray design element (feature or group of features);
- bioassay (i.e., hybridization (for data coming out of feature extraction software) or data transformation (for derived data));
- quantitation type (i.e., intensity, ratio, present/absent call, and others).

Expression Profiler operates with data as two-dimensional matrices. The design element dimension is transferred one-to-one from ArrayExpress to Expression Profiler. In order to specify the other dimension, users can select which bioassays they want to analyze and which quantitation types should be used. The simplest case is to select all bioassays in the experiment and just one quantitation type, typically log ratio (see Figure 6.7). However, it is possible to select more than one quantitation type and not all bioassays.

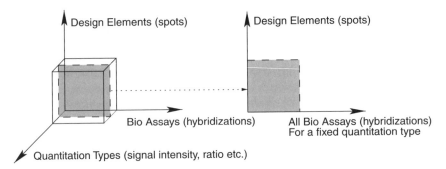

FIGURE 6.7. The schematic view of selecting a subset of data from ArrayExpress into Expression Profiler for the analysis.

Future releases of Expression Profiler will try to integrate better with ArrayExpress, possibly to allow direct queries into the database for more precise data retrieval.

6.9 Conclusions

There are many stand-alone programs that do some combination of what Expression Profiler offers. Expression Profiler stands out from these and other similar packages by providing a simple Web-based interface to a so-phisticated, integrated collection of modules that span the needs of expression data analysis from initial data retrieval and filtering to data manipulation and various means of analysis and cross-comparison.

An important distinguishing feature of this collection of tools is their flexibility: the tools are implemented as server-side components, which means that they can easily be scaled up to suit heavier computational means, they can be used concurrently by multiple users in a collaborative fashion, and, perhaps most significantly, every tool in Expression Profiler was designed with extensibility in mind. When new algorithms and databases become available, they can be added to Expression Profiler seamlessly, fitting neatly into the existing framework.

Furthermore, as we are designing Expression Profiler for future use, it is becoming more and more apparent that there is a need for the support of a canonical, open interface for the exchange of experimental expression data between software tools. We tackle this by providing a number of simple ways for data import/export and also by working closely with the ArrayExpress database team on implementing the support for the MAGE-ML data format and linking directly to this standard repository of expression experiment data.

It is already evident that expression data analysis is not a simple matter of following a cookbook recipe in a step-by-step fashion. There are no algorithms that lead to a full, unambiguous, analytical description of the experiment. Instead, the user should be able to choose from an overwhelmingly large number of methods and specialized algorithms and, importantly, often by way of experiment, arrive at conclusions and then validate them through still further analysis. We hope to continue to extend Expression Profiler in such a way that the number of available approaches will be accessible to users at every level of bioinformatical sophistication and that it will guide them through the data analysis maze. Future releases of Expression Profiler will not only concentrate on improving and adding analytical algorithms but also on improving the user interface, developing a collaborative environment, and offering ways to integrate disparate but complementary data for analysis.

Mailing List

The mailing list `ep-users@ebi.ac.uk` has been set up for the users of Expression Profiler. Update announcements are forwarded via the mailing list as well as answers to questions that users may have are provided. To subscribe to the mailing list, please send the command "subscribe ep-users" in the body of an e-mail to the address `majordomo@ebi.ac.uk`.

References

Brazma A, Hingamp P, Quackenbush J, Sherlock G, Spellman P, Stoeckert C, Aach J, Ansorge W, Ball CA, Causton HC, Gaasterland T, Glenisson P, Holstege FC, Kim IF, Markowitz V, Matese JC, Parkinson H, Robinson A, Sarkans U, Schulze-Kremer S, Stewart J, Taylor R, Vilo J, Vingron M (2001). Minimum information about a microarray experiment (MIAME) toward standards for microarray data. *Nature Genetics* 29(4):365–371.

Brazma A, Jonassen I, Vilo J, Ukkonen E (1998). Predictin gene regulatory elements *in silico* on a genomic scale. *Genome Research* 8(11):1202–1215.

Eisen M, Spellman PT, Botstein D, Brown PO (1998). Cluster analysis and display of genome-wide expression patterns. *Proceedings of the National Academy of Science USA* 95:14863–14867.

Everitt BS, Landau S, Leese M (2001). *Cluster Analysis*. London: Arnold, 4th edition.

Herrero J, Valencia A, Dopazo J (2001). A hierarchical unsupervised growing neural network for clustering gene expression patterns. *Bioinformatics* 17:126–136.

Hughes JD, Estep PW, Tavazoie S, Church GM (2000). Computational identification of *cis*-regulatory elements associated with groups of functionally related genes in *saccharomyces cerevisiae*. *Journal of Molecular Biology* 296(5):1205–1214.

Kemmeren P, van Berkum NL, Vilo J, Bijma T, Donders R, Brazma A, Holstege FC (2002). Protein interaction verification and functional annotation by integrated analysis of genome-scale data. *Molecular Cell* 9(5):1133–1143.

Kohonen T (1997). *Self-Organizing Maps*. Berlin, Heidelberg: Springer. (Second Extended Ed. 1997).

Kruglyak S, Tang H (2001). A new estimator of significance of correlation in time series data. *Journal of Computational Biology* 8(5):463–470.

Legendre P, Legendre L (1998). *Numerical Ecology*. Developments in Environmental Modelling. Elsevier.

Pilpel Y, Sudarsanam P, Church GM (2001). Identifying regulatory networks by combinatorial analysis of promoter elements. *Nature Genetics* 29:153–159.

The Gene Ontology Consortium (2000). Gene ontology: Tool for the unification of biology. *Nature Genetics* 25:25–29.

van Helden J, André B, Collado-Vides J (1998). Extracting regulatory sites from the upstream region of yeast genes by computational analysis of oligonucleotide frequencies. *Journal of Molecular Biology* 281(5):827–842.

Vilo J, Brazma A, Jonassen I, Robinson A, Ukkonen E (2000). Mining for putative regulatory elements in the yeast genome using gene expression data. In: *Proceedings of Eighth International Conference on Intelligent Systems for Molecular Biology (ISMB-2000)*, volume 8, 384–394. La Jolla, California: AAAI Press.

Vilo J, Kivinen K (2001). Regulatory sequence analysis: Application to interpretation of gene expression. *European Neuropsychopharmacology* 11(6):399–411.

Winzeler EA, Shoemaker DD, Astromoff A, Liang H, Anderson K, Andre B, Bangham R, Benito R, Boeke JD, Bussey H, Chu AM, Connelly C, Davis K, Dietrich F, Dow SW, Bakkoury M, Foury F, Friend SH, Gentalen E, Giaever G, Hegemann JH, Jones T, Laub M, Liao H, Davis RW, et al. (1999). Functional characterization of the *S. cerevisiae* genome by gene deletion and parallel analysis. *Science* 285(5429):901–906.

7

An S-PLUS Library for the Analysis and Visualization of Differential Expression

Jae K. Lee
Michael O'Connell

Abstract

This chapter describes a genomics library for S-PLUS® 6. One focus of the work involves the testing of hypotheses regarding differential expression. In this area, we provide methods and S-PLUS functions for expression error estimation based on pooling errors within genes and between duplicate arrays for genes in which expression values are similar. This is motivated by the observation that errors between duplicates vary as a function of the average gene expression intensity and by the fact that many gene expression studies are implemented with a limited number of replicated arrays (Lee, 2002).

Our clustering and visualization methods take advantage of S-PLUS Graphlets™, lightweight applets that are simply created using the Java and XML-based graphics classes and the java.graph graphics device that are new to S-PLUS 6. In addition to providing interactive graphs in a Web browser, the Graphlets enable connection to gene-information databases such as NCBI GenBank. Such connectivity facilitates incorporation of additional annotation information into the graphical and tabular summaries via database querying on the URL. The S-PLUS 6 genomics library is available at www.insightful.com/arrayAnalyzer. This site also provides updates regarding ongoing work in genomics and related areas at Insightful.

7.1 Introduction

This chapter describes a genomics library for S-PLUS® 6. Many of the functions currently in use support various microarray data analyses at the University of Virginia School of Medicine. One focus of our work involves the testing of hypotheses regarding differential expression. In this area, we provide methods and S-PLUS functions for expression error estimation based on pooling errors within genes and between replicate arrays for genes in which expression values are similar. This is motivated in part by the observation that variability between replicates often varies as a nonlinear function of the average gene expression intensity and by the fact that most gene expression studies are implemented with a limited number of replicated arrays (Chen et al., 1997; Lee, 2002).

Our clustering and visualization take advantage of S-PLUS Graphlets, lightweight applets that may be created with Java and XML-based graphics classes using the `java.graph` graphics device that is new to S-PLUS 6. In addition to providing interactive graphs in a Web browser, the Graphlets enable connection to gene-information databases such as NCBI GenBank. Such connectivity facilitates incorporation of additional annotation information into the graphical and tabular summaries via database querying on the URL.

The local pooled error (LPE) and clustering approaches are described in Section 7.2 and a detailed analysis is presented in Section 7.3 using data from a 4-chip oligonucleotide microarray study of a melanoma cell line. The cluster methods are also illustrated in Section 7.3 using the Alizadeh et al. (2000) lymphoma data. The chapter concludes with a summary of the approaches presented and future plans for the software.

7.2   Assessment of Differential Expression

It is common practice to assess each gene's differential expression intensities by examination of (typically pairwise) contrasts among experimental conditions. Such comparisons have been historically assessed as fold changes whereby genes with greater than twofold or threefold changes are highlighted for interpretation. A number of authors (Dudoit et al., 2000; Jin et al., 2001) have pointed out the basic flaw in this approach i.e., that genes with high fold change might also exhibit high variability, and hence their differential expression between experimental conditions may not be significant. Similarly, genes with less than twofold changes may have highly reproducible expression intensities and hence significant differential expression across experimental conditions.

In order to assess differential expression in a way that controls both false positives (genes declared to be differentially expressed when they are not) and false negatives (genes declared to be not differentially expressed when they are), the standard approach emerging is one based on statistical significance and hypothesis testing, with careful attention paid to multiple comparisons issues. One issue with this approach is that it relies on reasonable estimates of reproducibility or within-gene error to be constructed. Since microarray experiments are expensive (and/or the RNA sample is limited), they are typically performed with a limited number of replicates

(Eisen et al., 1998; Scherf et al., 2000). As such, error estimates constructed solely within genes may result in underpowered differential expression comparisons. For example, two-sample t-tests and permutation-based tests require estimation of a gene's within-condition errors, and such estimates may not be accurate given typical experimental replication.

In many microarray experiments, the variability of individual log-intensity measurements has been found to decrease as a nonlinear function of that gene's expression intensity (Lee, 2002). This is likely due to common background-noise processes at each spot of the microarray. At high levels of expression intensity, this background noise is dominated by the expression intensity, while at low levels the background-noise process is a larger component of the observed expression intensity. Note that log intensity, rather than original intensity, is directly relevant to the discovery of differential expression because we are interested in the genes with high proportions of change, rather than absolute-magnitude change, under varying experimental conditions.

Motivated by the issues described above, namely that

- the hypothesis testing framework for assessing differential expression is appropriate,
- microarray experiments typically have a limited number of replicates, and tests using error estimates constructed solely within genes may be underpowered, and
- variability of individual gene expression intensity measurements has been found to vary as a (nonlinear) function of a gene's expression intensity,

we construct local pooled error (LPE) estimates for within-gene expression intensities, whereby variance estimates for genes are formed by pooling variance estimates for genes with similar expression intensities.

Note that even though a gene may show quite distinctive expression patterns under different biological conditions, its expression under a particular condition is generally quite consistent, provided the quality of array instrumentation and experimental conditions have been well-controlled (Lee, 2001). Therefore, the error of a gene's expression intensity among replicated arrays can be safely considered to be a result of experimental and/or individual sampling factors. Such an error needs to be estimated reliably in order to test the significance of differential expression patterns.

Once a subset of genes is identified from the statistical comparisons described above, these genes may then be investigated by additional statistical and graphical methods such as clustering and classification (Golub et al., 1999; Scherf et al., 2000, Vilo et al., Chapter 6, this volume).

7.2.1 Local Pooled Error

Expression intensities are usually transformed as log base 2. This allows a natural interpretation of differential expression as fold changes and makes the right-skewed intensity distribution symmetric and closer to a normal distribution.

For cDNA array data, let R_j and G_j be the two (red and green) observed fluorescence intensities at spot j for $j = 1, \ldots, J$ genes. As an initial exploratory plot of expression intensities within a chip, most analysts prefer to plot the log intensity ratio $d = \log_2 R/G$ versus the mean log intensity $a = \log_2 \sqrt{RG}$ rather than plotting $\log_2 R_j$ vs. $log_2 G_j$ (see, for example, the following chapters in this volume: Dudoit and Yang, Chapter 3, Irizarry et al., Chapter 4, Colantuani et al., Chapter 9 and Wu et al., Chapter 14). The d vs. a plot provides a raw look at the data and is useful in detecting outliers and patterns of intensity variation as a function of mean intensity, whereas the $\log_2 R$ vs. $\log_2 G$ plot sometimes gives the illusion of better reproducibility than is actually present. The d vs. a plot is also useful for implementing within-chip normalization methods (Yang et al., 2001; Dudoit and Yang, Chapter 3, this volume).

Similarly for oligonucleotide data (e.g., Affymetrix GeneChip^{TM} or Amersham CodeLink^{TM}arrays), let x_{ijk} be the observed expression intensity at gene j for array k and sample i. One may examine a plot of the log intensity ratio between arrays and within genes. For duplicate arrays, $k = 1, 2$, plots of $d = \log_2 x_{ij1}/x_{ij2}$ vs. $a = \log_2 \sqrt{x_{ij1}x_{ij2}}$, $j = 1, \ldots, J$, can facilitate the investigation of between-duplicate variability in terms of overall intensity. For different samples or experimental conditions, $i = 1, 2$, plots of $d = \log_2 x_{1jk}/x_{2jk}$ vs. $a = \log_2 \sqrt{x_{1jk}x_{2jk}}$, $j = 1, \ldots, J$, show differential expression between samples and within genes.

Although the $\log_2$ transformation enables a convenient interpretation of differential expression as fold changes, it is not a transformation that typically stabilizes variance. Figure 7.1 shows plots of the log intensity ratio $d = \log_2 x_{ij1}/x_{ij2}$ between duplicate oligonucleotide arrays for RNA samples from a melanoma cell line (MM5) run under two sample conditions versus the overall intensity $a = \log_2 \sqrt{x_{ij1}x_{ij2}}$, $j = 1, \ldots, 12568$ genes; $i = 0, 24$ samples or experimental conditions (start and 24 hours later of a gelatin matrix condition); $k = 1, 2$ duplicates for each experimental condition. In this particular case, a more appropriate transformation from a variance stabilization perspective would be the square root, also shown in Figure 7.1.

Based on the considerations and observations outlined above, we propose pooling error estimates within local regions of expression intensity, where baseline variance functions are obtained from replicate arrays run under a particular experimental condition. Our local pooled error (LPE) estimates are obtained by fitting a nonparametric local regression to estimated variances of pairwise contrasts d, constructed for percentiles of a using available replicates of experimental conditions as detailed below.

The local pooled error (LPE) is derived by first estimating a baseline variance function for each of the compared experimental conditions, say A and B. For example, when duplicated arrays $(A1, A2)$ are used for condition

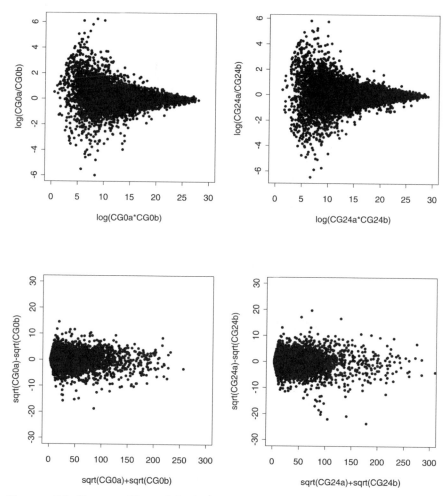

FIGURE 7.1. Top row: Plots of the log intensity ratio $\log_2(x_{ij1}/x_{ij2})$ between duplicate samples under the same (treatment) condition versus the overall intensity $\log_2(x_{ij1}x_{ij2})$, $j = 1, \ldots, 12568$ genes; $i = 1, 2$ (treatment) conditions, $k = 1, 2$ duplicates. Bottom row: Plots of the square root intensity ratio $\sqrt{x}_{ij1} - \sqrt{x}_{ij2}$ between duplicate samples under the same (treatment) condition versus the overall intensity $\sqrt{x}_{ij1} + \sqrt{x}_{ij2}$, $j = 1, \ldots, 12568$ genes; $i = 1, 2$ (treatment) conditions, $k = 1, 2$ duplicates. Data are the melanoma data described in Section 7.3; the left-hand column shows the samples at 0 hours and the right-hand column after 24 hours of a gelatin matrix condition.

A, the variance of d ($= A1 - A2$ and $A2 - A1$) on each percentile range of $a = A1 + A2$ is evaluated. When there are more than duplicates, all pairwise comparisons of (d, a) are pooled together for such estimation. A nonparametric local regression curve is then fit to the variance estimates on

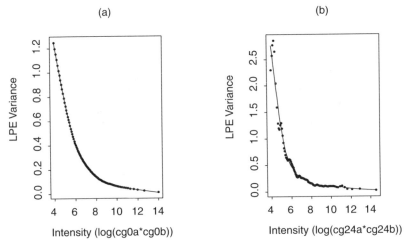

FIGURE 7.2. Plots of the local pooled error variance of the log intensity ratio d within treatments versus the overall intensity a within treatments. Data are the melanoma data described in Section 7.3: (a) start and (b) 24 hour. The LPE estimates decay exponentially in both cases. The 24-hour array replicates show larger LPE variances than the start replicates.

the percentile subintervals (refer to Figure 7.2 as an example; Cleveland, 1979). The baseline variance function for condition B is similarly derived.

Therefore, each baseline-error distribution is derived in two steps from all the replicated arrays under each condition: (1) estimation of the error at subintervals of equal percentiles and (2) a nonparametric local regression fit to the subinterval errors. This two-stage variance function estimation is adopted to facilitate the hypothesis test development and because direct non-parametric estimation can result in inappropriately small or large estimates of error in the subinterval at the right-hand end of the expression intensity domain.

Others have also recommended using pooled error methods in the analysis of differential expression. Nadon et al. (2001) constructed an ad-hoc regression of standard deviation estimates $\hat{\sigma}_j$ on the observed mean log intensities $\bar{y}_j$ over the K replicates using a B-spline. Rocke and Durbin (2001), developed a two-component variance model that captures most of the mean-variance structure. Durbin et al. (2002), Huber et al. (2002), and Munson (2001) described a family of transformations that stabilize the variance of microarray data.

Our approach is not only motivated by our observation that the variance of the log intensity ratio typically decreases as a function of the overall intensity but that biologically genes are thought to have their own inherent variability. This has led some analysts to define and fit models to individual genes and to estimate expression variability within genes (e.g., Jin et al.,

2001). Although this is a safe approach biologically, the resulting tests for
differential expression may be underpowered since researchers typically cannot replicate many conditions, and thus within-gene variability estimates
are based on very few degrees of freedom (from a small number of replicated
arrays).

On the other hand, global estimates of variability and global variance
function estimation (e.g., Carroll and Ruppert, 1988; O'Connell et al.,
1993) are not plausible biologically or even empirically. However, array
instrumental/experimental error, which can be observed from duplicated
arrays, is relatively homogeneous within a local intensity region.

As a result of the considerations described above, we believe the best
estimates of variability lie between the extremes of estimating error solely
within genes and estimating error as some global function of the mean.
We believe that within-gene comparisons based on our local pooled error method are faithful to within-gene mechanisms while simultaneously
borrowing strength from genes with similar expression intensities for the
experimental conditions under consideration.

7.2.2   Tests for Differential Expression

Two-sample t-statistics can be used for testing each gene's differential expression. However, the p-value associated with this one-step statistic must
be adjusted for the number of possible comparisons (Milliken and Johnson, 1984). Most of these adjustments (e.g., Bonferroni or Fisher's LSD
corrections) often lead to very conservative tests so that many meaningful
biological expression patterns can be missed with a high false-negative error
rate. Using a permutation sampling strategy, Dudoit et al. (2000) suggested
Westfall and Young's step-down method to remedy this problem. However,
this method requires a large number of replicated arrays (e.g., ten or more
replicates for all conditions). When a very small number of replicates are
available per condition (e.g., duplicate or triplicate), this approach does
not provide a reliable hypothesis-testing framework.

We evaluate the significance of our LPE statistics with different strategies: (1) when three or more replicated arrays (per condition) are available
and (2) when only duplicates are available. In the first case, each gene's
median expression log intensity for the two compared experimental conditions is calculated to derive a z-statistic of median difference assuming
the normal approximation of medians. Note that this test is different from
a nonparametric median test, which typically has low statistical power in
small sample cases (David, 1981). Such a normal approximation is justified
by approximate normality of the array log intensity values within local intensity ranges. Even though the distribution of log intensity values on the
whole range often has heavy tails (compared to a normal distribution), it is
found to closely follow a normal distribution within a local intensity range
(data not shown).

Following the approach outlined in Mood et al. (1974), let $\{Y_{(k)}\}_{k=1}^{n}$ be the order statistics of a random sample $\{X_i\}_{i=1}^{n}$ i.i.d. from a distribution F (or density function f). Then $Y_{(k)}$ is asymptotically distributed as normal with mean ξ and variance $\{p_\xi(1 - p_\xi)\}/\{nf^2(\xi)\}$, where $p_\xi = k/(n + 1)$ and $\xi = F^{-1}(p_\xi)$.

Based on this result, the LPE test for the median log intensity difference is approximated as a z-statistic,

$$z = \frac{M_1 - M_2}{\sigma_{pooled}},$$

where M_i, $i = 1, 2$, is the median intensity of the ith sample, $\sigma_{pooled}^2 = 1.57\{\sigma_1^2(M_1)/n_1 + \sigma_2^2(M_2)/n_2\}$, n_1 and n_2 are the number of replicates in the two array samples being compared, and $\sigma_i^2(M_i)$, $i = 1, 2$, is the error estimate from the ith LPE baseline-error distribution at median M_i.

The constant factor 1.57 is obtained if ξ and f in the asymptotic variance expression above are replaced by the median and a standard normal density function. Note that even though the LPE pooled variance is 1.57 times larger than that of the sample-mean-based z-statistic, this LPE statistic (based on medians) is robust to outliers if there are three or more replicates.

When only duplicates per condition are available for two comparing conditions A and B (say, A_1, A_2 arrays and B_1, B_2 arrays), one can still calculate the LPE test statistics for the median difference as before. However, in this case medians are the same as means, and the LPE test is no longer robust. Therefore, we identify differentially expressed genes slightly differently in this case. First, the LPE test becomes a sample-mean-based z-test so that the scaling factor 1.57 for the pooled variance is not needed. Next, because this significance evaluation can be considerably affected by an outlier, we also use an LPE-based test for outlier detection and employ a procedure comparing both pairs $(A_1 - B_1, A_2 - B_2)$ and $(A_1 - B_2, A_2 - B_1)$ for reliable discovery. Genes that show significant differences in either of these two pairs are flagged as outliers.

We are currently extending our LPE analysis to small, saturated experimental designs in which just one or two of the experimental conditions are replicated and these are used to estimate a baseline variance function for all the chips. We are also exploring the incorporation of error estimation approaches for saturated designs into the variance function such as described in Haaland and O'Connell (1994). In this setting, a biologist may examine the effects of multiple experimental factors using a small number of chips and obtain tests and estimates of differential expression across various combinations of the experimental conditions.

The difference between our LPE method and other approaches, such as Bonferroni-corrected t-test and Westfall and Young step-down methods, is illustrated below for an oligonucleotide microarray study comparing two mouse immune environments. This array study was done in triplicate for two immune environments: no exposure and 48 hour exposure to an allergen. Table 7.1 shows the t-test and Westfall and Young p-values along with LPE p-values for nine genes randomly chosen from the 365 genes exhibiting significant differential expression by the LPE test (out of 12,488 spots on the array).

We note that the p-values of the LPE-based test require a less conservative adjustment for multiple comparisons because the baseline error is evaluated using all the genes in each local intensity range, that is, the p-value is already based on a (local) family-wise error rate. Nonetheless, the Bonferroni-corrected p-values for the LPE method are shown for comparison with other methods.

Table 7.1 provides some examples where the Westfall and Young method and Bonferroni-corrected t-test were unable to identify genes that were detected as differentially expressed with the LPE method. For example, the immune response E-alpha gene, which is known to be relevant to the mouse immune mechanism, is highly significant by LPE but insignificant by the other methods. Transplantation antigen (LPE p-value is 0.03) has a 2.6 fold change between the two conditions, and the observed intensities are consistently found to be around 1200 and 4000 under the two conditions, respectively. This gene shows no significance using Westfall and Young and Bonferroni-adjusted t-test methods.

We have done a simulation study to examine error rates for the LPE test and compare the approaches presented in Table 7.1. Results show that the two-sample t-test and Westfall and Young approaches are not able to identify differentially expressed genes with statistical significance, especially when only duplicates or triplicates per condition are available, while the LPE method is able to identify such genes (data not shown).

7.2.3   Cluster Analysis and Visualization

A variety of partitioning and agglomerative cluster analysis methods are available in S-PLUS 6, including a library of algorithms described in Kaufman and Rousseeuw (1990). The partitioning methods include K-means (kmeans()), partitioning around medoids (pam()), and a fuzzy clustering method (fanny()) in which probability of membership of each class is estimated. A method for large datasets, clara(), is also included; this is based on pam(). The hierarchical methods include agglomerative methods, which start from individual points clustering and successively merge clusters until one cluster representing the entire dataset remains, and divisive methods, which start by considering the whole dataset and split it until each object is separate. The available agglomerative methods are agnes(), mclust(), and hclust(). The available divisive methods are diana() and mona().

The agglomerative hierarchical methods agnes() and hclust() use several measures for defining *between-cluster* dissimilarity, including a group average, nearest neighbor (or single linkage), and furthest neighbor (com-

TABLE 7.1. Fold change and statistical significance of the oligonucleotide microarray data on a mouse immune-response study. Log intensity values of nine genes are illustrated from triplicated arrays used for each of two mouse immune conditions. The p-values are reported by the Westfall and Young step-down method, two-sample t-test, and the LPE method (p-values of t-test and LPE are adjusted by a Bonferroni correction).

Gene	No exposure (log2)	48 hour (log2)	Fold change (log2)	W–Y p-value	t-test p-value	LPE p-value
Histocompatibility antigen	10.73, 10.62, 10.57	5.89, 5.36, 5.76	−4.87	0.10	0.09	$< 1.0e - 15$
Immune response E-alpha	9.94, 9.82, 9.54	6.45, 5.87, 6.42	−3.40	0.20	1.00	$< 1.0e - 15$
Interferon gamma	7.13, 7.43, 7.07	10.24, 10.51, 10.14	3.11	0.20	0.52	$2.15e - 07$
Interferon regulator	8.96, 8.55, 8.45	5.77, 5.96, 5.35	−2.78	0.40	1.00	$6.65e - 07$
Lymphopain mRNA	10.88, 10.70, 10.61	8.84, 9.01, 8.85	−1.85	0.20	0.59	$1.63e - 04$
CD19 antigen	8.13, 8.86, 8.68	6.23, 6.59, 7.16	−2.09	1.00	1.00	$4.01e - 04$
Topoisomerase inhibitor	11.42, 10.87, 11.69	9.73, 9.81, 9.63	−1.69	1.00	1.00	$2.25e - 03$
DNA binding inhibitor	11.56, 11.37, 11.29	10.12, 9.55, 9.73	−1.64	0.80	1.00	$5.40e - 03$
Transplantation antigen	9.89, 10.57, 10.46	11.96, 11.87, 11.78	1.40	1.00	1.00	$3.46e - 02$

plete linkage). These methods proceed by merging the two clusters with the smallest between-cluster variability, based on the chosen between-cluster dissimilarity measure, at each stage of the process. The `mclust()` method assumes that data are generated from an underlying mixture of probability distributions (e.g., Gaussian distributions) and provides insight into the number of clusters, a quantity that is derived from a model selection process in its probability framework. The divisive method `diana()` starts by finding the most disparate object and splitting it into a splinter group.

All cluster methods are very sensitive to the choice of distance or dissimilarity *between points* (i.e., samples or genes). S-PLUS includes two commonly used functions for creating distances or dissimilarities between points, namely `dist()` and `daisy()`. The correlation function, `cor()`, may also be used, and `1-cor(x)` produces a matrix representing the dissimilarities between columns (samples) of a matrix $\mathbf{x} = (x_{ijk})$. The `dist()` function simply constructs distances between rows as Euclidean, Manhattan, maximum, and binary. If the data are normalized with mean zero and variance one prior to calling `dist()`, the resulting matrix is equivalent to a dissimilarity matrix produced using `cor()`.

Hierarchical methods similar to `hclust()` and `agnes()` have been widely used for the cluster analysis of microarray data. Yeung et al. (2001) discuss the benefits of model-based clustering for microarray analysis. Results from the hierarchical methods are typically represented with a dendrogram showing the hierarchy from all samples to individual samples or from all genes to individual genes. Genes with obviously nonsignificant expression values should be omitted from the clustering analyses. Genes included in the clustering analyses may be chosen using the LPE-based tests for differential expression described earlier in Section 7.2.

The partitioning methods, primarily `kmeans()` and `pam()`, are appropriate when distinct sets of subpopulations are hypothesized. Results from using the partitioning methods are typically represented with cluster biplots and silhouette plots. Cluster biplots show the subpopulations separated in the first two principal component dimensions. Silhouette plots show how well individual samples are classified.

In silhouette plots, for each object i (a sample or experimental condition typically), the silhouette value is computed and then represented in the plot as a bar of length $s(i)$. If A denotes the cluster to which object i belongs, we define

$$a(i) = \text{average dissimilarity of } i \text{ to all other objects of } A.$$

Now consider any cluster C different from A and define

$$d(i, C) = \text{average dissimilarity of } i \text{ to all objects of } C.$$

After computing $d(i, C)$ for all clusters C not equal to A, we take the smallest of them:

$$b(i) = \min_{C \neq A} d(i, C).$$

The cluster B that attains this minimum, namely $d(i, B) = b(i)$, is called the neighbor of object i. This is the second-best cluster for object i. The value $s(i)$ can now be defined:

$$s(i) = \frac{b(i) - a(i)}{\max(a(i), b(i))}.$$

We can see that $s(i)$ always lies between -1 and 1. The value $s(i)$ may be interpreted as follows:

$$s(i) \sim 1 \longrightarrow \text{object } i \text{ is well classified,}$$

$$s(i) \sim 0 \longrightarrow \text{object } i \text{ lies between two clusters,}$$

$$s(i) \sim -1 \longrightarrow \text{object } i \text{ is badly classified.}$$

The silhouette of a cluster is a plot of the $s(i)$ ranked in decreasing order. The silhouette plot shows the silhouettes of all clusters next to each other so the quality of the clusters can be compared. The average silhouette width of a partitioning cluster analysis is the average of all the $s(i)$'s from every cluster. This is a measure of quality, or goodness, of the cluster analysis.

One typically runs pam() several times, using a different number of clusters within a specified range appropriate for the number of samples, and compares the resulting silhouette plots. One can then select the number of clusters yielding the highest average silhouette width. If the highest average silhouette width is small (e.g., below 0.2), one may conclude that no substantial structure has been found.

7.3   Analysis of Melanoma Expression

This section illustrates some of the methods described above by way of an example involving the MM5 melanoma cell-line for which expression was determined at the start and 24 hours later of a gelatin matrix condition that assimilates the in vivo cellular condition of melanoma (Fox et al., 2001). This simple experimental design thus involved one factor (matrix condition) at two levels (0 and 24 hours), with expression being measured twice (or duplicated arrays) for each time point. The main hypothesis of interest involves discovering genes showing differential expression at the two time points because these genes are believed to be relevant to tumor invasion and metastasis.

We first read in the data; we illustrate this for one of the two 24 hour samples:

```
import.data(FileName="data/24h A.xls",
   FileType = "EXCEL",ColNames = "",Format = "",
   TargetStartCol = "1", DataFrame = "cg24a",NameRow = "",
   StartCol = "1",EndCol = "END",StartRow = "1",EndRow = "END")
```

We use the average difference statistics reported by Affymetrix MAS 4.0 software for analysis in this case; this can be replaced by the signal intensity for Affymetrix MAS 5.0 software or by estimates from dChip software (Li and Wong, 2001). For quality-control purposes, we first omit spots for which less than seven perfect match (PM) and mismatch (MM) pairs were used in constructing the average difference statistics.

```
cg24a$avg.diff <- ifelse(cg24a$pairs.used<7,NA,cg24a$avg.diff)
cg24b$avg.diff <- ifelse(cg24b$pairs.used<7,NA,cg24b$avg.diff)
cga$avg.diff <- ifelse(cga$pairs.used<7,NA,cga$avg.diff)
cgb$avg.diff <- ifelse(cgb$pairs.used<7,NA,cgb$avg.diff)
OLIG <- data.frame(cg24a$avg.diff,cg24b$avg.diff,cga$avg.diff,
                   cgb$avg.diff)
```

We now adjust observed intensities so that interquartile ranges and medians on all chips are set to their widest range and highest median. The former is performed by multiplying by a scaling factor and the latter by adding a constant. Note that this is a simple constant-scale & location normalization step. At this point, we also filter out (or threshold) negative intensities and set them to 1 (0 on $\log_2$ scale). We note that for MAS 5.0 software, median-adjustment and threshold filtering may not be required due to the different software algorithm.

```
OLIG.NO <- between.norm(OLIG) OLIG.N <-
data.frame(logb(ifelse(OLIG.NO<=1,1,OLIG.NO),
    base=2))
names(OLIG.N) <- c("CG24a","CG24b","CGa","CGb")
```

7.3.1 Tests for Differential Expression

We now construct local pooled error estimates for each gene within treatments. This involves first estimating the baseline error between duplicates at subintervals of equal percentiles in overall intensity and then fitting a non-parametric regression function to the estimated subinterval errors. The S-PLUS function baseOLIG() is used for this two-stage procedure with the argument q set to 0.01 for percentiles. Note that the argument q acts as a smoothing parameter of the variance function and that a lowess smooth with appropriate degrees of freedom (default $1/10q$) is produced by this procedure. Figure 7.2 shows the LPE variance of the log intensity ratio d within treatments as a function of the overall intensity a within treatments for the two experimental conditions.

```
OLIG.cg24 <- baseOLIG(OLIG.N[,1:2], 0.01)
OLIG.cg0 <- baseOLIG(OLIG.N[,3:4], 0.01)
par(mfrow=c(1,2))
plot(OLIG.cg0, xlab = 'Intensity (log(cg0a*cg0b))',
    main='(a)',cex=0.8)
plot(OLIG.cg24, xlab = 'Intensity (log(cg024a*cg24b))',
    main='(b)', cex=0.8)
```

We now construct the tests for differential expression between the 0 and 24 hour conditions in the melanoma study using the lpeOLIG() function. We summarize results from the tests as a table with genes as rows (labeled by gene name) and columns comprising the medians of each condition (0 and 24 hours in this case), the LPE p-value for differential expression, an outlier flag, and the p-value for outlier detection (based on an LPE test within replicates). This summary table is sorted by LPE p-values, so that genes with the most significant differential expression are at the top of the table listing. The summary() method on the lpeOLIG() class also has an argument for truncating the table based on LPE p-values. A sample of differentially expressed genes from the tabular output for the melanoma study is provided in Table 7.2. The sample of genes shown is from rows $440-450$ of the summary output table (i.e., the genes that had the 440th to 450th lowest p-value based on the LPE test for differential expression).

```
LPE.cg <- lpeOLIG(OLIG.cg24, OLIG.cg0)
LPE.cg.summ <- summary(LPE.cg)
LPE.cg.summ[440:450,]
```

TABLE 7.2. Sample of output from the LPE analysis of the melanoma data showing median expression intensities for the cg24 and cg0 experimental conditions, LPE p-values, outlier flags, and p-values for some differentially expressed genes. * indicates a significant outlier at the 5% level, ** at the 1% level of significance.

Gene	cg24	cg0	LPE p-value	Outlier flag	p outlier
34777_at	8.2	5.4	$2.6e-6$	.	0.44
39286_at	8.1	4.9	$2.6e-6$	.	0.27
282_at	9.5	7.7	$2.7e-6$	.	0.47
508_at	8.1	9.6	$2.8e-6$	.	0.36
41732_at	9.8	11	$2.8e-6$	.	0.37
39967_at	8.2	5.3	$2.9e-6$	.	0.39
32116_at	3.2	8.9	$3.0e-6$	*	0.03
39385_at	6.4	8.6	$3.1e-6$	.	0.41
40234_at	4.1	9.5	$3.2e-6$	**	0.01
34104_i_at	9.9	8.3	$3.2e-6$	.	0.40
35022_at	9.1	7.3	$3.2e-6$	.	0.34

We now plot the p-values obtained from the LPE-based test versus the fold change as a volcano plot. The average fold change between conditions is used as the x-axis of this plot.

```
plot.lpe(LPE.cg, p = 0.05)
```

Figure 7.3 shows the differential gene expression as a volcano plot. In this plot, genes with significant expression are those below the horizontal line. This default horizontal line setting represents a Bonferroni cutoff value

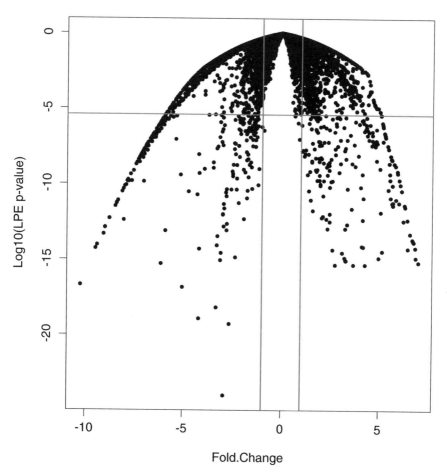

FIGURE 7.3. Plot of log10(p-value) versus Mean(Log2(Fold Change)), where the p-values are based on error estimates for within-gene expression comparisons based on the local pooled error approach. Data are the melanoma data described in Section 7.3.

of 0.05/12558, which is approximately −5.4 on the log10 scale. Note that genes below the horizontal line but between the vertical lines represent false negatives in a straight fold-change approach (i.e., these are significant using the LPE-based hypothesis test but have a fold change of less than 2). Also, points in the top-right and top-left sections of the plot represent false positives under a straight fold-change approach.

7.3.2 Cluster Analysis and Visualization

Once an interesting subset of genes are identified, one may choose to examine patterns within samples for these genes. In cancer diagnostics, there is considerable interest in subpopulations of cancer tissue samples. For example, distinct subpopulations identified within the collection of samples may have different etiologies and may be candidates for different clinical interventions. In the melanoma experiment described above, there were only four samples and two conditions. We therefore use cancer diagnostic data from Alizadeh et al. (2000) to illustrate the cluster analysis and visualization methods.

Alizadeh et al. (2000) characterized variability in gene expression among tumors in lymphoma patients using a customized cDNA lympho-chip. This chip included genes expressed in lymph cells and genes that play an important role in cancer. They ran samples from the three most common adult lymphomas on the lympho-chip: namely diffuse large B-cell lymphoma (DLBCL), follicular lymphoma (FL), and chronic lymphocytic leukaemia (CLL), and a variety of other lymphoma and leukemia cell lines. Each chip had a reference sample, with cy5 labeling used for the experimental samples and cy3 for the reference samples. Alizadeh et al. (2000) identified two distinct subtypes of DLCBL from a hierarchical cluster analysis of the resulting data; the relevant heat map and dendrogram from this analysis are given in Figure 3a of Alizadeh et al. (2000).

Note that Alizadeh et al. (2000) focused their attention on B-cell differentiation genes based on visual examination of hierarchical cluster analysis and heatmap visualization of 96 samples run on arrays of more than 10,000 genes. We do not recommend this qualitative approach; rather, we suggest genes be included in cluster analyses based on their differential expression according to a reliable hypothesis-testing procedure such as LPE-based discovery.

We provide two analyses of the subset of data presented in Figure 3a of Alizadeh et al. (2000) using hierarchical and partitioning cluster routines. We actually use the data as summarized by Cluster (Eisen et al., 1998) and prepared for viewing in TreeView (Eisen et al., 1998). Our hierarchical cluster method (hclust) uses a group average between-cluster dissimilarity measure. The partitioning method uses the partitioning around medoids method (pam).

```
aliz.cmat <- apply(mat3a,1,standard.norm)
# cluster rows
aliz.dist1 <- dist(t(as.matrix(aliz.cmat)))
aliz.hclust1 <- hclust(dist=aliz.dist1,method="average")
# cluster cols
aliz.dist2 <- dist(as.matrix(aliz.cmat))
aliz.hclust2 <- hclust(dist=aliz.dist2,method="average")
# color 6 = GC B like; color 5 = Activated B like;
# color 1 = GC centroblasts
array3a.colors <- c(rep(6,16),rep(1,2),rep(6,6),rep(5,23))
# plot heatmap and dendrograms
par(mai=c(0,0,0,0),omi=c(0,2.7,1.4,1.1))
image(aliz.cmat[aliz.hclust2$order,aliz.hclust1$order],
    axes=F,bty="n")
par(new=T,omi=c(6.55,2.75,0,1.15))
plclust2.fn(aliz.hclust2,cex=1,rotate.me=F,lty=1,
    colors=array3a.colors[aliz.hclust2$order])
par(new=T,omi=c(0.02,0.95,1.42,7.75))
plclust2.fn(aliz.hclust1,cex=0.1,rotate.me=T,lty=1)
# partitioning -- 2 classes (compare to 3 and up)
mat3a.2.pam <- pam(t(mat3a),2)
plot(mat3a.2.pam)
```

Results of the hierarchical cluster analysis are presented in Figure 7.4 (see color insert). Note that we have ordered the columns based on the default rule in S-PLUS, namely that, at each merge, the subtree with the tightest cluster is placed to the left. This is the opposite of the ordering used by the package Cluster (Eisen et al., 1998), which was used by Alizadeh et al. (2000). The columns of our heatmap are thus in approximately reverse order to those presented in Alizadeh et al. (2000). Also note that individuals within nodes are paired by their original order in S-PLUS, while this ordering is at random in the package Cluster (Eisen et al., 1998). Further, Alizadeh et al. (2000) use a weighting function that is not well documented in the Cluster (Eisen et al., 1998) manual. Overall, we produce a clustering result similar to that of Alizadeh et al. (2000) but we do not get as deep a dendrogram split between the two populations, and some of the samples are classified differently.

Since Alizadeh et al. (2000) were interested in identifying two specific subpopulations within the DLBCL samples, they may have used a partitioning method. We use the partitioning around medoids method (pam()). This analysis provides some evidence for the existence of two subpopulations rather than three, four, or five subpopulations based on average silhouette width. However, absolute values of the average silhouette width are fairly small in all cases. A silhouette plot for two subpopulations is provided in Figure 7.5.

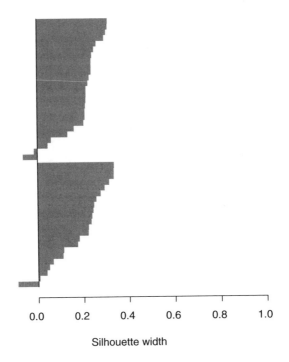

Silhouette width

Average silhouette width : 0.19

FIGURE 7.5. Silhouette plot for two subpopulations. Data are the Alizadeh et al. (2000) lymphoma data described above.

7.3.3 Annotation

The results of the hierarchical analysis of the Alizadeh et al. (2000) data may also be displayed as an S-PLUS Graphlet created through the S-PLUS 6 Java graphics (`java.graph`) device. This implementation produces a lightweight interactive applet (typically less than 30 kb in size) in a browser with mouseover metadata showing gene and sample information and expression intensity for each spot on the set of arrays. Genes are shown as rows and samples as columns. By clicking on a particular spot, the gene's accession number is sent to the NCBI UniGene database, and annotation information regarding that gene is returned in a lower browser frame. A screen shot showing the heatmap, dendrogram, and drilldown on the gene TNFRSF7 is shown in Figure 7.6 (see color insert).

S-PLUS Graphlets are typically deployed using a simple Web user interface with the S-PLUS engine on a server (e.g., the Insightful Analytic Server™ on UNIX platforms and the Insightful StatServer® on Windows platforms). In these server solutions, the data are read from a database or some other source (e.g., a Microsoft Excel file). Code snippets for StatServer and Analytic Server deployment (e.g., StatServer entry point function for reading data from a database) are available from the Web site.

Additional annotation of summary results from the LPE analysis of differential expression is also readily achieved from inside S-PLUS or through a web interface to the Analytic Server or StatServer by using functions from the Bioconductor annotation package (Gentleman and Carey, Chapter 2 this volume). For example, annotation for the sample of differentially expressed genes provided in Table 7.2 from GenBank may be obtained in S-PLUS 6 as follows:

```
hgu95Aid <- read.annotation("hgu95Aid")
hgu95All <- read.annotation("hgu95All")
names.genes <- dimnames(LPE.cg.summ)[[1]][440:450]
# look up genes by accession number
names.ACC <- unlist(hgu95Aid[name.genes])
genbank(names.ACC, disp = "browser")
# look up genes by LocusLink ID
name.LL <- unlist(hgu95All[names.genes])
locuslinkByID(as.numeric(name.LL))
# create page for LocusLink IDs
ll.htmlpage(name.LL, "main.html", "side.html")
```

7.4   Discussion

Gene discovery based on fold-change values is often misleading because variability in a gene's expression is often quite heterogeneous under different biological conditions and on different intensity ranges of array measurements (Yang et al., 2001). For example, larger variability between a pair of replicated arrays is found in genes with low-intensity expression values. Therefore, discovery using fold-change values is often misleading, and it is highly desirable to identify genes based on more rigorous measures of differential gene expression patterns.

In order to address this issue, several statistical methods for assessing the significance of differential gene expressions have been suggested based on estimating the expression variance of each gene under each experimental condition using replicated arrays (Dudoit et al., 2000; Kerr and Churchill, 2001; Wu et al., Chapter 14, this volume). For example, Dudoit et al. (2000) adjust p-values by permutation sampling as suggested by Westfall and Young. Bayesian mixture and regression models are also used to evaluate the statistical significance of differential expression (Lee et al., 2000; West et al., 2001). All of these methods have limitations, including: (1) the accuracy of the variance estimation for an individual gene is poor because the number of replicated arrays is often very small; and (2) the statistical significance of each gene's differential expression should be adjusted to avoid numerous false-positive findings. The first limitation results in most statistical tests being underpowered for genome-wide discovery. The adjustments referred to in the second limitation are generally performed assuming independence among all genes, which often results in a very conservative statistical significance measure and a high false-negative error rate.

Using our LPE approach, the sensitivity of detecting subtle expression changes can be dramatically increased and differential gene expression patterns can be identified with both small false-positive and small false-negative error rates. This is because, in contrast to the individual gene's error variance, the local pooled error variance can be estimated very accurately.

Clustering approaches have been widely applied to the analysis of gene expression data (Eisen et al., 1998; Scherf et al., 2000). In particular, the method of visualizing gene expression data based on cluster order, or cluster image map analysis, has been found to be an efficient approach for summarizing thousands of gene expression values and assisting in the identification of interesting gene expression patterns. However, it is important to understand that these clustering approaches do not directly provide any reliable measure of confidence for clustered expression patterns. A clustering method heuristically reorganizes the genes based on its predefined association distance and allocation algorithm, which only aids us in discerning coexpression patterns visually. Therefore, a validation step is required for such clustering discoveries before further inference can be drawn. For example, a bootstrapping method can be used for assessing reliability of clustering classifications of a fixed, known number of groups (Kerr and Churchill, 2001). Additional gene information is also extremely useful for discovering meaningful clustering patterns. Prior to clustering, it is also recommended to triage genes based on their statistical significance in differential expressions and to confirm consistent expression patterns within replicates (e.g., Ross et al., 2000).

The S-PLUS 6 genomics library is available at `www.insightful.com/ arrayAnalyzer`. The site will be regularly updated with ongoing work in genomics and related areas at Insightful.

Acknowledgments. We wish to acknowledge the following colleagues: P. Aboyoun, J. Betcher, D Clarkson, J. Gibson, A. Hoering, N. Jain, S. Kaluzny, L. Kannapel, D. Kinsey, P. McKinnis, D. Stanford, S. Vega, and H. Yan. This study was partially supported by the American Cancer Society grant RSG-02-182-01-MGO.

References

Alizadeh AA, Eisen MB, Davis RE, Ma C, Lossos IS, Rosenwald A, Boldrick JC, Sabet H, Tran T, Yu X, Powell JI, Yang L, Marti GE, Moore T, Hudson

T Jr, Lu L, Lewis DB, Tibshirani R, Sherlock G, Chan WC, Greiner TC, Weisenburger DD, Armitage JO, Warnke R, Levy R, Wilson W, Grever MR, Byrd JC, Botstein D, Brown PO, Staudt LM (2000). Distinct types of diffuse large B-cell lymphoma identified by gene expression profiling. *Nature*, 403:503–511.

Carroll RJ, Ruppert D (1988). *Transformation and Weighting in Regression.* Chapman and Hall: New York.

Chambers JM (1998). *Programming with Data: A Guide to the S Language.* Springer: New York.

Chen Y, Dougherty ER, Bittner ML (1997). Ratio-based decisions and the quantitative analysis of cDNA microarray images. *Biomedical Optics*, 2:364–374.

Cleveland WS (1979). Robust locally weighted regression and smoothing scatterplots. *Journal of the American Statistical Association*, 74:829–836.

David HA (1981). *Order Statistics*, John Wiley & Sons, Inc.: New York.

Dudoit S, Yang YH, Callow MJ, Speed TP (2000). Statistical methods for identifying differentially expressed genes in replicated cDNA microarray experiments, Technical Report #578, Department of Statistics, University of California at Berkeley: Berkeley, CA.

Durbin B, Hardin J, Hawkins DM, Rocke DM (2002) A variance-stabilizing transformation for gene-expression microarray data. *Bioinformatics* 18:105–110.

Eisen MB, Spellman PT, Brown PO, Botstein D (1998). Cluster analysis and display of genome-wide expression patterns. *Proceedings of National Academic Sciences USA*, 95(25):14863–14868.

Fox JW, Dragulev B, Fox N, Mauch C, Nischt R (2001). Identification of ADAM9 in human melanoma: Expression, regulation by matrix and role in cell-cell adhesion. *Proceedings of International Proteolysis Society Meeting.*

Golub TR, Slonim DK, Tamayo P, Huard C, Gaasenbeek M, Mesirov JP, Coller H, Loh ML, Downing JR, Caligiuri MA, Bloomfield CD, Lander ES (1999). Molecular classification of cancer: Class discovery and class prediction by gene expression monitoring. *Science*, 286(5439):531–537.

Haaland P, O'Connell M (1994). Inference for effect saturated fractional factorial designs. *Technometrics*, 37:82–93.

Huber W, von Heydebreck A, Sultmann H, Poustka A, Vingron M (2002). Variance stabilization applied to microarray data calibration and to the quantification of differential expression. *Bioinformatics*, 18:96S–104S.

Jin W, Riley R, Wolfinger RD, White KP, Passador-Gurgel G, Gibson G (2001). Contributions of sex, genotype and age to transcriptional variance in *Drosophila melanogaster*. *Nature Genetics*, 29:389–395.

Kaufman L, Rousseeuw PJ (1990). *Finding Groups in Data: An Introduction to Cluster Analysis.* John Wiley & Sons: New York.

Kerr MK, Churchill GA (2001). Bootstrapping cluster analysis: Assessing the reliability of conclusions from microarray experiments. *Proceedings of National Academic Sciences USA*, 98:8961–8965.

Lee MT, Kuo FC, Whitemore GA, Sklar J (2000). Importance of replication in microarray gene expression studies: Statistical methods and evidence from repetitive cDNA hybridizations. *Proceedings of National Academic Sciences USA*, 97:9834–9839.

Lee JK (2001). Analysis Issues for Gene Expression Array Data. *Clinical Chemistry* 47:1350–1352.

Lee JK (2002). Discovery and validation of microarray gene expression patterns, *LabMedica International* 19(2):8–10.

Li C, Wong WH (2001). Model-based analysis of oligonucleotide arrays: expression index computation and outlier detection. *Proceedings of National Academic Sciences USA*, 98:31–36.

Milliken GA, Johnson DE (1984). *Analysis of Messy Data* (Volume I). Van Nostrand Reinhold: New York.

Mood AM, Graybill FA, Boes DC (1974). *Introduction to the theory of statistics*, 3rd ed. McGraw-Hill, Inc.: New York.

Munson P (2001) A consistency test for determining the significance of gene expression changes on replicate samples and two convenient variance-stabilizing transformations. GeneLogic Workshop of Low Level Analysis of Affymetrix GeneChip Data.

Nadon R, Shi P, Skandalis A, Woody E, Hubschle H, Susko E, Rghei N, Ramm P (2001). Statistical inference methods for gene expression arrays, *Proceedings of SPIE, BIOS 2001, Microarrays: Optical Technologies and Informatics*, 4266:46–55.

O'Connell M, Belanger B, Haaland P (1993). Calibration and assay development using the four-parameter logistic model. *Chemometrics and Intelligent Laboratory Systems*, 20:97–114.

Rocke DM, Durbin B (2001). A model for measurement error for gene expression arrays. *Journal of Computational Biology*, 8:557–569.

Ross DT, Scherf U, Eisen MB, Perou CM, Rees C, Spellman P, Iyer V, Jeffrey SS, Van de Rijn M, Waltham M, Pergamenschikov A, Lee JC, Lashkari D, Shalon D, Myers TG, Weinstein JN, Botstein D, Brown PO (2000). Systematic variation in gene expression patterns in human cancer cell lines. *Nature Genetics* 24(3):227–235.

Scherf U, Ross DT, Waltham M, Smith LH, Lee JK, Kohn KW, Reinhold WC, Myers TG, Andrews DT, Scudiero DA, Eisen MB, Sausville EA, Pommier Y, Botstein D, Brown PO, Weinstein JN (2000). A cDNA microarray gene expression database for the molecular pharmacology of cancer. *Nature Genetics* 24(3):236–244.

Venables WN, Ripley BD (2000). *S Programming*. Springer: New York.

West M, Blanchette C, Dressman H, Huang E, Ishida S, Sprang R, Zuzan H, Olson J, Marks J, Nevins J (2001). Predicting the clinical status of human breast cancer by using gene expression profiles. *Proceedings of National Academic Sciences USA*, 98:11462–11467.

Yang YH, Dudoit S, Luu P, Speed TP (2001). Normalization for cDNA microarray data, *Proceedings of SPIE, BIOS 2001, Microarrays: Optical Technologies and Informatics*, 4266:141–152.

Yeung, KY, Fraley C, Murua A, Raftery A, Ruzzo WL (2001). Model-Based Clustering and Data Transformations for Gene Expression Data. Technical Report #396, Department of Statistics, University of Washington: Seattle, WA.

8

DRAGON and DRAGON View: Methods for the Annotation, Analysis, and Visualization of Large-Scale Gene Expression Data

CHRISTOPHER M.L.S. BOUTON
GEORGE HENRY
CARLO COLANTUONI
JONATHAN PEVSNER

Abstract

Database Referencing of Array Genes ONline (DRAGON) is a database system that consists of information derived from publicly available databases, including UniGene, Swiss Prot, Pfam, and the Kyoto Encyclopedia of Genes and Genomes (KEGG). Through a Web-accessible interface, DRAGON rapidly supplies information pertaining to a range of biological characteristics of all the genes in any large-scale gene expression dataset. The subsequent inclusion of this information during data analysis allows for deeper insight into gene expression patterns. A related set of visualization tools called DRAGON View has been developed to allow for the analysis of large-scale gene expression datasets in relation to biological characteristics of gene sets.

8.1   Introduction

Currently, microarrays are one of the primary technologies used in the generation of large-scale gene expression data. Microarrays allow for the parallel analysis of the mRNA expression for thousands of genes. A microarray is a solid support, such as a nitrocellulose membrane or glass slide, that has cDNAs or oligonucleotides corresponding to specific genes robotically spotted on it in a regular pattern. Radioactively or fluorescently labeled cDNAs converted from mRNAs extracted from biological samples such as cell culture or tissue samples are hybridized to the microarray. Detection of the relative amounts of specific cDNAs that hybridize to complementary spots on the microarray is accomplished with a phosphorimager or fluorescence reader (Bowtell, 1999; Cheung et al., 1999; Duggan et al., 1999; Lipshutz et al., 1999). Differential gene expression patterns between samples such as drug-treated versus untreated cell cultures or diseased versus control tissues are determined by comparing relative amounts of cDNA signal. Due to the amount of data that can be generated in a microarray experiment, computational methods are required for multiple tasks, including reading the data from the microarray, comparison of data between two or more experiments, analysis of data with appropriate statistical methods, and storage of the data in databases (Bassett et al., 1999; Brown and Botstein, 1999; Vingron and Hoheisel, 1999).

Microarray datasets are typically analyzed using two general approaches (Figure 8.1). First, inferential statistical methods can be applied to draw conclusions about which genes are significantly regulated in a comparison of two conditions. For example, a t-test may be applied to accept a hypothesis that a particular gene is differentially regulated at the 95% ($p < 0.05$) confidence level or a fold-difference value will be generated to evaluate relative expression of a gene in one sample versus another. A second approach is to apply descriptive or exploratory statistics to microarray datasets. A variety of methods can be applied to the entire set of gene expression data to describe patterns or signatures within the data, including K-means and hierarchical clustering algorithms (Michaels et al., 1998; Wen et al., 1998), principal component analysis (Tibshirani, 1999), genetic network analysis (Somogyi and Sniegoski, 1996; Somogyi et al., 1997; Liang et al., 1998; Szallasi, 1999), and self-organizing maps (Tamayo et al., 1999; Toronen et al., 1999). These methods identify similarity in the expression patterns of groups or clusters of genes across time or sample. When gene expression information is used to cluster samples, microarrays provide a powerful tool for medical diagnosis (Golub et al., 1999; Alizadeh et al., 2000; Colantuoni et al., 2000), toxicity analysis (Afshari et al., 1999; Nuwaysir et al., 1999), and drug screening (Debouck and Goodfellow, 1999; Fuhrman et al., 2000; Somogyi, 1999; Wilson et al., 1999). When expression information is used to cluster genes, groups of coordinately expressed (or coregulated) genes are classified into waves of expression over time or groups of similarly expressed genes across samples. By grouping genes according to their expression patterns, inferences can be made about their functional similarity. Previous studies have found that characteristics such as promoter elements, transcription factors, chromosomal loci, or cellular functions of encoded proteins have been associated with the coordinate expression of genes (Carr et al., 1997; Chu et al., 1998, Eisen et al., 1998; Gawantka

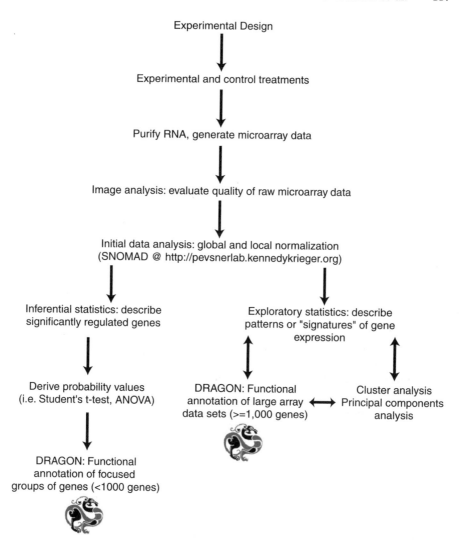

FIGURE 8.1. Overview of gene expression data collection and data analysis using microarrays. After choosing an appropriate experimental design, control and experimental biological samples are collected (e.g., cell lines +/− drug treatment). RNA is purified, labeled with fluorescence or radioactivity, and hybridized to a microarray typically representing thousands of genes. Image analysis is used to quantitate the amount that each gene is expressed. Preprocessing of the raw data may include global and local normalization procedures. Data analysis then consists of inferential statistics (to identify regulated genes with associated probability values) or exploratory statistics (to find patterns of meaning in the data, such as clustering of samples and/or genes). The DRAGON database described in this chapter automatically annotates gene expression datasets and provides visualization tools to provide insight into the biological relevance of genes whose expression is measured in microarray experiments.

et al., 1998; Heyer et al., 1999; Wen et al., 1998; Zhang, 1999; Spellman and Rubin, 2002).

Allowing for the rapid and comprehensive identification of the coordinate regulation of functionally related genes in large-scale gene expression datasets is one of the primary objectives of the development of DRAGON (Bouton and Pevsner, 2001). However, instead of starting with expression data and inferring similar biological characteristics through clustering, DRAGON allows the investigator to start with biological characteristics in order to identify which of those characteristics are associated with coordinate gene expression. This approach to analyzing expression data is not typically used because the task of understanding the biological characteristics of the thousands of genes typically present in a microarray dataset is usually left to the investigator's knowledge of the system in question, literature searches, and the tedious process of researching individual genes in public databases via the World Wide Web (WWW). DRAGON addresses this by making it possible for any investigator with access to the WWW and an Internet browser to automatically, comprehensively, and simultaneously annotate all genes from a microarray experiment with a range of biological characteristics. The information provided by DRAGON can then be integrated with expression data gained from the microarray to draw biological inferences. Thus, DRAGON acts as a tool for the biological annotation of genes in a microarray dataset. The information provided by DRAGON can be integrated with other sorts of analysis of the microarray expression data in order to gain a better understanding of how biological characteristics are related to gene expression patterns. For example, the suite of DRAGON View tools have been developed for this purpose and will be discussed in greater detail below. To summarize, DRAGON has been designed to include four criteria relevant to the analysis of microarray data. First, DRAGON simultaneously annotates large lists of microarray genes with several different types of biological information. Second, the types of information that DRAGON provides are relevant to the study of gene expression. Third, it is easy to update DRAGON as new data from genome projects and other bioinformatic sources are added to public databases. Finally, the information DRAGON provides is suitable for the grouping of microarray genes into biologically relevant categories so that the expression patterns of the annotated genes can be analyzed with reference to these categories.

In this chapter, we discuss the development and implementation of DRAGON and DRAGON View. First, we describe the means by which DRAGON is derived from publicly available databases. Then, we address the ways in which information provided by DRAGON can be integrated with other microarray data-analysis methodologies. Finally, we have applied DRAGON to two publicly available microarray datasets in order to demonstrate the types of analysis that can be performed with DRAGON and DRAGON View.

8.2   System and Methods

8.2.1   Overview of DRAGON

DRAGON is accessible via the World Wide Web (`http://pevsnerlab.kennedykrieger.org/dragon.htm`). There are three main components to DRAGON: (1) A search tool allows the user to study one gene at a time. (2) The annotate tool allows a user to annotate any number of genes (e.g., > 10,000) from a microarray experiment. Any other list of genes may be annotated as well, such as serial analysis of gene expression (SAGE) data. This annotation returns information generated by DRAGON concerning each DNA or corresponding protein in a list. (3) DRAGON View is a suite of three data visualization tools (DRAGON Families, Path, and Order) that allow for the analysis of the annotated gene expression datasets.

The DRAGON annotate tool, the centerpiece of DRAGON, uses information extracted from a number of publicly available databases accessible via the WWW (summarized in *Nucleic Acids Research* Volume 28, 1, 2000). These databases include UniGene (`http://www.ncbi.nlm.nih.gov/UniGene/`), SwissProt (`http://www.expasy.ch/sprot/`), and Pfam (`http://www.sanger.ac.uk/Software/Pfam/`). Figure 8.2 provides an overview of a subset of the tables, data types, and potential connections between data presently available in DRAGON. Each of these databases uses accession numbers to categorize its entries. In some databases, such as SwissProt, GenBank, and UniGene, accession numbers are unique to each gene or protein. In other databases, accession numbers represent families of hundreds to thousands of genes or proteins that are related by shared features such as sequence homology or similarity of cellular function. For example, accession numbers from the Pfam (Bateman et al., 2000) database classify hundreds to thousands of protein sequences into families that share homologous protein domains. Alternatively, the SwissProt database assigns a controlled vocabulary of keywords to its entries based on their known functions.

In a typical microarray experiment, the investigator generates a list of gene expression values that are ordered from those most up-regulated to those most down-regulated. Each gene expression value is associated with a GenBank accession number corresponding to the DNA element immobilized on the microarray. DRAGON uses a relational database system to associate additional features to each gene represented on the microarray, including UniGene, SwissProt, and Pfam accession numbers, chromosomal locations, keyword classification and cellular pathway participation. Figure 8.3 (see color insert) illustrates the assignment of a set of accession numbers to a given gene and its encoded protein and how the accession numbers identify biological characteristics of the gene and protein.

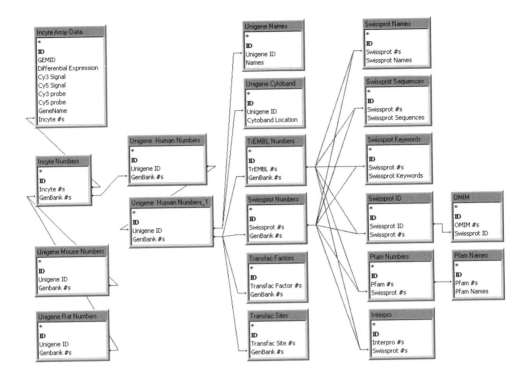

FIGURE 8.2. Overview of the information in DRAGON. This diagram represents a subset of the tables now available in DRAGON and the possible connections between them. Depending upon what type of information is desired, different sets of tables are joined with the table containing microarray gene expression data that is, for example, "Incyte Array Data" and "Incyte Numbers" in this diagram. Two "UniGene Human Numbers" tables are used to expand the "GenBank #s" from the "Incyte Numbers" table into all "GenBank #s" associated with each "UniGene ID," thereby providing a bridge between "GenBank #s" from the "Incyte Numbers" table and the "SwissProt Numbers," "TrEMBL Numbers," "Transfac Factors," and "Transfac Sites" tables. Further characterization of the proteins that genes from the microarray encode occurs by joining with tables derived from the SwissProt, Pfam, Interpro, and OMIM databases.

8.2.2 DRAGON's Hardware, Software, and Database Architecture

The DRAGON database architecture is outlined in Figure 8.4. The DRAGON database and DRAGON View tools are available for queries on the DRAGON Web site (http://pevsnerlab.kennedykrieger.org/dragon.htm). The files contained in the DRAGON database can also be

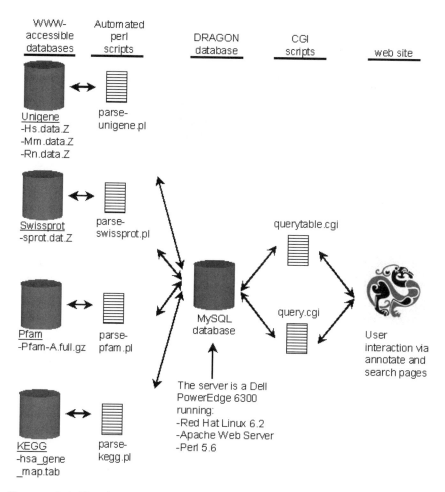

FIGURE 8.4. Database architecture for DRAGON. The data contained in the DRAGON database are derived from World Wide Web (WWW)-accessible databases that are downloaded by FTP, parsed using Perl scripts, and stored in tables in the MySQL relational database management system. The DRAGON database is housed on a Dell PowerEdge 6300 dual-processor server. The front end consists of a Web site that is searched using Perl (.cgi) scripts to allow for user-defined queries of the database.

downloaded (or ordered on CD-ROM) from the DRAGON Web site as tab-delimited text files. A Dell PowerEdge 6300, dual Xeon 550 MHz processor-based server running RedHat Linux 6.2 serves the DRAGON Web site with Apache (http://www.apache.org). MySQL (http://www.mysql.com) is used as the relational database management system. DRAGON is composed of multiple tables derived from flat files (indicated in parentheses following the names of each database) provided by the Uni-

Gene (`ftp://ftp.ncbi.nlm.nih.gov/repository/UniGene/Hs.data.Z`, `Rn.data.Z`, `Mm.data.Z`, and `Dr.data.Z`), SwissProt (`ftp://expasy.proteome.org.au/databases/swiss-prot/release/sprotXX.dat`, where XX is the current version number), Kyoto Encyclopedia of Genes and Genomes (KEGG; `http://www.genome.ad.jp/kegg/`), and Pfam (`ftp://ftp.sanger.ac.uk/pub/databases/Pfam/Pfam-A.full.gz`) databases. Further additions to DRAGON will include information derived from the Online Mendelian Inheritance in Man database (OMIM; `http://www.ncbi.nlm.nih.gov/omim/`), Transfac (`http://www.cbi.pku.edu.cn/TRANSFAC/`), Interpro (`http://www.ebi.ac.uk/interpro/`), the Biological and Biochemical Image Database (BBID; `http://www.grc.nia.nih.gov/bbid/bbid99.htm`), the Database of Interacting Proteins (DIP; `http://dip.doe-mbi.ucla.edu/`), and multiple yeast databases.

In order to create the DRAGON database tables, a "cron" process is used to automatically run a set of Perl scripts (`http://www.perl.com`) that use the Net::FTP module to check for updated versions of each database flat file at the respective Web sites for each of these databases on a daily basis. If a database file has been updated (based on the timestamp of each file), the Perl script downloads the new file, extracts specific information from it, and saves the information into a set of output files. Continual updating of DRAGON minimizes errors in the cross-referencing of the databases that can occur due to inconsistencies such as retired accession numbers. The output files are imported into a MySQL database using the DBI module. The MySQL database is queried via common gateway interface (`.cgi`) scripts also written in Perl. As the user selects the types of information with which they wish to annotate their dataset, an SQL statement is dynamically constructed that selects the appropriate tables in the database. When the query is submitted, the constructed SQL statement is passed to the MySQL database, processed, and returned in one of three possible output formats from which the user can choose. The query can be returned as either an HTML file viewable in a Web browser, a tab-delimited text file that is returned to the Web browser, or as a tab-delimited text file that is e-mailed to one or more e-mail addresses supplied by the user.

The DRAGON View tools are implemented as common gateway interface (`.cgi`) scripts written in Perl that run on the same server as the DRAGON database. Each tool provides comprehensive instructions related to the format and options required for proper data submission and analysis (`http://pevsnerlab.kennedykrieger.org/dragon.htm`).

8.2.3 Cross-Referencing Information in DRAGON

Two issues had to be resolved in the cross-referencing of database information performed by DRAGON. The first was that microarray gene lists typically have one GenBank accession number associated with each gene represented on the array. The SwissProt/TrEMBL databases only list a few GenBank accession numbers for each SwissProt/TrEMBL entry. As a result, it is rare for the GenBank accession number provided on a microarray gene list to match its corresponding SwissProt/TrEMBL entry. This poses a problem when attempting to provide a bridge between these two data sources. In DRAGON, cross-referencing via two redundant "UniGene Human Numbers" tables solves this bridging problem (Figure 8.2). This expands one GenBank number into all GenBank numbers associated with a given UniGene ID number so that the maximal number of microarray GenBank numbers can be associated with SwissProt GenBank numbers. Presently, DRAGON can query human, rat, mouse, and zebrafish (*Brachydanio rerio*) genes via UniGene ID numbers. We plan on adding capability for *D. melanogaster* and *C. elegans* queries soon.

A second issue that had to be resolved was that some GenBank accession numbers span large chromosomal regions and therefore can contain multiple genes. Because of this, these GenBank numbers can be included in multiple clusters in the Unigene database. Therefore, cross-referencing via a UniGene ID numbers table that contains these GenBank numbers can lead to the improper joining of UniGene ID and SwissProt/TrEMBL numbers in DRAGON. The problem of improper joining is solved in DRAGON by "filtering" out GenBank accession numbers that are included in more than one Unigene cluster. For example, in the build of Unigene current with this writing, 398 GenBank accession numbers that were associated with more then one UniGene ID number were identified and deleted from the "UniGene Human Numbers" table. Filtering is a routine part of the DRAGON build process every time it is updated and prevents improper joining.

Because these issues have been resolved, the DRAGON database schema (Figure 8.2) and Web-accessible user interface allow a user to traverse a large set of databases in order to provide a list of GenBank accession numbers with information derived from a range of publicly available databases. The benefit to this type of integration of databases within DRAGON is that large lists of genes from large-scale systems such as microarrays can be rapidly annotated simultaneously.

8.2.4   The DRAGON Search and Annotate Tools

The DRAGON search tool is used to gain information on one gene at a time. A typical search and its result are shown in Figure 8.5. As described above, the DRAGON database includes UniGene files that are downloaded as flat files and parsed. UniGene is an NCBI database that is intended to consist of unique genes (from 11 species), each with a unique accession number (e.g., Hs.12345) that is represented by a list of all known associated expressed sequence tags (ESTs). UniGene describes the location from which cDNA libraries encoding a particular EST are constructed. A query with the DRAGON search tool can thus be restricted to a particular tissue.

Instructions
1) Decide which database you would like to search by clicking on the radio button next to its name. **Note**: You can only search one database at a time.
2) Choose the types of information you would like provided by checking the appropriate checkboxes on the left.
3) Define the criteria for your search by typing them into the text boxes on the right. **Note**: You can check certain attributes on the left and not provide criteria for them on the right. If you do so, your search will be performed based only on your criteria, but will return all the different types of information you requested for the genes or proteins matching your criteria.
4) Choose whether you would like to limit your search to a certain number of returned genes.
5) Press "Submit Query" in order to generate your search.

⦿ **Unigene:**

☑ Find gene by name:	Example: keratin	keratin
☑ Find gene by cytoband:	Example: Xq28	
☑ Find gene by locuslink:	Example: 3846	
☑ Find gene by expression area:	Example: brain	brain
☑ Find gene by accession #:	Example: L24158	

○ **Swissprot:**

☐ Find protein by subcellular location:	Example: peroxisome	
☐ Find protein by description:	Example: synaptotagmin	
☐ Find protein by GenBank number:	Example: L24158	
☐ Find protein by function:	Example: DNA binding	
☐ Find protein by keywords:	Example: Signal	
☐ Find protein by medline #:	Example: 94237164	
☐ Find protein by amino acid seq:	Example: MSTNENANT	

○ **Pfam:**

☐ Find protein family by description:	Example: EF-hand	
☐ Find protein family by Swissprot ID:	Example: P26371	

Limit the number of genes that are returned: [ ]

⦿ **Output as an html table:**
(You can copy and paste this type of output into a spreadsheet program such as MS Excel.)

○ **Output to browser or save as tab-delimited text file:**
(If your browser asks you whether you would like open in the browser or save to disk, you can either open in the browser, select it all and paste it into a spreadsheet program such as MS Excel or save it to a folder on your computer and then open it later in a spreadsheet program.)

○ **Output to Email Address:**
(You can enter multiple addresses separated by commas. A tab-delimited text file will be emailed to you when it is ready. This option is good for large queries such as those that contain more than 1000 genes.)

[Submit Query]

			9 Records Found:	
Unigene-ID	Title	Cytoband	Locuslink	Expression Area
Dr.10665	type II cytokeratin	LG 23	30392	0 day fin regenerates;1 day fin regenerates;26 somite embryos, adult livers, shield stage embryos;3 day fin regenerates;brain;embryo, 14 somite;embryo, day 3;fin;fin, 8-day regeneration;heart;myocardium, endocardium, vessel;olfactory rosettes;ovary (pooled);pectoral fin;pooled 15-19 hour zebrafish embryos;pooled 15-19 hr zebrafish embryos;pooled 26-somite embryos;testis (pooled);whole body;whole embryo
Dr.780	type I cytokeratin, enveloping layer	LG 19	30327	0 day fin regenerates;1 day fin regenerates;26 somite embryos, adult livers, shield stage embryos;3 day fin regenerates;brain;embryo;embryo, 14 somite;embryo, day 3;fin;heart;kidney pooled from 300 wild type adults;myocardium, endocardium, vessel;olfactory rosettes;ovary (pooled);pectoral fin;pooled 15-19 hour zebrafish embryos;pooled 15-19 hr zebrafish embryos;pooled 26-somite embryos;testis (pooled);whole body;whole embryo
Hs.164568	fibroblast growth factor 7 (keratinocyte growth factor)	15q15-q21.1	2252	fetal heart;infant brain;fetal cochlea;embryonic lung
Hs.2785	keratin 17	17q12-q21	3872	HeLa (cell line);ovary (tumor);skin--epidermis;epididymus;skin;fibroblast;breast;olfactory epithelium;placenta;fetal heart;brain;pancreas (tumor)

FIGURE 8.5. The DRAGON search tool (upper panel) allows the user to input a query (such as "keratin" expressed in the region "brain") and then request associated information from databases such as UniGene, SwissProt, and Pfam. In this example, boxes in the UniGene category are checked, and the program returns a result (lower panel) including UniGene identifier, gene name, cytoband, LocusLink ID, and regions in which the gene is expressed. Only a portion of the output from this query is shown.

For example, the query in Figure 8.5 is for keratin genes that have been expressed in the brain.

The DRAGON annotate tool is used to assign information from a variety of databases to a gene expression dataset (Figure 8.6). The ability of DRAGON to annotate gene expression data is useful in many ways.

- Information derived from the DRAGON annotate tool can be used to organize expression values from a microarray experiment into families that are related by biological characteristics of the genes or of their encoded proteins. These organized data can be visually and quantitatively analyzed for correlations between particular biological characteristics and

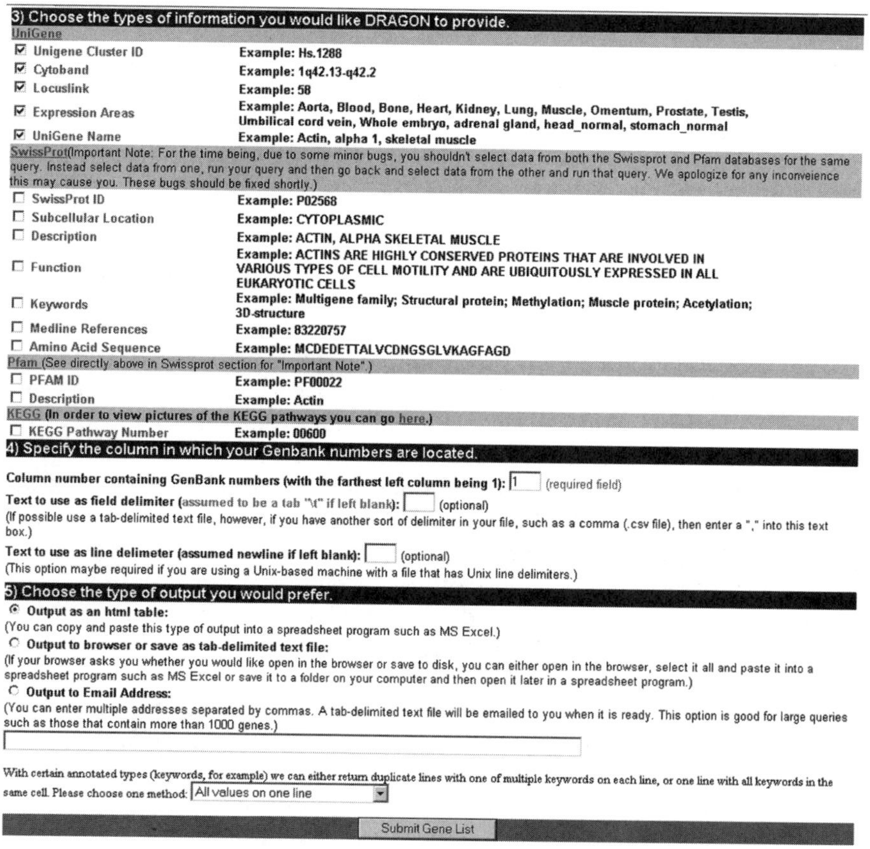

3) Choose the types of information you would like DRAGON to provide.

UniGene

☑ Unigene Cluster ID — Example: Hs.1288
☑ Cytoband — Example: 1q42.13-q42.2
☑ Locuslink — Example: 58
☑ Expression Areas — Example: Aorta, Blood, Bone, Heart, Kidney, Lung, Muscle, Omentum, Prostate, Testis, Umbilical cord vein, Whole embryo, adrenal gland, head_normal, stomach_normal
☑ UniGene Name — Example: Actin, alpha 1, skeletal muscle

SwissProt(Important Note: For the time being, due to some minor bugs, you shouldn't select data from both the Swissprot and Pfam databases for the same query. Instead select data from one, run your query and then go back and select data from the other and run that query. We apologize for any inconveience this may cause you. These bugs should be fixed shortly.)

☐ SwissProt ID — Example: P02568
☐ Subcellular Location — Example: CYTOPLASMIC
☐ Description — Example: ACTIN, ALPHA SKELETAL MUSCLE
☐ Function — Example: ACTINS ARE HIGHLY CONSERVED PROTEINS THAT ARE INVOLVED IN VARIOUS TYPES OF CELL MOTILITY AND ARE UBIQUITOUSLY EXPRESSED IN ALL EUKARYOTIC CELLS
☐ Keywords — Example: Multigene family; Structural protein; Methylation; Muscle protein; Acetylation; 3D-structure
☐ Medline References — Example: 83220757
☐ Amino Acid Sequence — Example: MCDEDETTALVCDNGSGLVKAGFAGD

Pfam (See directly above in Swissprot section for "Important Note".)

☐ PFAM ID — Example: PF00022
☐ Description — Example: Actin

KEGG (In order to view pictures of the KEGG pathways you can go here.)

☐ KEGG Pathway Number — Example: 00600

4) Specify the column in which your Genbank numbers are located.

Column number containing GenBank numbers (with the farthest left column being 1): [1] (required field)

Text to use as field delimiter (assumed to be a tab "\t" if left blank): [   ] (optional)
(If possible use a tab-delimited text file, however, if you have another sort of delimiter in your file, such as a comma (.csv file), then enter a "," into this text box.)

Text to use as line delimeter (assumed newline if left blank): [   ] (optional)
(This option maybe required if you are using a Unix-based machine with a file that has Unix line delimiters.)

5) Choose the type of output you would prefer.

⦿ Output as an html table:
(You can copy and paste this type of output into a spreadsheet program such as MS Excel.)
○ Output to browser or save as tab-delimited text file:
(If your browser asks you whether you would like open in the browser or save to disk, you can either open in the browser, select it all and paste it into a spreadsheet program such as MS Excel or save it to a folder on your computer and then open it later in a spreadsheet program.)
○ Output to Email Address:
(You can enter multiple addresses separated by commas. A tab-delimited text file will be emailed to you when it is ready. This option is good for large queries such as those that contain more than 1000 genes.)
[   ]

With certain annotated types (keywords, for example) we can either return duplicate lines with one of multiple keywords on each line, or one line with all keywords in the same cell. Please choose one method: [All values on one line ▾]

[Submit Gene List]

FIGURE 8.6. The DRAGON Annotate page allows a user to input data into a dialog box (not shown), or a tab-delimited text file can be uploaded from a local file. The user selects options (shown above) and then sends a request for annotation to the DRAGON database. Results may be returned as an HTML table, as a tab-delimited text file (suitable for import into a spreadsheet such as Microsoft Excel), or as an e-mail.

the coordinate expression of genes related by those characteristics. (See the discussion of DRAGON View tools below.) The relationship between gene expression values and each biological characteristic in turn can be analyzed for associations.

- The DRAGON annotate tool can rapidly compare the gene lists from two different microarray platforms to identify common genes, even if the GenBank accession numbers provided for the same gene on the two microarrays are different.

- The DRAGON annotate tool can also be used if an investigator has an interest in a particular gene or family of genes when designing an array experiment. DRAGON can survey different microarray systems to identify the system that best represents that particular gene or set of genes.

- The DRAGON annotate tool can be used to analyze data derived from other high-throughput gene expression analysis systems. For example, we have used DRAGON in the electronic differential analysis of Uni-Gene cDNA libraries from schizophrenic versus control human brain tissue (Johnston-Wilson et al., 2001).

8.2.5   The DRAGON View Data Visualization Tools

An essential attribute of the information derived through the use of DRAGON is that it can be incorporated into microarray data visualization and quantitative analysis methods. Large-scale, or "information-abundant," visualization methods (Shneiderman, 1999; http://lsr.ebi.ac.uk/%7Ealan/VisSupp/VisAware/VisAware.html), such as the color plots of clustered gene expression data produced by TreeView (Eisen et al., 1998; Alizadeh et al., 2000 http://rana.stanford.edu/software/) and TreeMaps (http://www.cs.umd.edu/hcil/treemaps/), are essential to the analysis of the large amounts of data generated by microarray experiments. The information provided by DRAGON aids in visualization of microarray data by providing an added level of structure to the data (i.e., biological characteristics) so that gene expression patterns can be viewed and analyzed in reference to biologically relevant characteristics. The suite of tools provided by DRAGON View is intended to aid in this type of visualization. In addition to the DRAGON View tools, the information provided by DRAGON can be integrated with clustering techniques such as K-means clustering (Tibshirani, 1999) in order to identify subsets of coregulated, functionally related genes. Software packages such as Partek 2000 (http://www.partek.com) and S-Plus 2000 (http://www.mathsoft.com/splus/) allow the incorporation of information provided by DRAGON into these techniques.

DRAGON View includes three data visualization tools: Families, Order, and Path. For each of these tools, the user provides information about the gene expression ratios derived from comparisons of a control and experimental condition and annotations such as Pfam numbers, SwissProt keywords, and KEGG pathway numbers obtained using the DRAGON annotate tool. The DRAGON View tools allow the user to visualize information about genes or their corresponding protein products. We are presently implementing a fourth visualization tool (DRAGON Gram).

A portion of the DRAGON Families output is shown in Figure 8.7a (see color insert) (Bouton and Pevsner, 2002). This tool sorts several hundred functional groups of genes to reveal families that have been coordinately up-regulated (e.g., genes encoding tRNA synthetases) or down-regulated (e.g., genes encoding collagens). DRAGON Families represents each gene in its corresponding protein family as a box that is clickable and hyperlinked to the NCBI LocusLink entry for that gene. Across each row, all the boxes correspond to genes in a given family (e.g., Aminoacyl-tRNA synthetases as defined in the SwissProt database and then annotated by DRAGON). Each box is also color-coded on a scale from red (up-regulated) to green (down-regulated). Furthermore, for all the functional families that are annotated (e.g., several hundred SwissProt keyword families), the program returns the families ranked in order according to the average ratio expression value for all of the genes in that group (see values in parentheses in Figure 8.7a).

A second tool in the DRAGON View suite is DRAGON Order (Figure 8.7b). This tool is similar to DRAGON Families in that it visualizes the expression data from a user's gene expression data as sorted into functional groups. DRAGON Order automatically presorts data based on ratio expression values. For each functional group (e.g., the "transmembrane" category in Figure 8.7b), the program generates a series of bars (vertical lines |) each of which represents a protein in that functional group. The position of the vertical bar indicates the extent to which that gene is up- or down-regulated. An equal distribution of vertical lines across the whole row means that there is no significant coexpression of a set of genes in that group. However, clusters of lines at either the far left or the far right of any given row are potentially interesting because they indicate that a set of related genes are all up- or down-regulated. For example, all five of the "Cell Adhesion" genes are clustered to the left of the row (Figure 8.7b). This kind of information would be difficult to detect by manual inspection of microarray datasets.

A third visualization tool is DRAGON Paths (Figure 8.7c). This tool relies on cellular pathways downloaded by file transfer protocol from the KEGG database (Kanehisa and Goto, 2000; Kanehisa, 2002). DRAGON Paths maps gene expression values onto cellular pathway diagrams. By viewing the expression levels derived from microarray data within the context of cellular pathways, the user might be able to detect patterns of expression that are not otherwise apparent. Conveniently, KEGG provides a coordinate file for every cellular pathway diagram in their database. These coordinate files provide the x, y coordinates on each diagram for every protein product in the diagram along with the LocusLink number of every corresponding gene. DRAGON Paths uses the LocusLink numbers (obtained using the annotate tool) to map the expression values for each of the user's LocusLink numbers with the corresponding coordinate location of that LocusLink number on the KEGG cellular pathway diagrams. The KEGG pathway diagrams also include enzyme commission (EC) numbers from the ENZYME database (Bairoch, 2000), offering DRAGON users another source of data on proteins corresponding to regulated genes. In DRAGON Paths, EC numbers on the pathways are hyperlinked to the NCBI LocusLink resource.

8.2.6 DRAGON Gram: A Novel Visualization Tool

After a dataset is annotated (e.g., by Pfam families), we implement a number of tools to visualize the data, including DRAGON Gram. The DRAGON Gram algorithm generates a plot of the number of genes (y-axis) versus the degree to which they are up- or down-regulated (Figure 8.10b; see color insert) (x-axis). For each functionally annotated gene subset, we calculate a histogram of differential gene expression values in local Z-score (standard deviation) units (see Chapter 9) or fold-difference values. A differential gene expression value of -2 represents a decrease in expression two standard deviations below the mean change. In a Gaussian distribution, such a value occurs in about 2.5% of the genes assayed. We use Microsoft Excel and/or Perl scripts to calculate histogram counts for each of the gene families using a histogram comprised of 13 bins (from < -2 to > 2)(Figure 8.10, x-axis). This density plot resembles a normal distribution ($n = 17678$ gene expression measurements). DRAGON Gram systematically evaluates the gene expression profile of each of several hundred functionally annotated gene families. For example, the actin family ($n = 44$ genes) includes actin, actin-binding proteins, and other actin-related molecules as defined by keywords in the SwissProt database. As a group, these genes are not significantly different from the majority of genes ($n = 17, 678 - 44$).

The statistical significance is estimated using a χ^2 analysis to measure the departure of the distribution (histogram) for each annotated functional group relative to the set of all genes (Figure 8.10, gray bars in each panel). According to the null hypothesis, gene expression in a functional group is irrelevant to the expression levels of each gene in that group. The χ^2 statistic is calculated for each functional subset by summing D values across the 13 histogram bins, where for each bin $D = (O - E)^2/E$ (we define O as observed counts and E as expected counts). D measures the deviation of a functional group from the expected distribution, and the χ^2 statistic is the sum of the D values across all bins. The χ^2 p-value is used to test the null hypothesis for large functional group sizes (e.g., $E > 5$ per bin). For small group sizes ($E < 5$), we perform a Monte Carlo test by generating random gene subsets drawn from the entire gene expression dataset. The Monte Carlo p-value is the fraction of random subsets whose χ^2 statistic exceeds the one obtained for the original functional group gene subset of interest. The appropriate p value based on Monte Carlo analysis is adjusted for the sample size tested.

For some functional groups shown in Figure 8.10, such as tRNA synthetases, DRAGON Gram shows that this group of genes has a distribution of expression values that is significantly altered ($p < 0.0001$). In general, a set of strict criteria is used to identify statistically significantly differentially regulated gene families using DRAGON Gram. A significant family will require (i) expression measurements for at least 20 genes, (ii) at least three of those measurements will have to have an absolute expression ratio value of > 1.8, (iii) at least ten of those measurements will have to be in the same direction, and (iv) the gene family will have to have a χ^2 p value of < 0.0001 as compared to the distribution of all gene measurements.

8.3   Implementation

To demonstrate the use of DRAGON and DRAGON View, we have applied them to two publicly available microarray datasets. The first dataset is the tab-delimited text file "Spo Spreadsheet" that provides the data for the Chu et al. (1998) study and is downloadable from the Brown Laboratory Web site (`http://cmgm.stanford.edu/pbrown/sporulation/additional/`). The second dataset was derived from the lead-treated astrocyte cell cultures experiment as described previously (Bouton and Pevsner, 2001) and is available from the Pevsner Laboratory Web site (`http://pevsnerlab.kennedykrieger.org/dragon.htm`).

Unicellular eukaryotic yeast (*Saccharomyces cerevisiae*) produces haploid spores via a process of meiosis and spore morphogenesis. Chu et al. (1998) monitored the expression profiles for 97% of all yeast genes during the process of sporulation. They clustered gene expression patterns according to the method of Eisen et al. (1998) in order to identify groups of genes related by their gene expression patterns over time. They then discuss certain genes within each phase of expression that relate to the processes involved in sporulation, including chromosomal segregation, DNA replication, spore morphogenesis, and transcriptional regulation. To further define functionally related genes within these phases of expression, we used to DRAGON to classify the yeast genes by the Pfam families of their encoded proteins. As described in the "Transcription factors" sector of the Chu et al. (1998) paper, the current understanding of transcriptional regulation of sporulation is limited. In particular, the transcription factors that control the middle and late phases of sporulation are not yet fully identified. In an attempt to provide information related to this, we selected all genes classified by DRAGON as having various DNA-binding domains in their encoded proteins. This set of genes was then clustered with Cluster software (`http://`

`rana.stanford.edu/software/`) and visualized with TreeView software (`http://rana.stanford.edu/software/`) (Figure 8.8; see color insert). Four subsets of genes were identified: up-regulated during early sporulation (Figure 8.8a), up-regulated during middle/late sporulation (Figure 8.8b), down-regulated during middle/late sporulation (Figure 8.8c), and down-regulated during early sporulation (Figure 8.8d). The expression profiles of certain genes in the up-regulated during middle/late sporulation group (Figure 8.8b), including YGL081W, YNL116W, YJR119C, YMR072W, YKR099W, YIR013C, and YFL003C, suggest their encoded proteins as potential mediators of the middle and late phases of transcriptional regulation during sporulation. This kind of analysis would have been very difficult to perform without a computational tool such as the DRAGON annotate tool since it would have required the investigator to either know about the biological characteristics of every gene in the dataset or methodically and painstakingly search for the biological characteristics of each of the genes in the dataset in turn.

Expression data from the Chu et al. (1998) study along with Pfam family information provided by DRAGON were imported into Partek 2000. Genes were clustered by expression values into four groups with a K-means clustering algorithm. Genes related by both Pfam family and cluster were identified. Figure 8.9 displays examples of genes related by both the domains of their encoded proteins and their expression profiles over time. Three members of the "Kinesin motor domain" family (PF00225) were steadily up-regulated during the course of sporulation, as mentioned by Chu et al. (1998). Forty members of various ribosomal subunit families (these proteins are represented by numerous Pfam numbers) are down-regulated early in the sporulation process but return to higher levels of expression at the 11 hour time point. This may be because spores are packed with ribosomes for subsequent germination and independent survival following their production, as mentioned by Chu et al. (1998). Sixteen members of the "Seripauperin and TIP1 family" (PF00660) were coordinately up-regulated during sporulation. Interestingly, Interpro documentation for this Pfam family revealed that the genes encoding these proteins are located at the extremities of the yeast chromosomes. Genes classified into the "WD domain G beta repeat" (PF00400) and "DEAD/DEAH box helicase" (PF00270) families shared a very similar expression profile, with early repression followed by a middle and late return to presporulation expression levels. All three isoforms of the glyceraldehyde-3-phosphate dehydrogenases (GAPDHs; PF00044) were coordinately down-regulated during the course of sporulation. Four minichromosome maintenance proteins or MCM 235 proteins (PF00493) were up-regulated early in sporulation during DNA replication and then down-regulated following DNA replication. MCM proteins are DNA-dependent ATPases that initiate chromosomal DNA replication through an interaction with autonomously replicating sequences (ARS). Additional information concerning all of the protein fami-

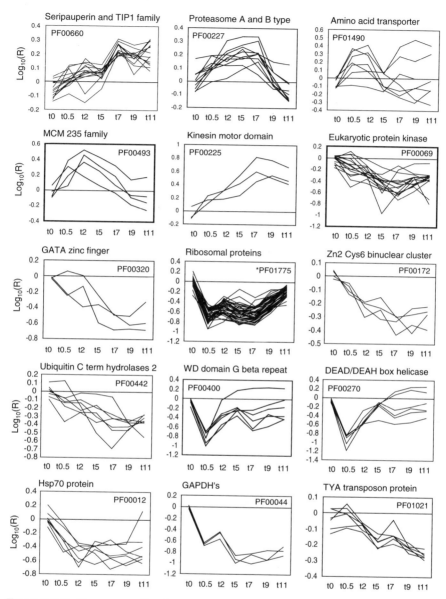

FIGURE 8.9. Genes encoding functionally related proteins are coordinately regulated. DRAGON classified genes in a yeast sporulation dataset by Pfam family. The data were then clustered with a K-means algorithm according to expression pattern similarity over time. Sets of genes related by both expression similarity and Pfam family were identified. Fifteen families are presented as an example. Each family is indicated by its title and Pfam accession number. The expression profiles for each gene in a given family are provided across seven time points (x-axis): zero hours ($t0$), 30 minutes ($t0.5$), two hours ($t2$), five hours ($t5$), seven hours ($t7$), nine hours ($t9$) and eleven hours ($t11$). The y-axis represents the inverse $\log_{10}(R)$, where R is the ratio of Cy3 and Cy5 intensities for each gene.

lies represented in Figure 8.9 can be obtained on the WWW at the Pfam Web site (http://dip.doe-mbi.ucla.edu/0). Once again, it would be very difficult to detect this type of coordinate regulation of functionally related genes without a means of annotating all of the genes on a microarray simultaneously.

In summary, through the use of DRAGON, we were able to identify a number of transcription factors potentially involved in the control of sporulation by combining information concerning gene expression and their encoded protein's functional domains (Figure 8.8). Additionally, in Figure 8.9 we have provided evidence for the coordinated expression of genes that encode functionally similar proteins. A number of the coordinately regulated protein families can be understood within the context of the cellular processes occurring during sporulation.

As a second example, the DRAGON and DRAGON View techniques were used in the analysis of a microarray dataset in which astrocyte cell cultures were treated with lead. Incyte cDNA microarrays were employed to detect differential gene expression in astrocyte cell cultures when treated with 10 μM lead for 24 hr (Bouton et al., 2001). Astrocytes, along with endothelial cells, comprise the blood–brain barrier. Lead is known to increase the permeability of the blood–brain barrier and can inhibit in vitro astrocyte–endothelial interactions relevant to brain microvessel development (Laterra et al., 1992). These effects may contribute to the encephalopathy seen at a clinical level following high-level lead exposure, especially in children (Clasen et al., 1974; Goldstein et al., 1974; Press, 1997). Therefore, astrocytes are an important cellular target of lead (Tiffany-Castiglioni, 1993).

In an attempt to identify the coordinate regulation of functionally related gene groups in this dataset, we used DRAGON and DRAGON View. The coordinate regulation of groups of functionally related genes has been identified in numerous studies (Chu et al., 1998; Eisen et al., 1998; Spellman et al., 1998; Wen et al., 1998; Heyer et al., 1999). Often, the identification of such groups can provide insight into the types of cellular processes that have been perturbed in the experimental state. We employed DRAGON and DRAGON View to screen our entire Incyte dataset for the coordinate regulation of functionally related genes. The two Incyte datasets, comprising 17, 678 genes, were combined and annotated with SwissProt database keywords using the DRAGON annotate tool. Using the SwissProt keywords, the DRAGON View analysis indicated that the tRNA synthetase and collagen families were the two largest coregulated gene groups in the entire combined dataset (Figure 8.10a; see color insert).

The DRAGON Gram χ^2 analysis was used in order to determine the statistical significance of the compared distribution histograms of all genes on the two Incyte arrays against the distribution histograms of the tRNA synthetase and collagen groups. Both groups were calculated to have expression profiles that were significantly different from that of the combined dataset (Figure 8.10b) according to the strict criterion of a $p < 0.0001$. The altered distribution of the tRNA synthetase and collagen families is not accounted for by the size of the family. Many other gene families of similar size such as the actins and neuropeptides (Figure 8.10b) have distributions that are not significantly different from the distribution of all genes.

The identification of the tRNA synthetases as a group of coordinately regulated genes was unexpected given our experimental system. However, a review of the literature indicated one potential cause of this phenomenon. Studies five decades ago first reported the cleavage of ribonucleic acids by lead (Dimroth et al., 1950; Farkas, 1968). Since then, lead has been routinely used as a tool for the study of the secondary structure of tRNA (Farkas et al., 1972; Krebs et al., 1972; Otzen et al., 1994; Pan et al., 1994; Ciesiolka et al., 1998; Perreau et al., 1999), mRNA (Farkas, 1975) and rRNA molecules (Winter et al., 1997). Lead is capable of cleaving RNA at specific sites through a mechanism of nucleophilic attack (Brown et al., 1983). Low micromolar concentrations of lead can cause RNA cleavage (Werner et al., 1976). Lead-induced RNA cleavage can occur both in vitro and in vivo (Kennedy et al., 1983). Under physiological conditions (pH 7.4 and 37°C), lead is 28 times more potent than the next most effective divalent cation, zinc, in cleaving RNA (Farkas, 1968; Otzen et al., 1994). Kennedy et al. (1983) prepared postmitochondrial fractions from the brains of rat pups treated with 4% lead carbonate maternal milk for 2 days. Lead-treated fractions displayed significantly lower amounts of protein synthesis activity than controls. A significant reduction in the level of tRNA synthetase activity accounted for most of the reduction in protein synthesis activity. Our microarray results, when analyzed in relation to these previous findings, indicate that the protein synthesis machinery is a target of lead treatment. This is suggested by the coordinate differential regulation of the amino acid biosynthesis-related genes (Figure 8.10b). Furthermore, it seems that tRNA synthetase activity in particular is disrupted by lead and that this ability to cleave tRNA molecules may be the cause of these disruptions. A reduction in the amount of tRNA available for the tRNA synthetases would cause a reduction in the synthetase activity and a subsequent up-regulation of the synthetases as a compensatory mechanism. Exportin-t is the major nuclear exporter of tRNA molecules (Arts et al., 1998a,b; Kutay et al., 1998; Lipowsky et al., 1999; Grosshans et al., 2000) and was also found to be significantly up-regulated in lead-treated astrocytes.

Through the use of DRAGON and DRAGON View we were able to identify the coordinate regulation of a group of functionally related molecules in lead-treated astrocytes. To our knowledge, this is the first time that the coordinate regulation of tRNA synthetases has been observed in any lead-treated biological tissue. It was because of our ability to annotate the entire dataset with functional information and then view the expression data in relation to those functional data that we were rapidly able to identify this as a significant finding within the dataset.

No investigator can be expected to remember, let alone know, all of the various biological characteristics of all of the thousands of genes in any large-scale gene expression dataset. By providing investigators with as much of this biological information as possible through automated computational approaches and visual analysis tools, one enhances their ability to identify the important biological processes occurring as a result of some experimental perturbation of the system being studied.

8.4   Discussion and Conclusion

The DRAGON annotate tool associates biologically relevant information derived from numerous public databases with gene expression information from microarray experiments. The subsequent analysis process includes the association of relevant information with microarray data, the visualization of microarray data in the context of associated biological characteristics, and the quantitative analysis of the association of certain biological characteristics with gene expression patterns. To illustrate the use of DRAGON, we have applied it to two microarray datasets available via the WWW. During the analysis of these datasets, we used visual and quantitative analysis methods to examine the correlation between gene expression patterns and biological characteristics such as membership in protein families and description by keywords. For example, in the Chu et al. (1998) dataset, we identified a number of transcription factors that may play a role in the regulation of the middle and late stages of sporulation. DRAGON accomplished this by integrating information concerning gene expression and biological characteristics of the proteins encoded by those genes. Furthermore, we identified a number of instances of the coordinate regulation of genes that are functionally related by the domains that they contain. Microarrays provide information concerning genome-scale expression patterns. This information can be used to better understand how cellular systems respond to certain stimuli. With the broad window through which we are now able to view gene expression patterns with microarrays, we may find that coordinate expression of functionally related gene products is a common attribute of the reaction of cellular systems to changing external and internal conditions.

By integrating biologically relevant information with the analysis of microarray data, certain types of gene expression phenomena can be discerned more easily and examined in light of the experimental paradigm being tested. A comprehensive definition of biological data regarding each gene on a microarray list through the interconnection of as many public databases as possible is an eventual goal in the development of DRAGON. DRAGON would then be able to supply a multidimensional network of information related to the expression patterns and biological characteristics of all genes on a microarray. This growth of DRAGON is dependent upon the continued integration of public databases (Robbins, 1993; Frishman et al., 1998; Spence and Aurora, 1999). Along with the question of database integration comes the crucial matter of data integrity within and across databases (Macauley et al., 1998).

In conclusion, DRAGON is a database that can be used to annotate microarray data with a broad range of biologically relevant characteristics. Multiple types of visual and quantitative analyses, such as the DRAGON View tools, can make use of the information provided by DRAGON in order to gain a deeper understanding of expression patterns observed with microarrays. DRAGON is freely available for queries via the WWW. Biology has entered a new era where computers are required in order to collect, store, and analyze experimental data. DRAGON is one such tool that provides access to in silico biological information in order to enhance the analysis of "wet-laboratory"-generated data.

References

Afshari CA, Nuwaysir EF, Barrett JC (1999). Application of complementary DNA microarray technology to carcinogen identification, toxicology, and drug safety evaluation. *Cancer Research*, 59:4759–4760.

Alizadeh AA, Eisen MB, Davis RE, Ma C, Lossos IS, Rosenwald A, Boldrick JC, Sabet H, Tran T, Yu X, Powell JI, Yang L, Marti GE, Moore T, Hudson J Jr, Lu L, Lewis DB, Tibshirani R, Sherlock G, Chan WC, Greiner TC, Weisenburger DD, Armitage JO, Warnke R, Staudt LM, et al. (2000). Distinct types of diffuse large B-cell lymphoma identified by gene expression profiling. *Nature*, 403:503–511.

Arts GJ, Fornerod M, Mattaj IW (1998a). Identification of a nuclear export receptor for tRNA. *Current Biology*, 8:305–314.

Arts GJ, Kuersten S, Romby P, Ehresmann B, Mattaj IW (1998b). The role of exportin-t in selective nuclear export of mature tRNAs. *EMBO Journal*, 17:7430–7441.

Bairoch A (2000). The ENZYME database in 2000. *Nucleic Acids Research*, 28:304-305.

Bassett DE, Eisen MB, Boguski MS (1999). Gene expression informatics—it's all in your mine. *Nature Genetics Supplement*, 21:51–55.

Bateman A, Birney E, Durbin R, Eddy SR, Howe KL, Sonnhammer ELL (2000). The Pfam protein families database. *Nucleic Acids Research*, 28:263–266.

Bouton CM, Pevsner J (2001). DRAGON: Database Referencing of Array Genes Online. *Bioinformatics*, 16:1038–1039.

Bouton CM, Pevsner J (2002). DRAGON View: Information visualization for annotated microarray data. *Bioinformatics*, 18:323–324.

Bouton CM, Hossain MA, Frelin LP, Laterra J, Pevsner J (2001). Microarray analysis of differential gene expression in lead-exposed astrocytes. *Toxicology and Applied Pharmacology*, 176:34–53.

Bowtell DDL (1999). Options available—from start to finish—for obtaining expression data by microarray. *Nature Genetics Supplement*, 21:25–32.

Brown PO, Botstein D (1999). Exploring the new world of the genome with DNA microarrays. *Nature Genetics Supplement*, 21:33–37.

Brown RS, Hingerty BE, Dewan JC, Klug A (1983). Pb(II)-catalysed cleavage of the sugar-phosphate backbone of yeast tRNAPhe—implications for lead toxicity and self-splicing RNA. *Nature*, 303:543–546.

Carr DB, Somogyi R, Michaels G (1997). Templates for looking at gene expression clustering. *Statistical Computing and Statistical Graphics Newsletter*, April, 20–29.

Cheung VG, Morley M, Aguilar F, Massimi A, Kucherlapati R, Childs G (1999). Making and reading microarrays. *Nature Genetics Supplement*, 21:15–19.

Chu S, DeRisi J, Eisen M, Mulholland J, Botstein D, Brown PO, Herskowitz I (1998). The transcriptional program of sporulation in budding yeast. *Science*, 282:699–705.

Ciesiolka J, Michalowski D, Wrzesinski J, Krajewski J, Krzyzosiak WJ (1998). Patterns of cleavages induced by lead ions in defined RNA secondary structure motifs. *Journal of Molecular Biology*, 275:211–20.

Clasen RA, Hartmann JF, Starr AJ, Coogan PS, Pandolfi S, Laing I, Becker R, Hass GM (1974). Electron microscopic and chemical studies of the vascular changes and edema of lead encephalopathy. A comparative study of the human and experimental disease. *American Journal of Pathology*, 74:215–240.

Colantuoni C, Purcell AE, Bouton CML, Pevsner J (2000). High throughput analysis of gene expression in the human brain. *Journal of Neuroscience Research*, 59:1–10.

Debouck C, Goodfellow PN (1999). DNA microarrays in drug discovery and development. *Nature Genetics Supplement*, 21:48–50.

Dimroth K, Jaenicke L, Heinzel D (1950). Die spaltung der pentose-nucleinsaure der hefe mit bleihydroxyd. *Liebigs Annalen der Chemie*, 566:206–210.

Duggan DJ, Bittner M, Chen Y, Meltzer P, Trent JM (1999). Expression profiling using cDNA microarrays. *Nature Genetics Supplement*, 21:10–14.

Eisen MB, Spellman PT, Brown PO, Botstein D (1998). Cluster analysis and display of genome-wide expression patterns. *Proceedings of the National Academy of Sciences, USA.*, 95:14863–14868.

Farkas WR (1968). Depolymerization of ribonucleic acid by plumbous ion. *Biochimica et Biophysica Acta*, 155:401–409.

Farkas, W.R. (1975) Effect of plumbous ion on messenger RNA. Chem. Biol. Interact. 11, 253-263.

Farkas WR, Hewins S, Welch JW (1972). Effects of plumbous ion on some functions of transfer RNA. *Chemica-Biological Interactions*, 5:191–200.

Frishman D, Heumann K, Lesk A, Mewes H-W (1998). Comprehensive, comprehensible, distributed and intelligent databases: Current status. *Bioinformatics*, 14:551–561.

Fuhrman S, Cunningham MJ, Wen X, Zweiger G, Seilhamer J, Somogyi R (2000). The application of Shannon entropy in the identification of putative drug targets. *Biosystems*, 55:5–14.

Gawantka V, Pollet N, Delius H, Vingron M, Pfister R, Nitsch R, Blumenstock C, Niehrs C (1998). Gene expression screening in *Xenopus* identifies molecular pathways, predicts gene function and provides a global view of embryonic gene expression. *Mechanisms of Development*, 77:95–141.

Goldstein GW, Asbury AK, Diamond I (1974). Pathogenesis of lead encephalopathy. Uptake of lead and reaction of brain capillaries. *Archives of Neurology*, 31:382–389.

Golub TR, Slonim DK, Tamayo P, Huard C, Gaasenbeek M, Mesirov JP, Coller H, Loh ML, Downing JR, Caliguri MA, Blommfield CD, Lander ES (1999). Molecular classification of cancer: Class discovery and class prediction by gene expression monitoring. *Science*, 286:531–537.

Grosshans H, Hurt E, Simos G (2000). An aminoacylation-dependent nuclear tRNA export pathway in yeast. *Genes and Development*, 14:830–840.

Heyer LJ, Kruglyak S, Yooseph S (1999). Exploring expression data: Identification and analysis of coexpressed genes. *Genome Research*, 9:1106–1115.

Johnston-Wilson NL, Bouton CM, Pevsner J, Breen JJ, Torrey EF, Yolken RH (2001). Emerging technologies for large-scale screening of human tissues and fluids in the study of severe psychiatric disease. *International Journal of Neuropsychopharmacology*, 4:83–92.

Kanehisa M, Goto S (2000). KEGG: kyoto encyclopedia of genes and genomes. *Nucleic Acids Research*, 28:27–30.

Kanehisa, M (2002). The KEGG databases at GenomeNet. *Nucleic Acids Research*, 30:42–46.

Kennedy JL, Girgis GR, Rakhra GS, Nicholls DM (1983). Protein synthesis in rat brain following neonatal exposure to lead. *Journal of the Neurological Sciences*, 59:57–68.

Krebs B, Werner C, Dirheimer G (1972). Action of divalent lead on phenylalanine tRNA of brewer's yeast. *European Journal of Toxicology, Hygiene, Environment* 5:337–342.

Kutay U, Lipowsky G, Izaurralde E, Bischoff FR, Schwarzmaier P, Hartmann E, Gorlich D (1998). Identification of a tRNA-specific nuclear export receptor. *Molecules and Cells*, 1:359–369.

Laterra J, Bressler JP, Indurti RR, Belloni-Olivi L, Goldstein GW (1992). Inhibition of astroglia-induced endothelial differentiation by inorganic lead: A role for protein kinase C. *Proceedings of the National Academy of Sciences, USA*, 89:10748–10752.

Liang S, Fuhrman S, Somogyi R (1998). Reveal, a general reverse engineering algorithm for inference of genetic network architectures. *Pacific Symposium on Biocomputing*, 3:18–29.

Lipowsky G, Bischoff FR, Izaurralde E, Kutay U, Schafer S, Gross HJ, Beier H, Gorlich D (1999). Coordination of tRNA nuclear export with processing of tRNA. *RNA*, 5:539–549.

Lipshutz RJ, Fodor SPA, Gingeras TR, Lockhart DJ (1999). High density synthetic oligonucleotide arrays. *Nature Genetics Supplement*, 21:20–24.

Macauley J, Wang H, Goodman N (1998). A model system for studying the integration of molecular biology databases. *Bioinformatics*, 14:575–582.

Michaels GS, Carr DB, Askenaki M, Fuhrman S, Wen X, Somogyi R (1998). Cluster analysis and data visualization of large-scale gene expression data. *Pacific Symposium on Biocomputing*, 3:42–53.

Nuwaysir EF, Bittner M, Trent J, Barrett JC, Afshari CA (1999). Microarrays and toxicology: the advent of toxicogenomics. *Molecular Carcinogenesis*, 24:153–159.

Otzen DE, Barciszewski J, Clark BF (1994). Dual hydrolytic role for Pb(II) ions. *Biochimie*, 76:15–21.

Pan T, Dichtl B, Uhlenbeck OC (1994). Properties of an in vitro selected Pb2+ cleavage motif. *Biochemistry*, 33:9561–5.

Perreau VM, Keith G, Holmes WM, Przykorska A, Santos MA, Tuite MF (1999). The *Candida albicans* CUG-decoding ser-tRNA has an atypical anticodon stem-loop structure. *Journal of Molecular Biology*, 293:1039–1053.

Press MF (1977) Lead encephalopathy in neonatal Long–Evans rats: Morphologic studies. *Journal of Neuropathology and Experimental Neurology*, 36:169–193.

Robbins RJ (1993). Report of the Invitational DOE Workshop on Genome Informatics, 26–27 April 1993. www.bis.med.jhmi.edu/Dan/DOE/whitepaper/inf_rep2.HTML. US Department of Energy: Washington, DC.

Shneiderman B (1999). Supporting creativity with advanced information-abundant user interfaces. www.spotfire.com/pdf/info_rich.htm.

Somogyi R (1999). Making sense of gene-expression data. In: *Pharma Informatics, A Trends Guide*, Elsevier Trends Journal, pp. 17–24. Elsevier: Amsterdam.

Somogyi R, Fuhrman S, Askenazi M, Wuensche A (1997). The gene expression matrix: towards the extraction of genetic network architectures. *WCNA96*, 30:1815–1824.

Somogyi R, Sniegoski CA (1996). Modeling the complexity of genetic networks: understanding multigenic and pleiotropic regulation. *Complexity*, 19:45–63.

Spellman PT, Sherlock G, Zhang MQ, Iyer VR, Anders K, Eisen MB, Brown PO, Botstein D, Futcher B (1998). Comprehensive identification of cell cycle-regulated genes of the yeast *Saccharomyces cerevisiae* by microarray hybridization. *Molecular Biology of the Cell* 9:3273–3297.

Spellman PT, Rubin GM (2002). Evidence for large domains of similarly expressed genes in the *Drosophila* genome. *Journal of Biology*, 1:5.1–5.8.

Spence P, Aurora R (1999). From reductionist to constructionist, but only if we integrate. In: *Pharma Informatics, A Trends Guide*, Elsevier Trends Journal, pp. 37–39. Elsevier: Amsterdam.

Szallasi Z (1999). Genetic network analysis in light of massively parallel biological data acquisition. *Pacific Symposium on Biocomputing*, 4:5–16.

Tamayo P, Slonim D, Mesirov J, Zhu Q, Kitareewan S, Dmitrovsky E, Lander ES, Golub TR (1999). Interpreting patterns of gene expression with self-organizing maps: Methods and applications to hematopoetic differentiation. *Proceedings of the National Academy of Sciences*, 96:2907–2912.

Tibshirani R, Hastie T, Eisen M, Ross D, Botstein D, Brown P (1999). Clustering methods for the analysis of DNA microarray data. Technical report. Department of Statistics, Stanford University: Standord, CA.

Tiffany-Castiglioni E (1993). Cell culture models for lead toxicity in neuronal and glial cells. *Neurotoxicology* 14:513–536.

Toronen P, Kolehmainen M, Wong G, Castr'en E (1999). Analysis of gene expression data using self-organizing maps. *FEBS Letters*, 451:142–146.

Velculescu VE, Zhang L, Vogelstein B, Kinzler KW (1995). Serial analysis of gene expression. *Science*, 270:484–487.

Vingron M, Hoheisel J (1999). Computational aspects of expression data. *Journal of Molecular Medicine*, 77:3–7.

Wen X, Fuhrman S, Michaels GS, Carr DB, Smith S, Barker JL, Somogyi R (1998). Large-scale temporal gene expression mapping of central nervous system development. *Proceedings of the National Academy of Sciences*, 95:334–339.

Werner C, Krebs B, Keith G, Dirheimer G (1976). Specific cleavages of pure tRNAs by plumbous ions. *Biochimica et Biophysica Acta*, 432:161–175.

Wilson M, DeRisi J, Kristensen H-H, Imboden P, Rane S, Brown PO, Schoolnik GK (1999). Exploring drug-induced alterations in gene expression in Mycobacterium tuberculosis by microarray hybridization. *Proceedings of the National Academy of Sciences, USA*, 96:12833–12838.

Winter D, Polacek N, Halama I, Streicher B, Barta A (1997). Lead-catalysed specific cleavage of ribosomal RNAs. *Nucleic Acids Research* 25:1817–1824.

Zhang MQ (1999). Large-scale gene expression data analysis: A new challenge to computational biologists. *Genome Research*, 9:681–688.

9

SNOMAD: Biologist-Friendly Web Tools for the Standardization and NOrmalization of Microarray Data

CARLO COLANTUONI
GEORGE HENRY
CHRISTOPHER M.L.S. BOUTON
SCOTT L. ZEGER
JONATHAN PEVSNER

Abstract

The use of DNA microarrays and other gene expression analysis techniques throughout the biological sciences has put extremely large, complex datasets in the hands of biologists who, for the most part, are not formally trained in computational or statistical methods. The majority of gene expression datasets have extensive artifactual bias and/or noise, which are not apparent upon superficial inspection. The SNOMAD gene expression analysis tools are an effort to make important normalization and quality control methods available to a wide audience of biological scientists working with gene expression data. Methods available in the SNOMAD tools include background subtraction, global mean normalization, local mean normalization across absolute intensity, local variance correction across absolute intensity, and ratio correction across the physical surface of the microarray. The SNOMAD web-implementation, available free of charge to all researchers at http://pevsnerlab.kennedykrieger.org/snomad.htm provides these tools without the downloading or installation of additional software, and does not require users to have any statistical or computer programming expertise.

9.1 Introduction

Serial analysis of gene expression (SAGE), cDNA library sequencing, differential display, cDNA subtraction, multiplex quantitative RT-PCR, cDNA microarrays, and oligonucleotide microarrays are some of the most powerful tools for the high-throughput analysis of gene expression in biological systems. The power of these techniques lies not only in their ability to investigate the transcriptome on a pan-genomic scale but to do so across a large number of diverse biological systems and experimental paradigms.

Multiple sources of noise are inherent in all microarray technologies. The target of gene expression analysis (i.e., variation in actual gene expression levels) can be obscured by heterogeneity in biological sample acquisition, microarray platform technology, probe labeling, hybridization strategy, signal detection, and image analysis methods. Each of these sources introduces an amount of artifactual variance and/or bias into gene expression data that complicates the estimation of expression levels within a dataset as well as the comparison of expression changes between datasets. The importance of both validity within datasets and reliability between datasets is becoming more evident as the size, number, and quality of publicly available gene expression datasets increases and the potential yield of relational data integration across datasets becomes greater.

Several groups have addressed basic normalization processes, such as background subtraction and global mean normalization (Beissbarth et al., 2000; Hegde et al., 2000; Liao et al., 2000, Schuchhardt et al., 2000; Smid-Koopman et al., 2000; Tseng et al., 2001), as well as methods directed at the identification of differentially expressed genes (Claverie, 1999; Eickhoff et al., 1999; Hilsenbeck et al., 1999; Wittes and Friedman, 1999; Manduchi, 2000; Kadota et al., 2001; Park et al., 2001). Speed and collaborators have developed methods for the normalization of gene expression data both within and between hybridization experiments (Yang et al., 2002; Dudoit and Yang, Chapter 3, this volume; and http://www.stat.berkeley.edu/users/terry/zarray/html/papersindex.html), while others have addressed differences in the variance of gene expression observations at different absolute gene expression levels.

Many software packages, both public and commercial, have been directed at facilitating the comparison of multiple microarray datasets and the global analysis of expression data. The methods and algorithms in this report are directed at the normalization and standardization of microarray datasets in preparation for their input into such algorithms (i.e., the refinement of the differential gene expression metric). Findings generated with data-mining algorithms are only as reliable as the data used as their input. It is our intention that the final metric of differential gene expression generated by SNOMAD (Standardization and NOrmalization of MicroArray Data) aid in (1) the identification of differentially expressed genes, (2) in the assessment of the statistical significance of identified genes, and (3) in the comparison of entire expression profiles across diverse microarray technologies, experimental paradigms, and biological systems. The Web-implementation of the SNOMAD tools is designed for easy use by biologists with no programming or statistical expertise. See http://pevsnerlab.kennedykrieger.org/snomad.htm.

It is common practice in gene expression analysis to apply a global mean normalization to raw array-element intensities. This ensures that the mean signal intensity or ratio of all array-elements will be equivalent across multiple microarray experiments. Although this is useful, much artifactual variation present in gene expression data is not constant across the range of element signal intensities and hence cannot be addressed by global normalization processes.

Here, we present three local normalization processes that address biases and variances that are systematic but are distributed nonuniformly across distinct dimensions within gene expression data. The first of these procedures is a mean normalization employing a locally calculated mean in order to correct bias present in observed gene expression ratios. This bias varies systematically as array-element intensity changes; hence, the mean gene expression ratio used to normalize individual expression ratios is calculated locally across the range of array-element intensities. Just as the goal of a global mean normalization is to ensure that the mean of all observed expression ratios is equal to 1, the goal of this local mean normalization is to ensure that the mean gene expression ratio is approximately equal to 1 at all points across the range of array-element signal intensities.

Following this local mean normalization, we describe a local variance correction that rectifies inequality in the variance of observed gene expression ratios across this same dimension of element intensity. Variance in the measurement of expression ratios for genes with low expression levels is often very different from that for genes with higher expression levels. We standardize gene expression ratios to a standard deviation calculated locally across the range of element intensities in order to correct for this unequal variance.

The last of these local normalization methods is directed at the correction of artifacts that are spatially systematic across the physical surface of a microarray. Array-element intensities are normalized to a smooth mean intensity calculated locally across the surface of the microarray. This method is useful for detecting and/or correcting artifacts introduced during the robotic printing of microarray-elements or in the hybridization of the labeled cDNA to the array.

9.2   Methods and Application

9.2.1   Overview of Experimental and Data Analysis Procedures

The analysis of one-channel DNA microarray data is fundamentally different from the analysis of two-channel data. One-channel data are generally derived from technologies that employ the hybridization of individual labeled cDNA probes to a microarray, while two-channel data are derived from the simultaneous competitive hybridization of two distinct cDNA probes, each labeled with a different fluorophore. The primary distinction is that raw one-channel data reflect absolute intensities, while two-channel data contain intensities and ratio data. Although the local mean normalization across an array surface can be applied to either one- or two-channel data, the other local normalization processes detailed in this report include the use of ratio data, and are hence designed for application to two-channel data or paired one-channel data (Figure 9.1A and B). However, it is not always the case that datasets can be paired: experimental design of a gene expression experiment determines which specific algorithms should be applied to a dataset (Figure 9.1).

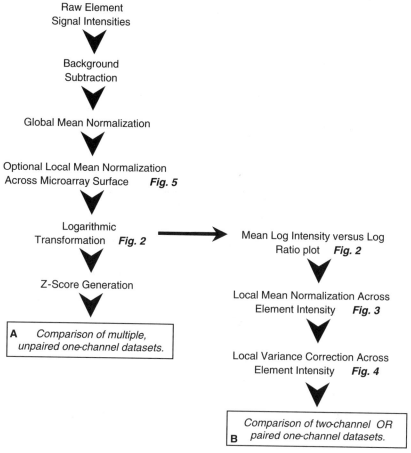

FIGURE 9.1. Overview of experimental and data analysis procedures (see the text).

9.2.2 Background Subtraction

Some microarray technologies use the subtraction of a local background intensity from each array-element intensity, while others use a low number of background measurements in order to calculate a mean background intensity, which is subtracted from many or all element intensities. The Web implementation of SNOMAD allows either (or neither) of these options to be applied to microarray data being uploaded for analysis.

9.2.3 Global Mean Normalization

Global normalization methods should be carried out on raw array-element intensities prior to the application of the local normalization detailed in this report. Although global mean normalization is rendered computationally obsolete by the local mean normalization detailed below, it is very useful in intermediate steps of analysis and in data visualization.

Several sources of variance may have a constant impact on all element signal intensities in a dataset, such as differential label incorporation into the probe, differing amounts of probe used, or differences in detection efficiency. In order to correct for this uniform variance, global normalization processes are applied to nearly all DNA microarray data. Both one- and two-channel array datasets require global normalization prior to analysis of expression values or ratios. In one-channel datasets, this entails the division of each element signal intensity by a correction factor. Two-channel data are best handled as two individual one-channel datasets in this respect, such that each element signal intensity is divided by a correction factor specific to the channel from which it was derived. This correction factor can be calculated in a number of different ways. Perhaps the most common is the normalization of all element signal intensities to the mean intensity of all elements contained within a single array (for one-channel data) or single channel (for two-channel data): element intensity/mean element intensity = global mean normalized intensity.

Alternatively, global normalization can also be achieved by normalizing all ratios in a two-channel dataset, or paired one-channel datasets, to a mean ratio of 1. This global normalization ensures that the mean gene expression ratio for each individual dataset will be equal to a value of 1.0 (zero on the log scale) and equivalent across datasets. When using this method of normalization, it is best that the number of genes involved be large (several hundred minimum) and that the samples to be compared be highly similar (e.g., from the same gross tissue source). Similarly, it is inadvisable to use global normalization methods with data derived from an array consisting entirely of genes that are expected to change (especially if expected to change in the same direction) or are all thought to be involved in the process being studied.

Normalization to a subset of genes designated as invariant, or "housekeeping genes," is also a common practice. This can lead to problems if any of the genes contained within this subset are regulated in the experimental paradigm under study. Without prior validation of gene subsets for the particular samples being studied it is not easy to determine which genes' expression will be invariant. Another alternative is to add known amounts of labeled cDNAs not otherwise present in the samples being studied, but included on the array. Such additions require precise quantification of RNA prior to addition and should be made at the earliest possible point in sample preparation. These intensity values can then be used for global normalization.

Although global mean normalization ensures that the mean gene expression value or ratio for individual datasets will be equivalent (mean of global mean normalized dataset = 1.0), global normalizations (any normalization applying the identical transformation to all element signal intensities) cannot address effects that are nonuniformly distributed through a dataset. Two examples of factors that introduce variance differentially across the span of array-element signal intensities include (1) differences in the dynamic range of dye detection between the two dyes used in a two-color hybridization, and (2) differences in the magnitude of variance among measures taken at varying levels within these dynamic ranges. In this report, we present two local normalization processes that address biases and variances that are nonuniformly distributed across absolute signal intensity and a third that examines artifacts introduced systematically across the physical surface of the microarray.

9.2.4 Standard Data Transformation and Visualization Methods

A number of commonly used transformations and scatterplots are invaluable during the qualitative assessment, quality control, and quantitative analysis of gene expression data. Figure 9.2 depicts the comparison of two sets of intensity values generated in a single two-channel hybridization experiment using the UniGEM V 1.0 array technology from Incyte Genomics, Inc. In these plots, each gene is plotted as a single point with an intensity derived from each of two microarray data channels. Even after global mean normalization, using a scatterplot of normalized raw intensities, it is difficult to gain insight into the nature of the distribution of data points in a large dataset (Figure 9.2A). The dominant feature of this display is the substantial correlation between expression levels across the two experiments being compared. However, it is the difference or change in expression levels, not the similarity, that is of greatest biological interest. Such differences are not easily seen in Figure 9.2A. One reason is that expression levels are strongly skewed to higher values (a preponderance of low intensity values), reflecting the fact that proportional rather than absolute change in expression is the more natural scale. To display proportional change, a logarithmic transformation is applied to all intensities from both fluorescence

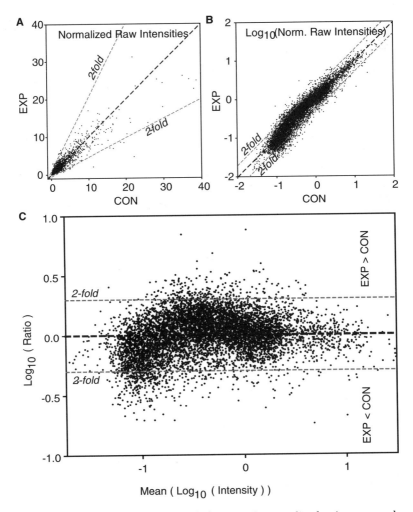

FIGURE 9.2. Standard scatterplots. (A) Plot of normalized microarray-element signal intensities. (B) Plot of $\log_{10}$ (normalized intensities). (C) Plot of mean ($\log_{10}$ intensity) vs. $\log_{10}$(ratio). As a reference, the 2-fold regulation value is depicted in each of the plots. Data depicted in Figures 9.2 and 9.3 were generated from human postmortem brain cDNA using the UniGEM V 1.0 array technology from Incyte Genomics, Inc. (Colantuoni et al., 2001). Abbreviations: CON, control; EXP, experimental. Each of the plots depicted in this figure can be created using the SNOMAD Web application.

channels (or both of the paired one-channel datasets). This results in a scatterplot that displays much more clearly many of the attributes of the data's distribution (Figure 9.2B). The dataset in this case appears to have a symmetrical distribution about a line of slope 1.

Even after logarithmic transformation, the scatterplot depicted in Figure 9.2B accentuates similarities rather than differences in expression levels across the two samples. In order to focus our analysis on differences, a third scatterplot (Figure 9.2C) is generated to directly reflect relative change in gene expression (y-axis) as a function of absolute level of intensity (x-axis). The mean gene expression level is plotted on the x-axis as the mean log intensity (averaged across the two array-element intensities being compared). The metric of differential gene expression is plotted on the y-axis as the log of the ratio of these same two intensities. The use of logarithmic transformations of values on both axes not only distributes values well across mean element signal intensity (x-axis) but provides a symmetrical scale (about zero) for ratio values (y-axis). This is not the case when true ratios are used. This scatterplot makes evident subtle imbalances in the distribution of observed gene expression ratios (i.e., deviations of the mean gene expression ratio from a value of 1—asymmetry about the zero value in the log scale on the y-axis).

9.2.5 Local Mean Normalization Across Element Signal Intensity

Here, we describe a local mean normalization that corrects for systematic bias in gene expression ratios measured between two samples, such as that which became apparent in Figure 9.2C. This procedure is "local" rather than "global" in that data are normalized not to the mean of all ratios but to the mean of ratios proximal (near in absolute element signal intensity) to the ratio being normalized. In gene expression datasets derived from many technologies, the mean expression ratio often deviates significantly from a value of 1 at points across the range of element signal intensities. These deviations are nonuniform across the range of microarray-element signal intensities and therefore require local rather than global mean normalization. Figure 9.3 demonstrates the use of a robust local regression ("loess" in the R statistical language) in order to ensure a "balanced" distribution of gene expression ratios across the entire range of element signal intensities (i.e., a mean gene expression ratio of 1 at all points across the range of element signal intensities). This local regression is fit to the differential gene expression ratios (y-axis) across the range of gene expression levels (x-axis). The "loess" function, or locally weighted regression (Cleveland, 1981; Hastie and Tibshirani, 1990), is a method for estimating the conditional mean (or typical value) of one variable (expression ratio) as a function of a second (expression level). It is a "robust" procedure in that the mean ratio estimate is insensitive to a small fraction of outlying or extreme values. In the Web implementation of SNOMAD, this fraction can be determined by the user via the setting of the "trim" variable. Additionally, the smoothness of the loess curve is controlled by a bandwidth parameter, termed "span"

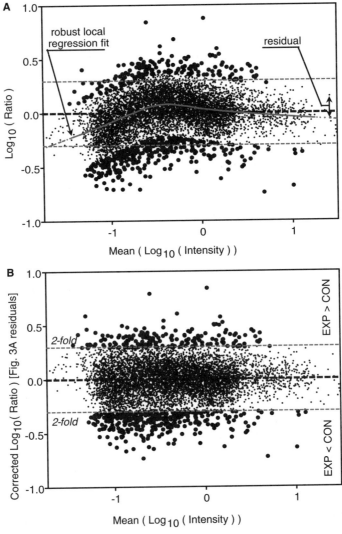

FIGURE 9.3. Local normalization across element signal intensity. (A) Plot of mean $\log_{10}$ intensity vs. $\log_{10}$ ratio with overlaid local mean log ratio ("loess," gray line). (B) Plot of the mean $\log_{10}$ intensity vs. the corrected $\log_{10}$ ratios (i.e., the residuals generated from the data and fit pictured in panel A). The highlighted points in each of the plots show the genes that have a twofold or greater difference between the two samples. This local normalization should be performed following the standard transforms detailed in Figure 9.2. Data depicted in this figure are the same data as those analyzed in Figure 9.2. This local mean normalization and its visualization are possible using the SNOMAD Web application.

(the "window" or fraction of data included in the locally calculated mean ratio), which can also be specified by the user.

Any deviation of the loess estimate of the mean $\log_{10}$ ratio from zero indicates a local imbalance in gene expression ratios among those genes with similar levels of expression (Figure 9.3A, y-axis: gray local regression fit line vs. dashed zero line). For example, at low expression levels (Figure 9.3A, leftmost portion of plot), the mean log ratio is below zero, indicating a preponderance of expression ratios less than 1. There is no biological reason to believe that genes expressed at a particular level would all show expression changes in the same direction. Rather, this "imbalance" is most likely an artifact of the experimental system.

We can remove this artifactual bias through the subtraction of the loess estimate of the mean expression ratio from each element's $\log_{10}$ ratio at that particular local element intensity level. This is equivalent to using the vertical distance (in the y-axis) of each $\log_{10}$ ratio from the smooth curve in Figure 9.3A as the adjusted $\log_{10}$ ratio values in Figure 9.3B. These values are termed "residuals" of the loess fit process. The y-axis values in this adjusted plot are still in the same $\log_{10}$ ratio units as the unadjusted values, however, this normalization has ensured a mean expression ratio of 1 at all points across the range of element signal intensities. It is significant that this adjustment is correcting for systematic deviation from a mean ratio of 1 that is nonuniformly distributed across the range of element intensities (x-axis). If the effect were uniform across this x-axis, global normalization methods would be sufficient for its correction.

Datasets such as that depicted in Figure 9.3A are quite common in the microarray field and will pass most quality-control measures in both the academic and commercial sectors (as did this dataset). It is important, however, that local imbalances in gene expression ratios be corrected before further analysis of ratios takes place. This also highlights the utility of visualization tools such as those in Figure 9.2 in the quality control of microarray data.

It should be noted that normalizing gene expression data with the goal of a zero mean log ratio can lead to problems if there are a very small number of array elements (> 100) involved in the analysis or if the vast majority of elements on an array represent genes whose expression has changed in the experiment under study.

9.2.6 Local Variance Correction Across Element Signal Intensity

Following the local mean normalization detailed in Figure 9.3, the mean corrected expression ratio has a value of zero at all points along the x-axis (this is a necessary result of the local mean normalization). Despite this fact, variance in these differential gene expression ratios varies independently of this mean ratio and is not equal across the range of gene expression levels. The unequal variance in expression ratios can be readily identified upon inspection of the same $\log_{10}$ ratio vs. mean $\log_{10}$ intensity scatterplot used previously (Figure 9.4A). We have observed such differences in the variance of expression ratios across expression levels in microarray data derived from a number of commercially available and custom DNA arrays, including Clontech, Research Genetics, Incyte, and Affymetrix microarray systems.

Figure 9.4A depicts the comparison of two paired one-channel datasets generated using the GeneChip array technology from Affymetrix, Inc. The data depicted in Figure 9.4 have already undergone the local mean normalization detailed in Figure 9.3. The variance in gene expression ratios is approximately one order of magnitude greater at low expression levels (Figure 9.4A, leftmost portion of plot) than at high expression levels (Figure 9.4A, rightmost portion of plot). A uniform cutoff applied to the differential expression ratios at this point in the analysis (e.g., the selection of all elements beyond "2-fold" cutoffs pictured in Figure 9.4A) would yield a list consisting primarily of elements at very low expression levels due to the large amount of variance at low expression levels. When variance in the observed gene expression ratios is unequal, it is inappropriate to apply uniform cutoff values for the selection of interesting expression changes (e.g., 2-fold cutoffs).

Here, we use a variance correction that entails the standardization of expression ratios to a standard deviation calculated locally across the range of element signal intensities. This standard deviation is calculated locally again using the "loess" function in the R statistical language. Because this transformation includes division of differential gene expression values by a locally calculated standard deviation, the resulting values are local Z-scores (in local standard deviation units). Figure 9.4 depicts the use of two independent robust local regressions in order to correct for unequal variance across the range of element signal intensities (x-axis). Because the distribution of gene expression ratios (y-axis) is not necessarily symmetrical about the zero value (i.e., a ratio of 1), it is useful to estimate separately the spread among positive and negative gene expression ratios. This allows for the independent correction of changes in variance across gene expression levels which are asymmetric about zero on the log ratio scale of the y-axis.

Variance is defined as the mean squared deviation of values from the mean value. Hence, the local estimates of variance are obtained by regressing the squared deviation of the adjusted $\log_{10}$ ratios from the mean $\log_{10}$ ratio $([Y - \text{local mean}(Y)]^2)$ at each gene expression level across the x-axis. Because the local mean(Y) is equal to zero (as a result of the local mean normalization described in Figure 9.3), the estimate of variance reduces to the regression of the squared $\log_{10}$ ratios (Y) across mean $\log_{10}$ intensity (X). Local estimates of the standard deviation (Figure 9.4A, gray curves) are simply the square root of the local estimates of the variance. As with the first local normalization (Figure 9.3), the "span" (which controls

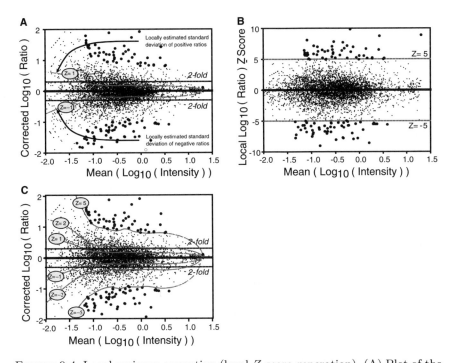

FIGURE 9.4. Local variance correction (local Z-score generation). (A) Plot of the mean $\log_{10}$ intensity vs. the corrected $\log_{10}$ ratios. This is the same plot that is depicted in Figure 9.3B but for a new dataset. Independent robust local regressions are generated to the square of the positive and negative ratios. The square root of these fit values (gray lines) is used to generate local Z-scores: corrected $\log_{10}$ ratio/local STDEV = Z-score. Local STDEV (locally estimated standard deviation) of gene expression ratios = [loess of (ratios)^2 across levels]$^{0.5}$. (B) Plot of the mean $\log_{10}$ intensity vs. the local Z-scores. (C) Plot of the mean $\log_{10}$ intensity vs. the corrected $\log_{10}$ ratios. This is the same plot that is depicted in panel A with the addition of curves corresponding to particular local Z-scores (gray lines). The points highlighted in each plot correspond to differential expression ratios that lie five local standard deviation units from the mean of 0. This variance correction should be performed following the standard transforms detailed in Figure 9.2 and the local normalization detailed in Figure 9.3. Data depicted here were generated from human leukocyte cDNA using the GeneChip photolithography technology from Affymetrix, Inc. (Golub et al., 1999). This dataset is publicly available from the Whitehead Institute via the Internet at http://waldo.wi.mit.edu/mpr/data_set_all_aml.html). Prior to all steps in the analysis of these Affymetrix data, all array-elements with an "average difference" value (expression level) less than or equal to zero in either one-channel dataset were removed from the analysis. Additional quality-control and analysis tools specifically designed for application to Affymetrix data can be found at http://biosun01.biostat.jhsph.edu/ ririzarr/affy/index.html, http://biosun1.harvard.edu/complab/computat.htm (Schadt et al., 2000), and http://www.egcrc.org/labs/miles/mm_analysis.html. This local variance correction and its visualization as seen in panels A and B are possible using the SNOMAD Web application.

the smoothness of the locally calculated standard deviation) and "trim" (which controls the fraction of most extreme values that do not influence the locally calculated standard deviation) parameters of the loess function can be determined by the user in the Web implementation of SNOMAD.

In order to correct for variance in gene expression ratios (y-axis) that is unequal across the range of gene expression levels (x-axis), each local mean adjusted log expression ratio (Y value) is standardized to (divided by) the estimation of the standard deviation of log ratio observations that share similar mean expression levels (i.e., that are proximal on the x-axis, as defined by the "span" parameter). This results in the generation of Z-scores in locally estimated standard deviation units (Figure 9.4B, local Z-scores on the y-axis).

The resulting set of local Z-scores are centered at zero (as a result of the local mean normalization) and have a variance and standard deviation of 1 at all points across the range of gene expression levels (as a result of this variance correction). This uniform variance across the x-axis makes appropriate the application of uniform gene expression ratio cutoffs (now expressed in local standard deviation units; Figure 9.4B). In Figure 9.4B, data points lying outside five local standard deviation units from zero are highlighted, as are their positions in the dataset prior to this local variance correction (Figure 9.4A). Without conversion of the adjusted $\log_{10}$ ratios to local Z-scores, this variance correction can be visualized on a y-axis that is unchanged from Figure 9.4A by indicating the $\log_{10}$ ratio values that correspond to particular Z-scores across the range of signal intensities (Figure 9.4C, gray lines).

By using Z-scores rather than intensities, intensity differences, ratios, or log ratios, the gene expression values (Y values) are no longer directly reflective of any absolute measure of change in gene expression. Z-scores, in standard deviation units, are directly reflective of a data point's position within the distribution of differential expression values. Locally calculated Z-scores take this one step further by indicating a data point's position within the distribution of expression ratios for genes at a particular expression level within a particular dataset. As part of a dataset containing several thousand gene expression ratios, this local Z-score becomes more meaningful in the identification of significantly differentially expressed genes than estimates of absolute expression or expression ratio. This is evidenced by the variability in estimates of expression ratio generated by multiple array technologies when evaluating identical samples (Butte et al., 2001; Colantuoni et al., 2001; Tsien et al., 2001).

Additionally, this variance correction reduces the need for arbitrary fold-induction cutoffs. For example, 2-fold is often applied to gene expression data as a standard cutoff for significant differential expression. In one dataset, this could lie within a single standard deviation unit of the mean, while in a dataset with less variance a 2-fold induced gene may lie many standard deviations from the mean. Similarly, within a single dataset, a ratio of 2 might lie within one local standard deviation unit at low expression levels (where variance tends to be higher) while this same ratio at a higher expression level may be beyond several local standard deviations. It is this situation of unequal variance that makes the application of standard ratio cutoffs to data (especially to data similar to those pictured in Figure 9.4A) inappropriate.

Local Z-scores provide a metric of differential gene expression that is ideal for comparison across multiple datasets. The comparability of differential expression values across many experiments is important for the selection of genes regulated across experiments as well as for input into other algorithms such as clustering of genes or expression profiles (Chu et al., 1998; Eisen et al., 1998), self-organizing maps (Toronen et al., 1999), relevance networks (Butte et al., 2000), support vector machines (Brown et al., 2000), principal components analysis (Alter et al., 2000; Holter et al., 2000; Raychaudhuri et al., 2000; Holter et al., 2001), multidimensional scaling (Khan et al., 1998), gene shaving (Hastie et al., 2000), and relevance networks (Lin et al., Chapter 13, this volume). The local Z-scores generated here are ideal for input into such algorithms.

9.2.7   Local Mean Normalization Across the Microarray Surface

The final local normalization procedure that we discuss is used to detect and correct artifacts that are systematically distributed across the physical surface of the microarray. These effects can be introduced during the production or use of arrays and are most often independent of the bias and variance effects that we have detailed in previous sections. These include artifacts that can be introduced during the robotic printing, hybridization, washing, or imaging of microarrays. Although this transformation is applied to data prior to the local normalizations discussed above (Figure 9.1), we describe it here because the other normalizations that we have discussed are necessary for the quantitative assessment of this procedure (Figure 9.5D and E; see color insert).

This methodology entails the use of spatial coordinates describing each element's position on the array in combination with the intensity value for each element in order to calculate a mean element intensity locally across the spatial surface of the microarray. When this smooth local mean is compared between duplicate microarrays that have undergone different experimental hybridizations, differences in this locally calculated mean indicate localized artifactual bias in the observed element intensities. These local means can then be used to correct for the very effects that they are used to detect.

Figure 9.5 (see color insert) depicts the comparison of two independent one-channel gene expression datasets derived from a radioactivity-based microarray system (Human GeneFilters, Research Genetics, Inc.). Images of the two arrays used in this analysis are pictured in Figure 9.5A (color insert). Using the intensity and position of each of the elements in an individual array, the "loess" function in R is employed to calculate a smooth local mean element intensity across the surface of each microarray (Figure 9.5B; color insert). SNOMAD allows the user to define both the smoothness ("span") and robustness ("trim") of this locally weighted regression. The difference between these two local mean intensities reveals localized bias in the measured element signal intensities (Figure 9.5C).

This localized bias in element signal intensities is also evident upon examination of the results of a nave (one that does not consider spatial position on the array) investigation of differentially expressed genes between the two microarray experiments. A scatterplot of globally normalized log element intensities from each of the arrays is depicted in Figure 9.5D (left panel). The center panel depicts the scatterplot of mean expression level vs. local Z-score following transformation with the two local normalizations detailed above (Figure 9.3 and Figure 9.4). A number of differentially expressed genes are highlighted in this plot. When represented as points in their original physical position (Figure 9.5D, right panel), it is clear that these array-elements localize to particular regions of the array surface (corresponding highlighted points and red and blue arrowheads). The regions to which these genes localize correspond to regions of the array surface where the difference in background signal intensity between the two arrays is especially high (Figure 9.5A). This is strong evidence that the detected bias in element intensities is due to hybridization artifacts present in the two microarray images used to generate these data (possibly due to uneven hybridization conditions, membrane drying, or uneven washing). Hence, the majority of array-elements identified in this analysis would be misidentified as being differentially expressed between the two samples under study.

To correct this artifactual bias in element signal intensities, we propose the normalization of all array-element signal intensities to the same smooth local mean intensities used to detect the bias. Using the locally estimated mean intensities (Figure 9.5B), we can now normalize all element intensities by dividing each individual intensity by the value of the local mean intensity at the identical physical position. Element intensities from the two array experiments are plotted against one another before (Figure 9.5D) and after (Figure 9.5E) the local mean normalization is applied to each of the individual one-channel datasets. The Pearson's correlation coefficient, r, is slightly greater after the local surface normalization (Figure 9.5D, left panel vs. Figure 9.5E, left panel), and perhaps more significant is the finding that differentially expressed array-elements no longer show systematic distribution across the array surface (Figure 9.5D, right panel vs. Figure 9.5E, right panel).

It is important to note that this algorithm assumes that array-elements are not spatially ordered with regard to sequence or biological function and requires that this spatial mapping be identical between the two sets of intensities to be compared. This requirement is automatically met when analyzing two-channel data because the two sets of element signal intensities are derived from two different scans of the identical microarray surface. When analyzing paired one-channel datasets using this local normalization, it is necessary that the two sets of intensities being compared be derived from duplicate microarray printings.

This normalization successfully removes the detected bias in element signal intensities. However, because the value of the local mean intensity at any one point is determined in large part by cDNA spot placement, the corrected element intensities resulting from this normalization are no longer a valid measure of absolute gene expression level. Hence, this normalization refines the metric of *differential* gene expression at the expense of the metric of *absolute* gene expression.

9.3   Software

SNOMAD has been integrated into an interactive, biologist-friendly Web application; see `http://pevsnerlab.kennedykrieger.org/snomad.htm`. The SNOMAD gene expression analysis tools are freely available to any researcher with Internet access and a standard HTML browser. No programming expertise or software download/installation is required. All of the data transformation and visualization tools detailed in this chapter are available through this web application. In addition, example datasets and tutorial instructions are available on the Web site as example applications of the SNOMAD transformations.

The freely available R statistical language (`www.r-project.org`) was used to develop the SNOMAD gene expression analysis tools. In order to make these tools available to biologists without statistical expertise or programming knowledge, an HTML form in combination with Perl scripts were used to create the Web implementation. The SNOMAD Web implementation includes all the data-visualization and transformation tools detailed in this report. Using the SNOMAD HTML form (Web page), users can upload their own gene expression data as a tab-delimited text file. With only a few mouse clicks, the user determines the specific transformations to be performed and the internal parameters controlling each transformation. Perl scripts then assemble and execute the appropriate R code without further input from the user. Depending on the option selected by the user, the results of the SNOMAD request are then returned via a new HTML page or as an e-mail attachment. Results include both a text file containing numeric values and image files depicting graphs of the data at all stages of transformation.

9.4 Discussion

We have described three normalization processes that approximate and then correct for biases and variances that are systematically distributed across different dimensions within gene expression data. Two of these procedures address effects that vary across the range of element signal intensities, while the third examines effects that are present across the spatial surface of the microarray. Terrence Speed and his collaborators have noted similar effects resulting from the robotic printing of microarrays and have proposed methods for their correction based on the normalization of array-elements derived from the same robotic arrayer print tip (`http://www.stat.berkeley.edu/users/terry/zarray/html/normspie.html`).

The utility of local normalizations such as those detailed and referenced here highlights the importance of the inspection of the distribution of expression ratios across systematic dimensions in gene expression datasets. Many unappreciated biases and sources of variance are introduced into gene expression data due to the complex nature of the experimental assay. Development of the biological paradigm, RNA extraction/preparation, probe generation, array hybridization, signal detection, and image analysis can all introduce artifactual variation in gene expression data. Other dimensions across which bias and/or variance could be systematically introduced include the GC content of DNA elements and array-element length in nucleotides. In the continuing effort to refine differential gene expression data, it will remain useful to uncover additional systematic sources of bias and variance and to describe the dimensions across which they vary, making it possible to identify and correct such artifacts.

References

Alter O, Brown PO, Botstein D (2000). Singular value decomposition for genome-wide expression data processing and modeling. *Proceedings of the National Academy of Sciences USA*, 97:10101–10106.

Beissbarth T, Fellenberg K, Brors B, Arribas-Prat R, Boer J, Hauser NC, Scheideler M, Hoheisel JD, Schutz G, Poustka A, Vingron M (2000). Processing and quality control of DNA array hybridization data. *Bioinformatics*, 16:1014–1022.

Brown MP, Grundy WN, Lin D, Cristianini N, Sugnet CW, Furey TS, Ares M, Haussler D (2000). Knowledge-based analysis of microarray gene expression data by using support vector machines. *Proceedings of the National Academy of Sciences USA*, 97:262–267.

Butte AJ, Tamayo P, Slonim D, Golub TR, Kohane IS (2000). Discovering functional relationships between RNA expression and chemotherapeutic susceptibility using relevance networks. *Proceedings of the National Academy of Sciences USA*, 97:12182–12186.

Butte AJ, Ye J, Haring HU, Stumvoll M, White MF, Kohane IS (2001). Determining significant fold differences in gene expression analysis. *Pacific Symposium on Biocomputing*, **X**:6–17.

Chu S, DeRisi J, Eisen M, Mulholland J, Botstein D, Brown PO, Herskowitz I (1998). The transcriptional program of sporulation in budding yeast. *Science*, 282:699–705.

Claverie JM (1999). Computational methods for the identification of differential and coordinated gene expression. *Human Molecular Genetics*, 8:1821–1832.

Cleveland WS (1981). Lowess: Program for smoothing scatterplots by robust locally weighted regression. *The American Statistician*, 35:54.

Colantuoni C, Jeon OH, Hyder K, Chenchik A, Khimani AH, Narayanan V, Hoffman EP, Kaufmann WE, Naidu S, Pevsner J (2001). Gene expression profiling in postmortem Rett Syndrome brain: Differential gene expression and patient classification. *Neurobiology of Disease* 8, 847-65.

Eickhoff B, Korn B, Schick M, Poustka A, van der Bosch J (1999). Normalization of array hybridization experiments in differential gene expression analysis. *Nucleic Acids Research*, 27:e33.

Eisen MB, Spellman PT, Brown PO, Botstein D (1998). Cluster analysis and display of genome-wide expression patterns. *Proceedings of the National Academy of Sciences USA*, 95:14863–14868.

Golub TR, Slonim DK, Tamayo P, Huard C, Gaasenbeek M, Mesirov JP, Coller H, Loh ML, Downing JR, Caligiuri MA, Bloomfield CD, Lander ES (1999). Molecular classification of cancer: Class discovery and class prediction by gene expression monitoring. *Science*, 286:531–537.

Hastie T, Tibshirani R (1990). Exploring the nature of covariate effects in the proportional hazards model. *Biometrics* 46:1005–1016,

Hastie T, Tibshirani R, Eisen MB, Alizadeh A, Levy R, Staudt L, Chan WC, Botstein D, Brown P (2000). 'Gene shaving' as a method for identifying distinct sets of genes with similar expression patterns. *Genome Biology*, 1:research0003.

Hegde P, Qi R, Abernathy K, Gay C, Dharap S, Gaspard R, Hughes JE, Snesrud E, Lee N, Quackenbush J (2000). A concise guide to cDNA microarray analysis. *Biotechniques*, 29:548–550, 552–554, 556 passim.

Hilsenbeck SG, Friedrichs WE, Schiff R, O'Connell P, Hansen RK, Osborne CK, Fuqua SA (1999). Statistical analysis of array expression data as applied to the problem of tamoxifen resistance. *Journal of the National Cancer Institute*, 91:453–459.

Holter NS, Maritan A, Cieplak M, Fedoroff NV, Banavar JR (2001). Dynamic modeling of gene expression data. *Proceedings of the National Academy of Sciences USA*, 98:1693–1698.

Holter NS, Mitra M, Maritan A, Cieplak M, Banavar JR, Fedoroff NV (2000). Fundamental patterns underlying gene expression profiles: Simplicity from complexity. *Proceedings of the National Academy of Sciences USA*, 97:8409–8414.

Iyer VR, Eisen MB, Ross DT, Schuler G, Moore T, Lee JC, Trent JM, Staudt LM, Hudson J, Boguski MS, Lashkari D, Shalon D, Botstein D, Brown PO (1999). The transcriptional program in the response of human fibroblasts to serum. *Science*, 283:83–87.

Kadota K, Miki R, Bono H, Shimizu K, Okazaki Y, Hayashizaki Y (2001). Pre-processing implementation for microarray (prim): An efficient method for processing cDNA microarray data. *Physiological Genomics*, 4:183–188.

Khan J, Simon R, Bittner M, Chen Y, Leighton SB, Pohida T, Smith PD, Jiang Y, Gooden GC, Trent JM, Meltzer PS (1998). Gene expression profiling of alveolar rhabdomyosarcoma with cDNA microarrays. *Cancer Research*, 58:5009–5013.

Lee CK, Klopp RG, Weindruch R, Prolla TA (1999). Gene expression profile of aging and its retardation by caloric restriction. *Science*, 285:1390–1393.

Liao B, Hale W, Epstein CB, Butow RA, Garner HR (2000). MAD: A suite of tools for microarray data management and processing. *Bioinformatics*, 16:946–947.

Manduchi E, Grant GR, McKenzie SE, Overton GC, Surrey S, Stoeckert CJ (2000). Generation of patterns from gene expression data by assigning confidence to differentially expressed genes. *Bioinformatics*, 16:685–698.

Park PJ, Pagano M, Bonetti M (2001). A nonparametric scoring algorithm for identifying informative genes from microarray data. *Pacific Symposium on Biocomputing*, **X**:52–63.

Raychaudhuri S, Stuart JM, Altman RB (2000). Principal components analysis to summarize microarray experiments: Application to sporulation time series. *Pacific Symposium on Biocomputing*, 11:455–466.

Schadt EE, Li C, Su C, Wong WH (2000). Analyzing high-density oligonucleotide gene expression array data. *Journal of Cellular Biochemistry*, 80:192–202.

Schuchhardt J, Beule D, Malik A, Wolski E, Eickhoff H, Lehrach H, Herzel H (2000). Normalization strategies for cDNA microarrays. *Nucleic Acids Research*, 28:E47.

Smid-Koopman E, Blok LJ, Chadha-Ajwani S, Helmerhorst TJ, Brinkmann AO, Huikeshoven FJ (2000). Gene expression profiles of human endometrial cancer samples using a cDNA-expression array technique: Assessment of an analysis method. *British Journal of Cancer*, 83:246–251.

Toronen P, Kolehmainen M, Wong G, Castren E (1999). Analysis of gene expression data using self-organizing maps. *FEBS Letters*, 451:142–146.

Tseng GC, Oh MK, Rohlin L, Liao JC, Wong WH (2001). Issues in cDNA microarray analysis: Quality filtering, channel normalization, models of variations and assessment of gene effects. *Nucleic Acids Research*, 29:2549–2557

Tsien CL, Libermann TA, Gu X, Kohane IS (2001). On reporting fold differences. *Pacific Symposium on Biocomputing*, **X**:496–507.

Wittes J, Friedman HP (1999). Searching for evidence of altered gene expression: A comment on statistical analysis of microarray data. *Journal of the National Cancer Institute*, 91:400–401.

Yang YH, Dudoit S, Luu P, Lin DM, Peng V, Ngai J, Speed TP (2002). Normalization for cDNA microarray data: A robust composite method addressing single and multiple slide systematic variation. *Nucleic Acids Research*, 30:e15.

10

Microarray Analysis Using the MicroArray Explorer

PETER F. LEMKIN
GREGORY C. THORNWALL
JAI EVANS

Abstract

The MicroArray Explorer (MAExplorer) is an open-source Java-based microarray data-mining tool that is available from the open source Web site as both a ready-to-run program and source code from `http://maexplorer.sourceforge.net/`. MAExplorer helps analyze expression patterns of individual genes, gene families, and clusters of genes. It is used as a stand-alone Java application and may be used for both ratio and intensity quantified array data (e.g., Cy3/Cy5, Affymetrix, and others). Data-mining sessions may be saved for continuation at later times or shared with collaborators; significant gene subsets, plots, and reports may be saved on the local disk. Extensions, called MAEPlugins, enable users to add new analysis methods and access to new genomic databases as they become available. MAExplorer was implemented in Java so that the same software could run on many platforms (e.g., Windows, MacOS 8/9 and X, Solaris, Linux, and Unix).

10.1   Introduction

The MicroArray Explorer (MAExplorer) is a user-friendly data-mining tool written in Java (Lemkin et al., 2000, 2001). It is available from the open source Web site (`http://maexplorer.sourceforge.net/`) as well as the original site (`http://www.lecb.ncifcrf.gov/MAExplorer`). The site at SourceForge.net was created to encourage collaborative community development of MAExplorer. MAExplorer was initially created to help analyze gene expression patterns of mammary enriched cDNA microarrays developed by one of our collaborators (L. Hennighausen, NIDDK, NIH). Expression patterns are of individual genes, gene families, and clusters of genes. The tool was first developed as a Web applet for the Mammary Genome

Anatomy Program (MGAP) Web site (http://mammary.nih.gov/) to publish ^{33}P labeled mouse clone expression data from early 1700 clones spotted membranes of normal and knockout breast tissue mouse models. The MAExplorer applet was subsequently converted for use as a stand-alone Java application and its capabilities greatly extended for use with ratio or intensity quantified array data (e.g., Cy3/Cy5, Affymetrix, and others). The stand-alone version has other advantages over the applet: it starts up faster, it makes use of more memory for clustering, data-mining sessions may be saved for continuation at later times or sharing with collaborators, and significant gene subsets, plots, and reports may be saved on the local disk. MAEPlugins extensions enable addition of new analysis methods and access to new genomic databases as they become available. MAExplorer was implemented in Java so that the same software could run on many platforms (e.g., Windows, MacOS 8/9 and X, Solaris, Linux, and Unix). Figure 10.1 (see color insert) shows the MAExplorer graphical user interface. We welcome developers who would like to contribute to this Open Source project.

10.1.1   Need for the Methodology

Traditionally, researchers have studied the regulation of biochemical and genetic pathways as well as developmental programs on a gene-by-gene basis. The development of large-scale gene expression profiling using DNA microarray chip technology holds promise as a powerful tool to help in the discovery of new genes and biochemical pathways involved in the development and regulation of mammalian cells and diseases such as cancer (DeRisi et al., 1996; Weinstein et al., 1997; Alizadeh et al., 2000; Cooper, 2001; Schulze and Downward, 2001; Sorlie et al., 2001). Although there is no simple 1:1 correspondence between mRNA expression and protein expression because of posttranscriptional and posttranslational processing (Ideker et al., 2001), understanding which mRNAs are expressed will help in understanding biological pathways. During the last half decade, the DNA microarray chip has become a major biological research tool. "At one end, the investigator is interested only in finding the single change in gene expression that might be the key to a given alteration in phenotype... At the other extreme, the aim is to look at overall patterns of gene expression in order to understand the architecture of genetic regulatory networks, a global approach that could ultimately lead to complete description of the transcription-control mechanism of a cell" (Schulze and Downward, 2001).

In general, data-mining is a collection of exploratory analysis methods for uncovering relevant patterns of interest in data from a particular problem domain (Tukey, 1977; Cleveland, 1985; Tufte, 1997) and through direct manipluation (Schneiderman, 1997). Typically, this involves using various methods, including database, statistical, data-filtering, graphical, and direct-manipulation user interface techniques. In the context of microarrays, the goal is to help identify genes having similar expression patterns associated with particular experimental conditions. Such subsets of genes and their associated patterns may be useful for helping uncover gene functions and genetic pathways (Schena et al., 1995; Strausberg and Austin, 1997; Ermolaeva et al., 1998; Eisen et al., 1998; Hughes et al., 2000).

10.1.2   Basic Ideas Behind the Approach

MAExplorer is an integrated program that includes a variety of data analysis methods for sample and gene-set management, data normalization, gene-set data filtering, direct-manipulation graphic displays (pseudoarray images, scatterplots, expression-level plots, histograms, cluster plots, clustergrams (i.e., heat maps), and dendrograms), and tabular reports for both genes and sample data. Gene data in plots and reports are hyperlinked to pop-up Web pages on genomic Internet database servers. This extensive set of features is fully described in the MAExplorer Reference Manual available for online use or download from the Web site. Because of space limitations, we cannot show examples for most of the concepts described in this chapter and will only touch on some of this functionality. Instead, we refer you to the Reference Manual, which has many examples of screen captures illustrating the functionality described in this chapter.

A sample in a MAExplorer database is the quantified data from a hybridized array. A MAExplorer database consists of a directory tree of files that may be generated by various conversion tools. These include: (1) our Cvt2Mae data converter "wizard"; (2) the Web-based NCI-CIT Micro Array Database (mAdb, http://nciarray.nci.nih.gov/) at NIH; or (3) data edited manually (e.g., with Microsoft-Excel) into MAExplorer format. The detailed formats are described in Appendices C and D in the Reference Manual. Cvt2Mae translates commercial (Affymetrix, GenePix, Scanalyze, and others) or non-standard academic user-defined chip tab-delimited data files to ready-to-analyze MAExplorer format data files; this surmounts the data conversion problem. For most non-NIH users, using the wizard (1) is the simplest method.

All data files are tab-delimited ASCII text. A database consists of a directory with at least three subdirectories: Config/, Quant/, and MAE/. The Config/ directory contains a Gene In Plate Order (GIPO) file mapping genes on the array with position of spots on the array as well as specifying genomic identifiers (such as GenBank, LocusLink, UniGene, and others) for each gene, a samples database listing the hybridized samples, and a configuration file describing the array layout and other information about the data. Quant/ contains separate quantified .quant data files for each hybridized sample with intensity, background, array location and good spot flag data for each spot on the array. MAE/ contains startup or .mae files. Additional directories are created as needed and include **State/** which holds named gene sets and named sample lists when you save a data-mining session, and **Report/**, which contains `.txt` or `.gif` files generated when you save tab-delimited text reports or pop-up graphics plots, respectively.

New bioinformatic analysis methods may be added to MAExplorer by using a dynamically loadable Java MAEPlugins programming facility. Because new genomic analysis methods and databases are constantly becoming available, this facility lets researchers add new methods and access these databases without having to modify the kernel MAExplorer software.

MAExplorer, Cvt2Mae, MAEPlugins, documentation, and sample data are freely available for download over the Internet from our Web site and are easily installed on common operating systems. We encourage authors of new plug-in methods to make them freely available for the research community either on the `http://maexplorer.sourceforge.net/` Web site or their own site.

10.2   Methods—Statistical and Informatics Basis

MAExplorer consists of methods for manipulating sets of gene data across sets of hybridized experiment samples. The general paradigm is to use gene data filters to define a *working set of genes*—a subset of all genes in the database based on various filter criteria. Having found interesting subsets of genes, the user can then visualize these subsets and make reports of those sets of genes. We now discuss some of the concepts and methods used to implement this paradigm.

Users first select a predefined database of multiple *hybridized samples* abbreviated as HPs, representing different sample conditions from either local files or Web databases. *Individual samples* may be assigned to separate X and Y variables referred to as HP-X and HP-Y. These are often used for comparing two samples, as, for example, comparing two samples in an HP-X vs. HP-Y scatterplot. For Cy3/Cy5 ratio data, one can compare any two channels between separate HP-X and HP-Y samples or compare Cy3 vs. Cy5 of the same sample. We defined the *current sample* as the last sample HP-X or HP-Y sample selected.

The user may also assign any number of *replicate samples* to *sets* called HP-X and HP-Y "sets." An *ordered list* of samples used for gene expression profiling is called the HP-E "list." and may be assigned by the user. Subsequent analyses then operate on samples in these sets and list. The X and Y sets may be used for pairwise analyses comparing mean values of replicate samples. For example, if one had replicate controls and replicate tumors, then one could compare the mean values in scatterplots, perform t-test gene data filtering, or apply other data filters using mean data.

The simplest database consists of a single sample assigned to HP-X. MAExplorer can report on these data to some extent, although more samples allow additional types of analysis. If that single sample contains Cy3/Cy5 data, you can do some limited analysis such as plotting Cy3 vs. Cy5, data filtering the Cy3/Cy5 ratio, viewing an intensity or ratio histogram, or interrogating individual genes for their ratios, and more. The default startup mode is a single HP-X sample and a single HP-Y sample, even if sample sets were assigned. You can toggle between single and replicate HP-X and HP-Y "sets" of samples using the (Samples | Use HP-X & HP-Y "sets" else single samples) menu command.

The HP-E ordered list of individual samples is generally assigned using the ordered conditions of a particular project experiment. Typical ordered conditions might be a time series, cell cycle, developmental stages, drug dose-response, and others. The HP-E samples list is used for expression profile plots (EP plots), clustering, and reports. Additional data structures and methods to manipulate ordered lists of condition sets of replicate samples are being added.

The hybridized arrays are scanned and the images quantified into tab-delimited intensity data files using programs such as Axon's GenePix^{TM}, Scanalyze, Molecular Dynamics ImageQuant^{TM}, Research Genetics' Pathways^{TM}, and others as well as Affymetrix software that generates tab-delimited data. Specific microarray image quantification characteristics are determined by the particular image analysis program being used to process the images into quantified spot data files. These quantified data files must then be converted to a specific set of MAExplorer tab-delimited data files using the Cvt2Mae data conversion wizard tool. Most spot intensity data should be able to be converted for use with MAExplorer.

When comparing data between samples, it is necessary to normalize the data between samples—even those generated under supposedly exactly the same conditions. This is critical because of differences in amount of sample, labeling efficiency, variations in scanner operation, including gain and baseline settings, and other systematic errors. There are a variety of data normalization methods built into MAExplorer. Other methods could be added using a normalization plug-in with the MAEPlugin facility.

All analyses operate on a subset of genes called the *working gene set* defined by the data filter and computed in real-time by the intersection of the results of gene data filter tests selected by the user. The working gene set is then used for plotting, clustering, or reporting. There are a number of predefined gene sets that are fixed for a particular array and include named genes, ESTs, and so forth. There are three special gene sets, including a user-defined *Edited Gene List* (EGL) that may be created or edited manually or that capture the current cluster(s) during clustering. The two other special gene sets are the normalization and user data filter gene sets. Gene sets may be saved in named gene sets and used for data filtering, reports, normalization, or to synthesize new gene sets using Boolean operations. Similarly, the HP-X, -Y, and -E sets and lists of sample conditions may be saved and restored as named sample lists as well as manipulated with Boolean list operations. The current data-mining session state is the collection of all of these named gene sets and sample sets, normalization options, data filter thresholds and options, and other settings. The state may be saved to the disk as a named startup file when the database is saved. Starting MAExplorer on this file at a later time restores the session state at the time it was saved.

Sets of genes or sample condition lists are useful for tracking intermediate results in complex data-mining sequences of analysis operations. For example, derived named gene sets may be used in successive data filters and for reports.

One could do the following experiment given four different types of samples (e.g., virgin, pregnancy, lactation, and involution). First, compare two HP sample sets using a statistical test such as a t-test. Then save the resulting set of genes under the name "virgin vs. pregnancy." Then, compare the next two HP sample sets and save the resulting genes under the name "lactation vs. involution." Finally, compute the difference of genes found in "virgin vs. pregnancy" that are not found in "lactation vs. involution." This resulting gene set could then be saved (e.g., with the name "Genes found in virgin vs. pregnancy but not in lactation vs. involution"). Similarly, taking the intersection of these two named sets shows genes that are common between the two sets. Taking the union shows genes found in either of the two named sets.

Data reports of sets of genes or lists of samples may be exported to Excel spreadsheets as tab-delimited data (by cut and paste or by saving to a text file) or as dynamic spreadsheets with clickable links to external databases such as genomic databases. Clicking on a link displays information for that gene in the related genomic database in a pop-up Web browser. Full resolution plots can be saved as GIF files for documenting results or for publication purposes.

The main MAExplorer window displays a pseudoarray image of the currently selected HP sample, individual HP-X and HP-Y samples, or sets of samples. It generates either an intensity image or a ratio image of various types of data (e.g., Cy3/Cy5, Cy3/Cy3, Cy5/Cy5 from the same sample or from different samples, HP-X/HP-Y of single samples, mean HP-X "set"/mean HP-Y "set" or others). Depending on the pseudoarray mode, different numeric data values are presented when one clicks on a spot in the image. This has the side effect of defining the corresponding gene as the *current gene*. Similarly, clicking on a point that corresponds to a gene in a scatterplot or other type of plot also defines that gene as the current gene. You may alternatively specify the current gene using a pop-up *gene name guesser* window where you can type a gene name or partial name. Changing the current gene in any display causes all of the active displays to be updated with the current data for that gene. Similarly, changing the normalization method or data filters selected or data filter parameters via threshold sliders also causes all active displays to be updated and the working set of genes defined by the data filter to be recomputed. A SaveAs button is available on the various plot and report windows that lets you save the plots as full-resolution GIF files and reports as tab-delimited text files.

Scatterplots are useful for comparing two samples, two mean samples composed of HP-X and HP-Y "sets," and channels from the same ratio (e.g., Cy3, Cy5) data or channels from different HP samples or duplicate spots (denoted F1, F2 as duplicate grid fields) in the sample. Changing the normalization automatically rescales and redraws the scatterplot. You may use scrollbars to zoom in on particular regions of the plot. Histogram plots of ratio and intensity data for individual samples or sets of samples can be generated. Then, a histogram bin can be selected and used to define a ratio range, intensity range, or functions of these ranges for the data filter.

We define the expression profile EP$_j$ of a given gene j as an ordered list of normalized intensity or ratio data values for each sample for that gene. This ordered list of samples used is specified by the HP-E list of samples. Then, the expression profile plot (EP plot) of a given gene is the intensity on the vertical axis and the sample number on the horizontal axis. You may generate EP plots for any selected gene, a scrollable list of EP plots for a set of genes, or an overlay plot of expression profiles for a set of genes. The scrollable list and overlay EP plots are useful for comparing expression profiles in gene sets.

MAExplorer clustering methods all work by clustering genes based on their expression profiles across an HP-E ordered list of samples. Only genes within the working set of genes are clustered. They use the concept of cluster distance d_{ij} between gene i and gene j computed according to a particular distance metric on the expression profile space. Gene-gene similarity s_{ij} is defined as $(1.0 - d'_{ij})$, where d'_{ij} is d_{ij} scaled to have a range of 0.0 to 1.0. The metrics currently available are Euclidean distance and Pearson correlation coefficient. The three cluster methods include: (1) clustering by expression profiles most similar to that of the currently selected gene (an example is shown in Figure 10.2; see color insert). This finds the subset of genes whose cluster distance is less than the adjustable cluster distance threshold; (2) K-means or K-median clustering with a user-adjustable number of clusters K; and (3) hierarchical clustering of genes as a function of the HP-E sample expression-list. The latter method results in a clustergram heat map (one gene per row, one HP sample per column), with an optional zoomable dendrogram. Dynamic cluster reports are generated for all of these methods. Gene subsets may be captured from all of these methods for subsequent operations. Sneath and Sokol (1973) describe these and other clustering methods.

10.2.1 Analysis Paradigm

Generally, an investigator has a number of objectives when analyzing a set of data. The types of analyses performed and how useful they are depends on what one wishes to get out of the analyses as well as the type and quality of the data.

Data mining is an exploratory data analysis for uncovering of relevant patterns of interest in data from a particular problem domain (Tukey, 1977). Typically, this activity involves using various statistical techniques to identify the patterns, including cluster analysis. Researchers across a wide range of fields have suggested that a major aspect of this problem is finding the correct means of graphical presentation to allow humans to be a part of the pattern recognition process (Cleveland, 1985; Tufte, 1997). Tufte (1997) argues that the proper display of quantitative data in the context of the problem domain can aid in understanding complex sets of data. This carries over to the analysis of microarrays using statistical and graphical methods as well as quantitative and genomic knowledge databases. Jagota describes a number of these methods and applications for microarray data analysis and visualization (Jagota, 2001).

To avoid biasing the results, one should approach the analysis of a set of data with minimal expectations of what the patterns might be. However, some idea of what genes you might be interested in helps focus the search. Beware of the trap of mining the data until you find the patterns you hope for since they may have occurred by chance.

Obviously, this introduction to data mining is a first approximation to what is eventually required to successfully explore one's data. However, it does capture the flavor of the data-mining process. Typically, user's would refine their search using variations of the data filters and might contrast results using Boolean operations on gene sets and hybridized sample condition lists found under one set of conditions with those found under other sets of conditions. Figure 10.3 illustrates this iterative process.

The user makes some initial decisions on the experimental design such as the hybridized samples to compare as well as the types and numbers of replicates. A guess can then be made as to the normalization method to use and the gene subset to concentrate on when setting the initial data filter. Additional data filters may be used to remove bad or noisy data. Data are viewed in various modalities to get a feeling for their inherent dynamic range and where interesting outliers might appear. Clustering and plots may help bring these differences into view. The results are then evaluated and either the process is stopped or the views are refined by adjusting data normalization, filter parameters, data subsets to be investigated, clustering methods, plots and so forth, and the process repeated until the user is able to see the differences between gene subsets more clearly or no significant differences appear to be found.

Because data mining is a pattern discovery activity, the researcher should try to make use of all the tools available. It is open-ended because of the variety of ways data may be partitioned, normalized, prefiltered, clustered, and viewed. Patterns that are apparent in one view may not be apparent in another. When data-mining microarray data, look at correlated genes from

Color Plates

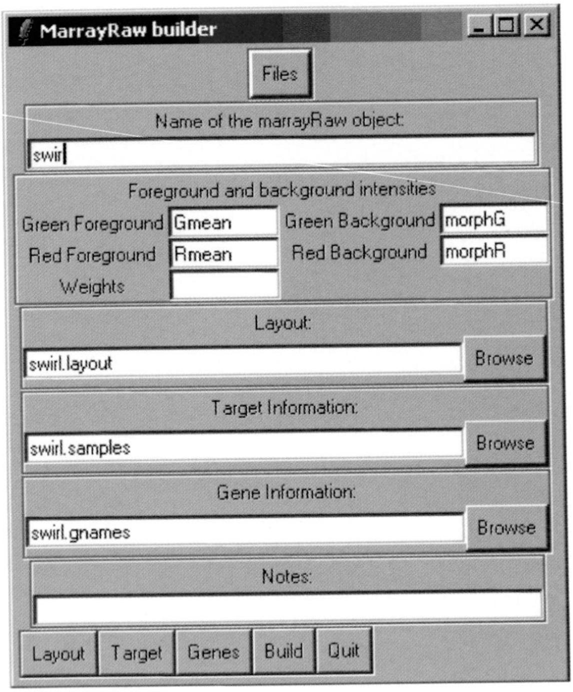

Color Plate 2

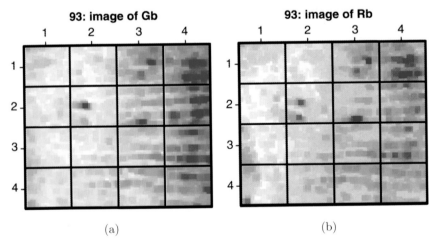

FIGURE 3.1. Screen shot of the widget for creating objects of class marrayRaw from image processing output file–widget.marrayRaw function.

FIGURE 3.2. 2D spatial images of background intensities for the Swirl 93 array. Panel (a): Cy3 background intensities using white-to-green color palette. Panel (b): Cy5 background intensities using white-to-red color palette.

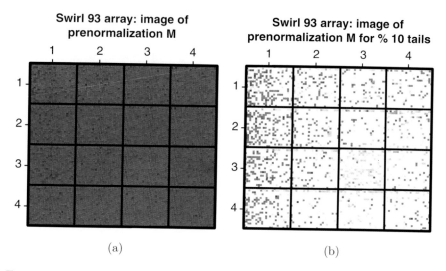

FIGURE 3.3. 2D spatial images of the prenormalization intensity log ratios M for the Swirl 93 array using a green-to-red color palette. Panel (a): All spots are displayed. Panel (b): Only spots with the highest and lowest 10% log ratios are highlighted.

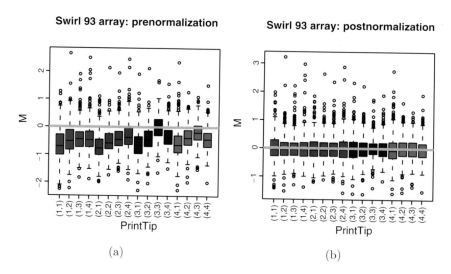

FIGURE 3.4. Boxplots by print-tip-group of the pre- and postnormalization intensity log ratios M for the Swirl 93 array.

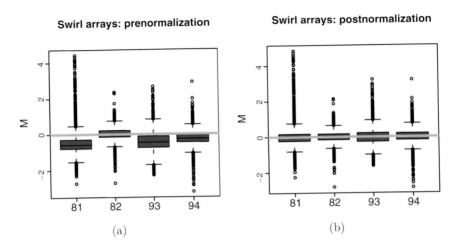

FIGURE 3.5. Boxplots of the pre-and postnormalization intensity log ratios M for the four arrays in the Swirl experiment.

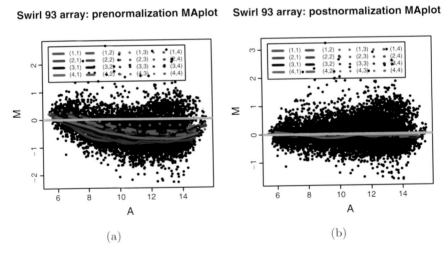

FIGURE 3.6. Pre- and postnormalization MA-plots for the Swirl 93 array, with the lowess fits for individual print-tip-groups. Different colors are used to represent lowess curves for print-tips from different rows of the print-head, and different line types are used to represent lowess curves for print-tips from different columns.

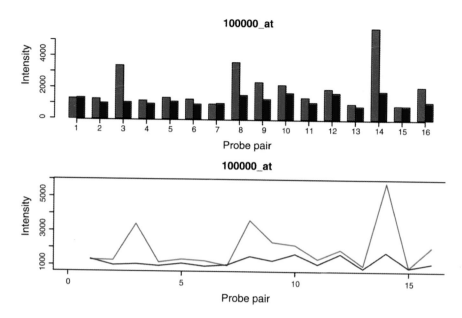

FIGURE 4.2. *PM* (red) and *MM* (blue) intensities for probes for gene with
`100000_at` ID shown as barplot and regular plot.

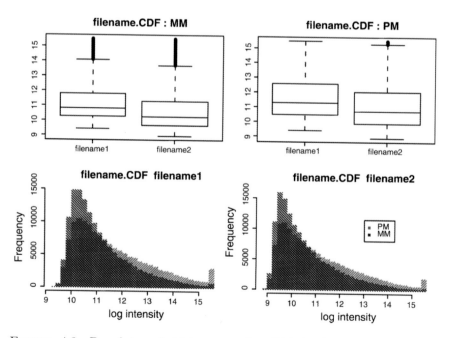

FIGURE 4.3. Boxplot and histograms for *PM* (red) and *MM* (blue)
intensities.

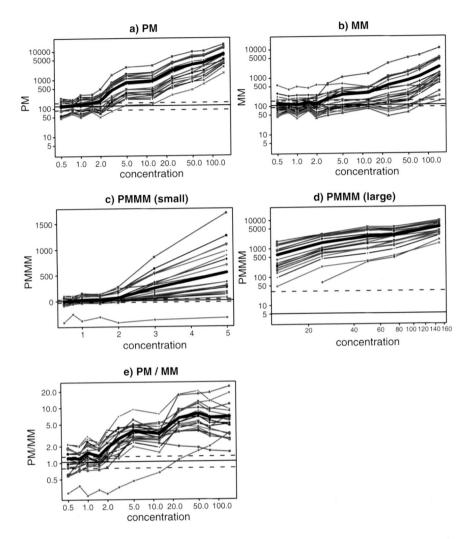

FIGURE 4.8. *PM*, *MM*, *PM-MM* for spike-in concentrations between 0 and 5, *PM-MM* for larger spike-in concentrations, and *PM/MM* values for each of the 20 probes representing spiked-in gene BioB-5 in the 12 spiked-in arrays from the varying concentration experiment plotted against concentration. The different probes are represented by the different colors and symbols. The horizontal line represents the median of the 20 BioB-5 probes for an array for which no spike-in was added (these nonspike-in data are not provided with the package) to the nonspiked-in array. The dashed lines are the 25th and 75th quantiles.

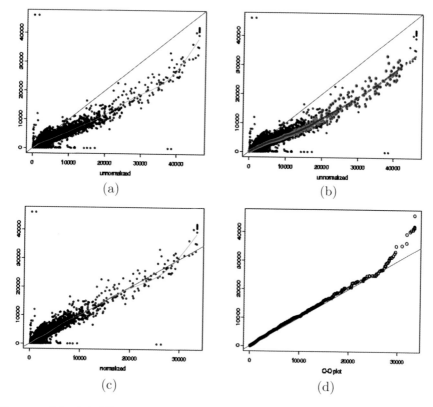

FIGURE 5.1. (a) The intensities of two arrays are plotted against each other. The baseline array shown on the probe y-axis is not as bright as the array shown on the x-axis. The smoothing spline (green curve) deviates from the diagonal line $y = x$, indicating the need for normalization. (b) The same plot as (a) with superimposed green circles representing the "Invariant set" based on which a piecewise-linear normalization curve is determined. Note that the smoothing spline in (a) is affected by several points at the lower-right-hand corner that might belong to differentially expressed genes, whereas the "Invariant set" does not include these points when determining the normalization curve (pink dots), leading to a different normalization relationship at the high end. Note that here we also use additional criteria to exclude the data points saturated in the array on the x-axis from the "Invariant set." (c) After normalization (y-axis is the baseline array, and x-axis is the normalized value of the array to be normalized), the scatterplot centers around the diagonal line, and the array to be normalized is adjusted to have a similar overall brightness as the baseline array. The smoothing spline (green curve) is also close to the diagonal line. (d) The Q–Q plot of probe intensities shows that the probes in the two arrays have similar distribution after normalization (adapted from Li and Wong, 2001b.)

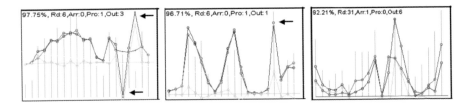

FIGURE 5.2. Single, probe, and array outliers. The pictures are for three different probe sets (adapted from Li and Wong, 2001b).

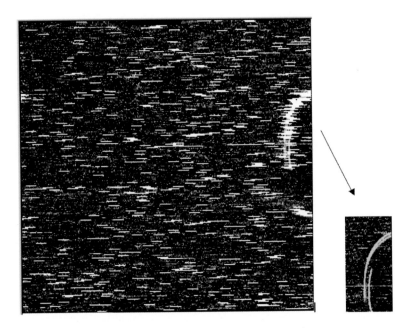

FIGURE 5.3. Outlier image. White bars indicate array outliers and pink dots indicate single outliers (adapted from Li and Wong, 2001b).

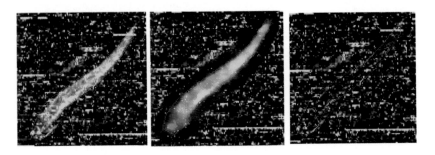

FIGURE 5.7. Image gradient correction.

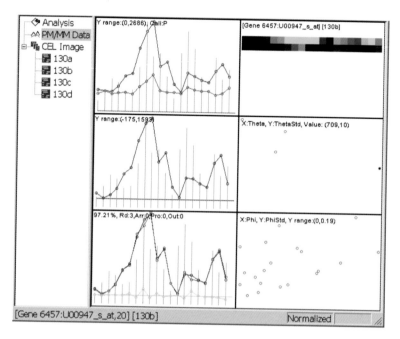

FIGURE 5.8. The "*PM/MM* Data View."

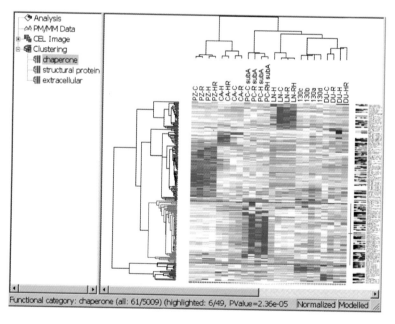

FIGURE 5.10. Hierarchical clustering and enriched gene clusters.

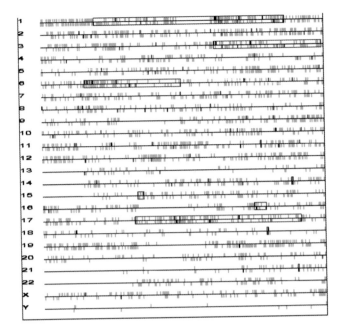

FIGURE 5.14. The "Chromosome view."

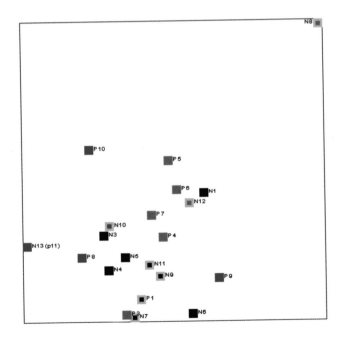

FIGURE 5.16. The "LDA classification view."

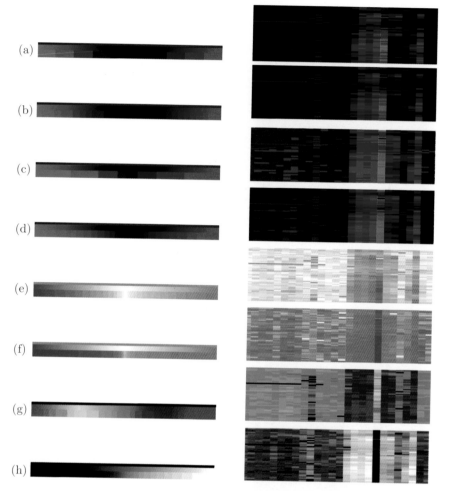

FIGURE 6.2. The effects of specifying different palettes and ways of choosing colors on visualization of microarray data. The color palettes on the left consist of three parts showing the color for the missing values (e.g., blue for a and gray for e), the continuous color palette, and the chosen colors themselves. Rows a, b, c, and d use the standard green-black-red palette with 10 and 30 colors using linear discretization, where colors are chosen at equal intervals, and 10 and 30 slots using harmonic discretization. Rows e and f: blue-white-pink palette with 30 harmonically and exponentially discretized slots. Row g: the color spectrum discretized linearly into 30 slots. Row h: black-and-white palette discretized harmonically into 30 slots. The blue-white-pink palette in rows e and f is particularly suitable for vision-impaired users, while the spectrum palette serves well to illustrate the data in a more colorful manner across smaller value ranges.

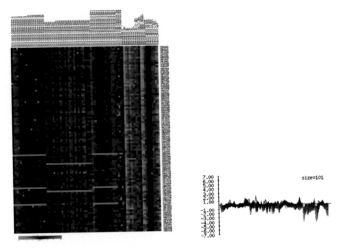

FIGURE 6.3. Expression profiles of 100 genes most similar to YGR128C as heat-map and line plot.

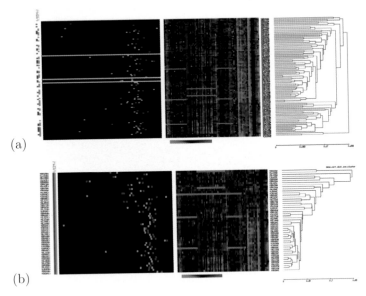

(a)

(b)

FIGURE 6.4. (a) A cluster of 98 genes based on the expression data (middle) and the two motifs G.GATGAG.T and TG.AAA.TTT found to be specific to the cluster by SPEXS (clustering on the right) matched to promoter regions of respective upstream sequences (left). (b) Hierarchical clustering (right) of all the yeast genes where the same two motifs occur within 40bp on their 600bp upstream sequences (left). It shows that all of the 52 genes where these two motifs are in close vicinity to each other have an almost unique expression response (middle).

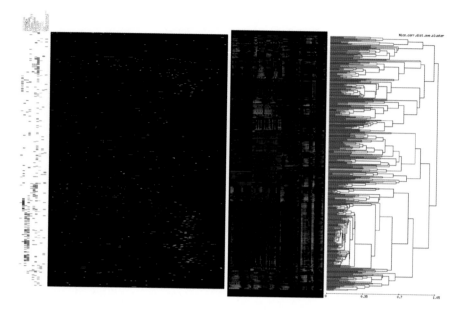

FIGURE 6.5. The combined visualization of microarray gene expression data clustering (right), the respective upstream sequences (middle), and several different patterns and their occurrences on these upstream sequences (left, middle). This illustrates that PATMATCH, in combination with EPCLUST and SPEXS, can uncover motifs that are specific to certain gene expression clusters and are often also conserved relative to the ORF start position. Note the grouping of motifs near the bottom-right-hand corner of the left images, corresponding to the highly coexpressed genes in the middle. Hierarchical clustering verifies these findings.

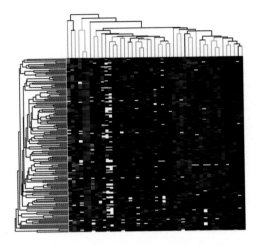

FIGURE 7.4. Dendrogram and heatmap for DLBCL and germinal center B-cells based on genes from germinal center B-cell expression; from Figure 3a of Alizadeh et al. (2000). Blue bars on the sample dendrogram represent Activated B-like DLBCL, and orange bars represent GC B-like DLBCL samples.

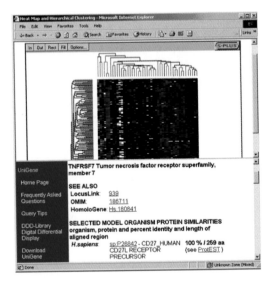

FIGURE 7.6. Heatmap and dendrogram for DLBCL samples corresponding to Figure 3a of Alizadeh et al. (2000). The gene highlighted is TNFRSF7, tumor necrosis factor receptor super family member 7. The heatmap and dendrogram are drawn using an S-PLUS graphlet produced using the java.graph device in S-PLUS 6. This features interactive metadata in the top right-hand corner of the graph and links to UniGene annotation information for genes (rows) via clicking on the heatmap.

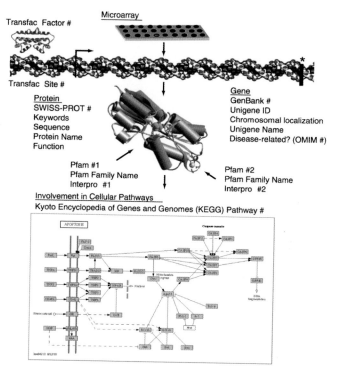

FIGURE 8.3. DRAGON uses accession numbers to define biological characteristics of genes and proteins. A microarray is a regular array of thousands of unique cDNAs or oligonucleotides spotted on a solid support. Each spot contains cDNA corresponding to a specific gene that encodes a protein. Accession numbers derived from publicly available databases provide information about the biological characteristics of both the gene and its corresponding protein. At the gene level, "Transfac Site" and "Transfac Factor" numbers indicate the presence of promoter regions on the gene and factors that bind to those promoter regions, respectively. The "GenBank "#"" and "UniGene ID" refer to EST sequences corresponding to fragments of the gene and a cluster of those EST sequences, respectively. The "UniGene Cytoband" indicates the chromosomal location of the gene. The "UniGene Name" is the name of the gene. The "OMIM "#"" indicates whether the gene is known to be involved in any human diseases. At the protein level, "Pfam "#"" and "Interpro "#"" indicate which functional domains the protein contains. The "SwissProt "#"" is a unique identifier for the protein and can be derived from either the SwissProt or TrEMBL databases. "SwissProt Keywords" are derived from a controlled vocabulary of 827 words that are assigned to proteins in the SwissProt database according to their function/s. "SwissProt Sequence" is the amino acid sequence for the protein. "SwissProt Name" is the SwissProt database name for the protein.

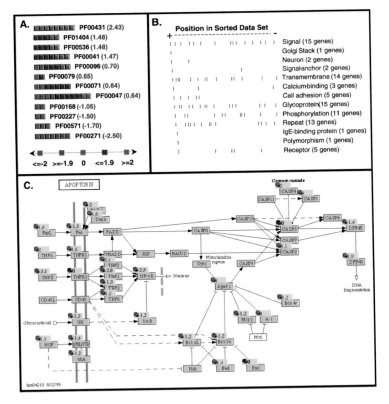

FIGURE 8.7. Examples of the graphical outputs of the three types of DRAGON View tools. (A) DRAGON Families produce rows of green (down-regulated), red (up-regulated) and gray (unchanged) boxes (see scale for the range of ratio values represented by each color). Each box represents one gene and is hyperlinked to its corresponding UniGene entry. Each row has a type identifier to its right that is hyperlinked to its description. To the far right is the average ratio expression value for all of the genes in that family. All rows are sorted from the most up-regulated family to the most down-regulated family. (B) DRAGON Order produces rows of black lines. Each line represents one gene, and its location in the row represents its position on a gene list sorted by ratio expression values. Lines at the far left represent the most up-regulated genes (+), and lines at the far right represent the most down-regulated (-). Each row's type (e.g., SwissProt keywords) is listed to the right. (C) DRAGON Paths maps the location and ratio expression value of genes from the submitted gene list onto KEGG cellular pathway diagrams. A green (down-regulated), red (up-regulated) or gray (unchanged) circle followed by the ratio expression value is mapped to the upper left corner of each corresponding protein box. Each protein box is hyperlinked to its corresponding LocusLink entry.

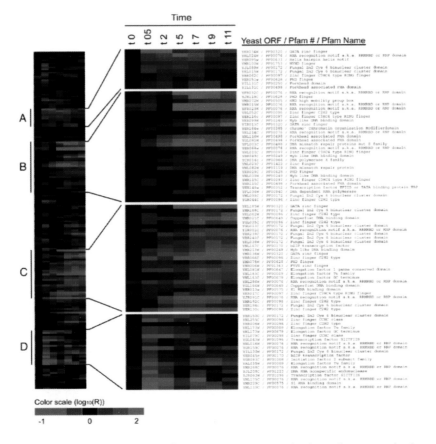

FIGURE 8.8. Expression of putative transcription factors during sporulation. The DRAGON annotate tool was first used to classify all genes according to Pfam family. A subset of those Pfam families was identified as classifying protein domains involved primarily in transcriptional regulation. This was performed by uploading a tab-delimited text file containing expression data and a list of GenBank accession numbers on the Annotate page. The resulting file was e-mailed due to the size of the file. It is better to e-mail larger files since they can take a while to annotate. The file was then sorted according to Pfam families, and genes in DNA-binding domain families were selected for clustering. The expression for all genes contained in these families was clustered with Cluster and then visualized with TreeView. Four expression patterns were identified as (A) up-regulated during early sporulation, (B) up-regulated during middle/late sporulation, (C) down-regulated during middle/late sporulation and (D) down-regulated during early sporulation. Specific genes in group B, including YGL081W, YNL116W, YJR119C, YMR072W, YKR099W, YIR013C, and YFL003C are candidates for regulation of the middle/late stages of sporulation.

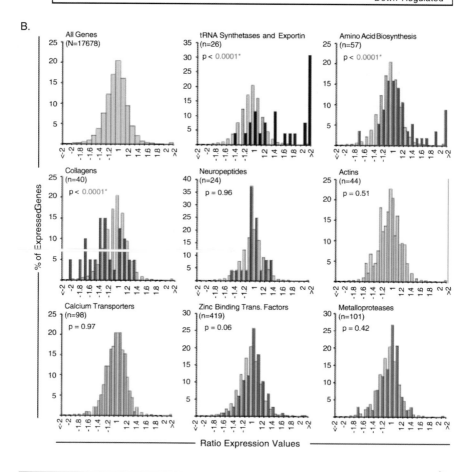

Color Plate 19

FIGURE 8.10. Functional group analysis of Incyte data. (A) The DRAGON Annotate tool was used to annotate all 17,678 genes analyzed on two Incyte microarrays with SwissProt keyword information in a manner similar to that described in Figure 8.8. The DRAGON Families information visualization tool was then used to visualize the ratio expression values of each keyword category across the entire dataset. This analysis was used to identify groups of biologically related genes that are coordinately regulated. In the entire Incyte dataset, one family ("Aminoacyl-tRNA synthetase") was significantly, differentially up-regulated by lead treatment and one was found to be significantly down-regulated ("Collagen"). The average expression value for each family is displayed in parentheses (blue text) to the right of each family. (B) The significance of the differential regulation of these two families was further investigated by comparing the distribution of all 17,678 gene expression values versus each gene family. Only three families were found to have significantly altered distributions: "tRNA synthetases and Exportin," "Amino Acid Biosynthesis," and "Collagens." The gray bars in each graph display the distribution of all gene expression values as shown in "All Genes." The colored bars in each graph display the distribution of all of the genes in that family (e.g., Amino Acid Biosynthesis), the number in parentheses provides the number of genes in that family (e.g., $n = 57$) and the p values indicate the probability of a given distribution happening by chance as determined by chi-square analysis. Significant p values ($p < 0.0001$) are displayed in red.

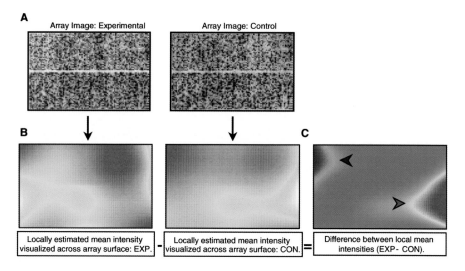

FIGURE 9.5. (A-C). Local mean normalization across microarray surface. The data depicted here were generated using mRNA derived from postmortem human brain analyzed using the Human GeneFilter microarray system from Research Genetics, Inc. (a) Images of the two arrays used to generate the data depicted here. (b) A robust local regression ("loess") is used to estimate the mean element signal intensity locally across the two-dimensional surface of each of the arrays. (c) The difference between these local mean estimates indicates localized regions where bias in element signal intensities is greatest (red and blue arrowheads).

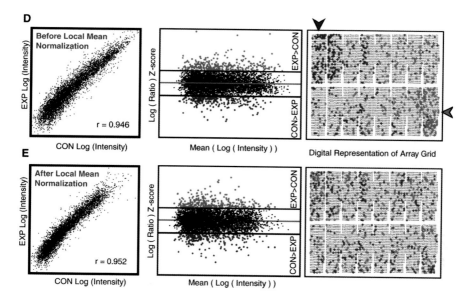

FIGURE 9.5. (D). Comparison of gene expression values prior to local mean normalization across the array surface. The left panel depicts a scatterplot of the global mean normalized $\log_{10}$ intensities for the two samples. The center panel shows an expression level vs. expression change scatterplot derived from the normalizations detailed above. Array elements identified as up- and down-regulated (center panel, red and blue highlighted points, respectively) localize to particular regions of the array surface (right panel, red and blue arrowheads). Array elements highlighted in this center panel correspond to those highlighted in the representation of the array grid surface (right panel). (E) Comparison of gene expression values after local mean normalization across the array surface. The amount of correlation between the two datasets has increased slightly (note the Pearson's correlation coefficient, or "r," in panel D vs. panel E). Array elements identified as differentially expressed (center panel) no longer show systematic distribution across the array surface (right panel). Abbreviations: CON, control; EXP, experimental. This local mean normalization and its visualization as seen in panels B and C are possible using the SNOMAD Web application."

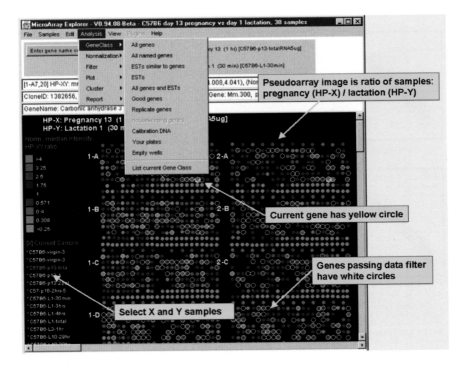

FIGURE 10.1. The main MAExplorer graphical user interface window
showing some of the pull-down menus, the pseudoarray image (from
MGAP duplicate spotted membranes (i.e., Grid 1-A is duplicated in Grid
2-A), and selectable individual samples on the left. The current gene is
selected by clicking on spots in the pseudoarray, points in plots, or
reports. Genes passing the data filter are indicated with white circles. The
user interacts directly with the program through pull-down menus, by
adjusting parameter sliders, and by clicking on spots in the pseudoarray,
points in plots, or cells in reports. Various pop-up scatterplots, histograms,
expression plots, cluster plots, dendrograms and clustergrams, and reports
may be generated. The set of all genes is restricted to a *working-genes*
subset computed by the data filter. Adjusting any of the *threshold state*
scrollbars causes the data filter to be recomputed. Any number of
orthogonal data filters may be combined to isolate a potentially
significant subset of genes.

Similar-Expression Cluster Report of Casein Beta

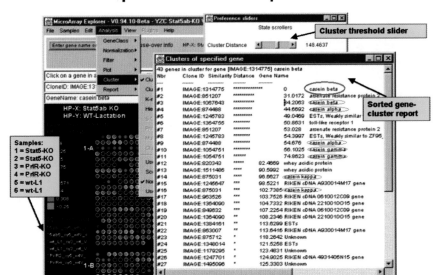

FIGURE 10.2. Example of expression profile similarity clustering of genes similar to Casein Beta (red circle) in an MAExplorer data-mining session on mouse mammary data. The data consist of 6 Cy5-labeled hybridizations (transplanted Stat5ab-KO, transplanted PrlR-KO, and wt-Lactation day (1) using wt-Virgin day 1 as the Cy3-labeled reference sample (Hennighausen et al., unpublished). The cluster distance threshold slider was adjusted to limit the number of genes passing the similar-cluster filter resulting in 43 genes. Passing genes are also saved in the Edited Gene List set that could be saved in a named gene set or further analyzed. Genes failing the cluster threshold distance test are not reported. These are presented in a dynamic cluster report of these genes sorted by similarity to the "seed" gene being tested. Selecting a new seed gene or adjusting the cluster threshold will recompute the cluster. Other useful options (not shown) include generating a cluster report (Cluster-Report), creating a popup list of expression profile plots (EP plot), and saving the cluster report in a tab-delimited text file (SaveAs).

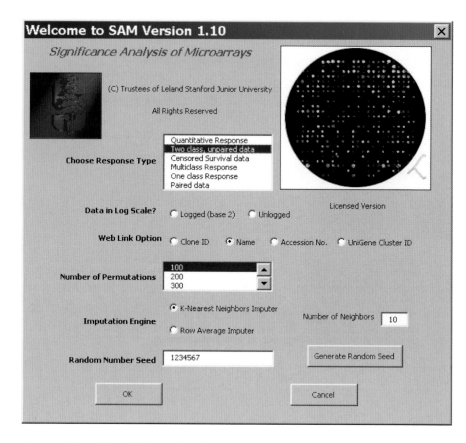

FIGURE 12.3. The SAM dialog box.

Color Plate 25

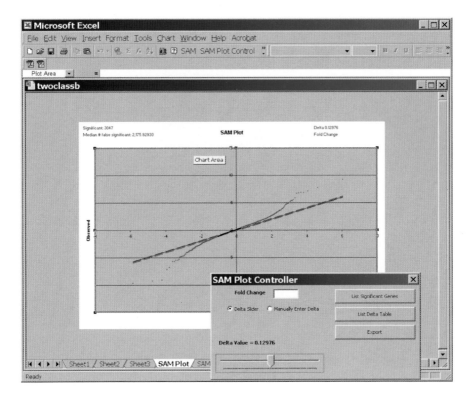

FIGURE 12.4. The SAM plot.

Color Plate 26

(a) methods with no noise

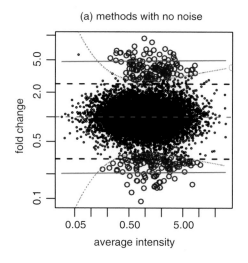

(b) methods with noise

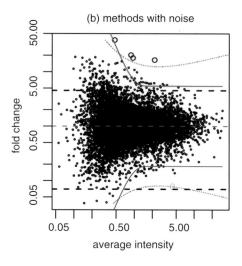

FIGURE 13.7. Effect of measurement error on shape of differential expression. Flat black dashed line for Chen et al. (1997); curved red dotted line for Newton et al. (2001) odds ratio of 1; smooth red solid line for our method. Blue circles are genes beyond odds ratio of 1. Horizontal red dashed line at 1 means no fold change. (a) no measurement error; (b) high measurement error ($\delta = 20$).

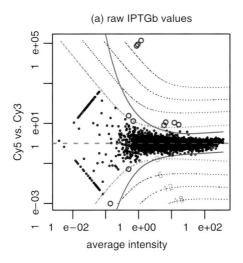

(a) raw IPTGb values

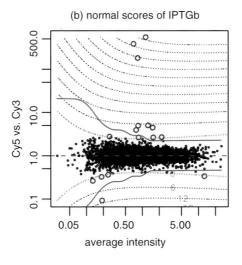

(b) normal scores of IPTGb

FIGURE 13.8. *Escherichia coli* IPTG-b analysis to compare Gamma—Gamma Bayesian method to normal-scores method. Each point is a gene, with blue circles for genes with odds ratio above 1 (red dotted lines at odds ratio of 1, black dotted lines at 10^6 increments in odds; see Newton et al., 2001). Solid purple line for our Bonferroni 5% criteria. (a) log–log scale of original expression levels, with negative values plotted along diagonals at far left; (b) normal-score-transformed data, antilogged and plotted on a log–log scale. Note three genes with huge IPTG-b signals. Our procedure is more conservative with no noise but detects patterns of variability when noise is present.

Color Plate 28

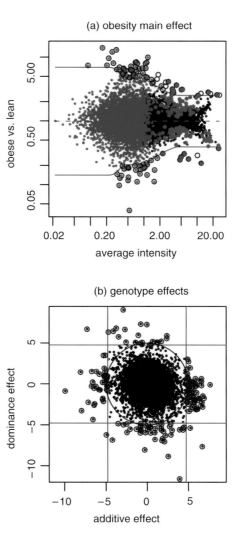

FIGURE 13.9. Diabetes and obesity study. Solid red line is our 5% Bonferroni limit. (a) Obesity main effect: green points have 1–6 negative adjusted values; purple points detected in Nadler et al. (2000); blue points beyond 5% line detected by our procedure. Clearly, methods based only on fold cannot detect all interesting patterns. (b) Genotype main effects: T scores scatterplot with blue circles on detected genes.

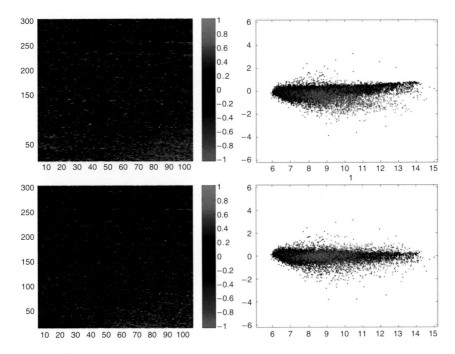

FIGURE 14.4. Preanalysis data transformation using the spatial lowess function. The upper left-hand panel shows log ratios of raw intensity data on a red—green color scale plotted on the spatial coordinates of the array. A bright red corner is apparent on the figure. The upper right-hand panel shows an RI plot of the same data. Spots from the bright corner of the array are highlighted in red. The lower two panels show the array view and the RI plots of data after transformations using the spatial lowess smoother.

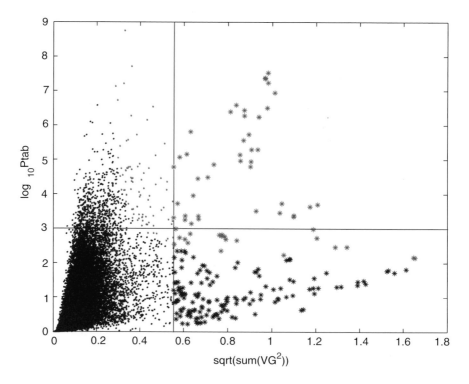

FIGURE 14.6. The volcano plot provides a summary of test statistics for differential expression of genes in the tumor survey. The null hypothesis is no differential expression (M0), and the alternative is differential expression across the groups defined by strains (M1). Under model M1, there are four relative expression values (VG) and the square root of the sum of squared VGs is shown on the x-axis of the figure. A vertical bar indicates the 0.05 multiple-test adjusted threshold for test F_3. The y-axis is the negative base 10 logarithm of the p-value for the per-gene F-test (F_1). A horiontal bar indicates the nominal 0.001 threshold for F_1. Genes that are significant at the 0.05 multiple-test adjusted level by the F_2 test are shown in red. In this analysis, each spot was treated as a separate gene, whereas each gene is actually represented by two spots on the array. Some pairing of points is apparent in the figure.

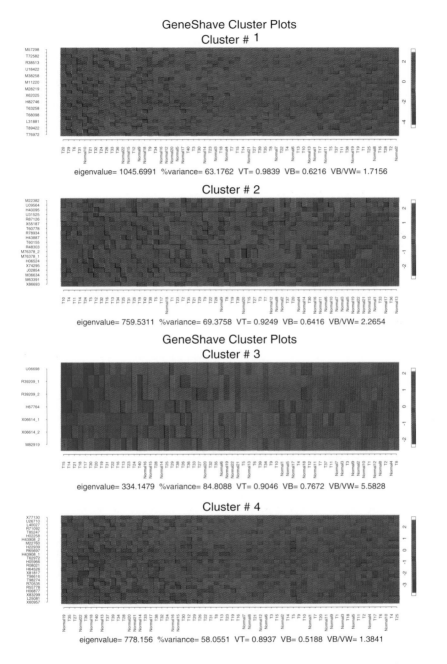

FIGURE 15.6. Heat maps of the first four unsupervised gene-shaving clusters for the reduced Alon colon data, sorted by the column mean gene.

Color Plate 32

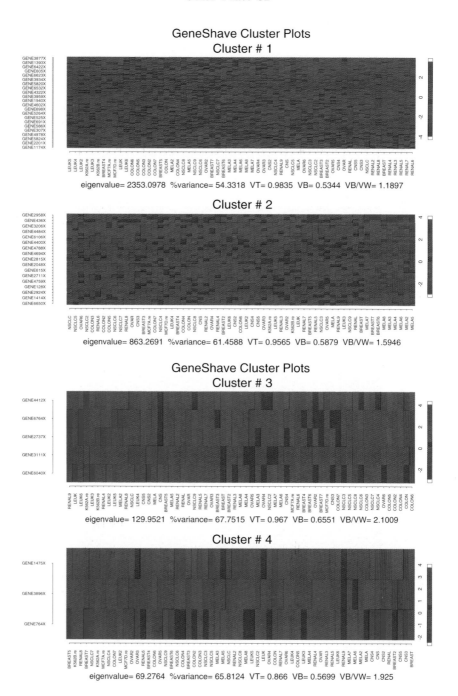

FIGURE 15.9. Heat maps of the first four supervised gene-shaving clusters for the NCI60 data, sorted by the column mean gene.

Color Plate 33

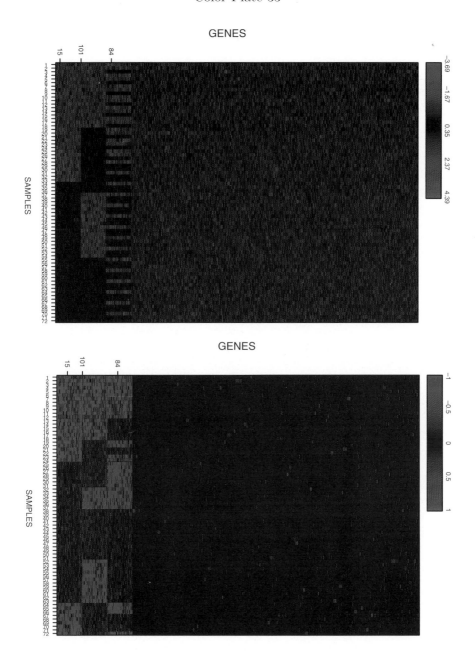

FIGURE 16.1. Simulated data. Raw data, x_{ji} (top) and P_{ji}^{*} (bottom), sorted by sample and gene. The POE scale is less noisy and provides improved resolution for the genes with signal.

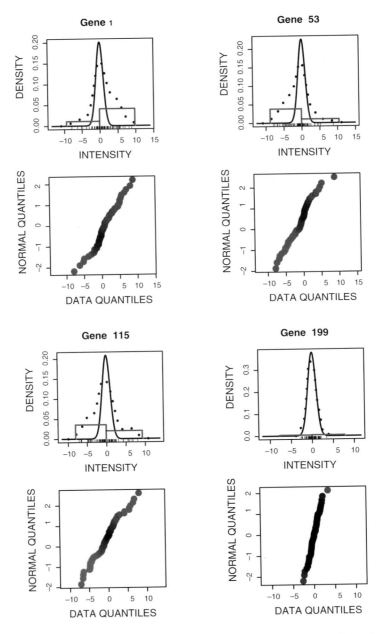

FIGURE 16.2. Fit of the mixture model on individual genes: Gene 1 is 1/3 overexpressed and 1/8 underexpressed; Gene 53 is 1/3 underexpressed, 1/8 overexpressed; Gene 115 is 1/4 overexpressed, 1/4 underexpressed; and Gene 199 is from the noise group. Color codes correspond to probability of expression. The fit of the normal component is judged by the linearity of the black portion.

GENES

GENES

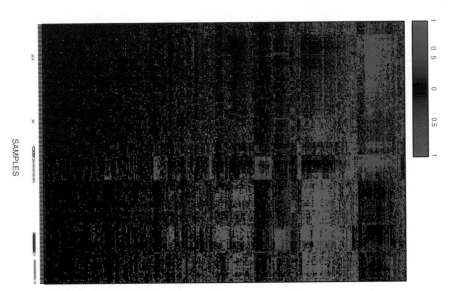

FIGURE 16.5. Lung cancer data. Expression values x_{ji} (top) and P_{ji}^{*} from POE (bottom). Genes and samples are ordered independently in the two panels to highlight the patterns that can be identified in the two scales.

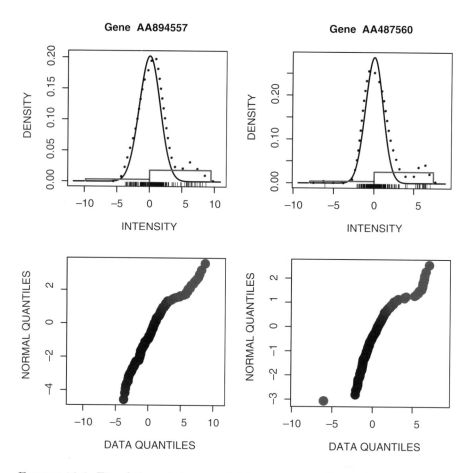

FIGURE 16.6. Fit of the mixture model in two genes from the lung cancer dataset: CKB (left) and CAV1 (right).

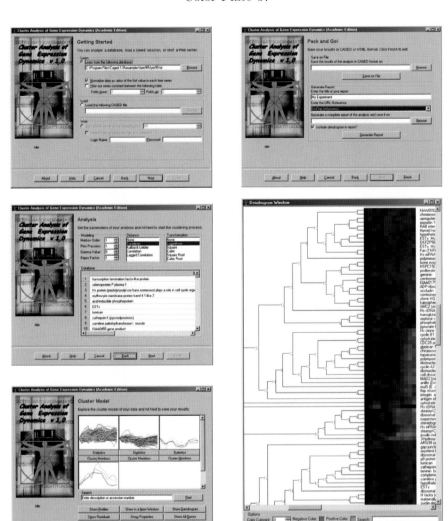

FIGURE 18.1. Snapshots of the main screens of CAGED 1.0.

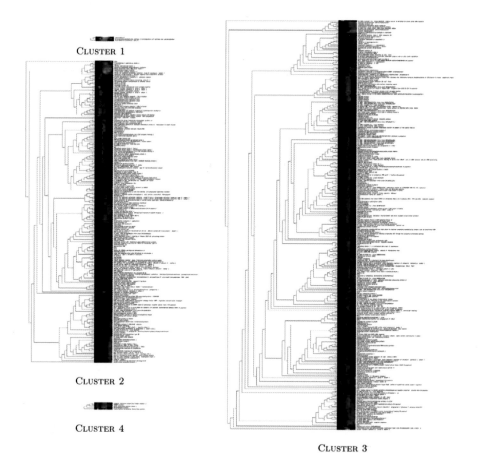

CLUSTER 1

CLUSTER 2

CLUSTER 4

CLUSTER 3

FIGURE 18.3. Binary tree (dendrogram) and labeled gene expression display showing the clustering model obtained by our method on the data reported in Iyer et al. (1999). The numbers on the branch points of the tree represent how many times the merging of two series renders the model more probable. The model identifies four distinct clusters containing 3 (cluster 1), 216 (cluster 2), 293 (cluster 3), and 5 (cluster 4) time series.

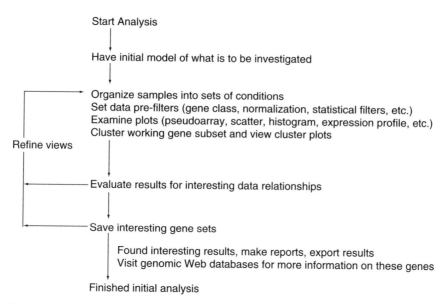

Start Analysis

Have initial model of what is to be investigated

Organize samples into sets of conditions
Set data pre-filters (gene class, normalization, statistical filters, etc.)
Examine plots (pseudoarray, scatter, histogram, expression profile, etc.)
Cluster working gene subset and view cluster plots

Refine views

Evaluate results for interesting data relationships

Save interesting gene sets

Found interesting results, make reports, export results
Visit genomic Web databases for more information on these genes

Finished initial analysis

FIGURE 10.3. Flow chart of a simplified view of the iterative data-mining process.

the point of view of what relationships might be interesting to a biologist when characterizing genes that cluster together. Additional information, such as knowledge of genes belonging to known pathways, may also be useful. Further investigate your results using various NCBI and PubMed database searches on the resulting genes, by designing other lab experiments to better uncover the causation, and other methods. Correlation does not imply cause and effect.

The process above must utilize an appropriate experimental design. Proper experimental design is critical for resolving significant differences in gene expression between experimental conditions and in making valid statistical inferences. Experimental design issues common to microarray experiments are discussed in Dudoit et al. (2000), Jagota (2001), Kerr and Churchill (2001), and Simon et al. (2001).

If users compared two different sample conditions, the type of analysis used would be different than if they were comparing an ordered sequence of sample conditions (e.g., time series, cell cycle, dose-response, tumor-stage, and so on). MAExplorer gives users the ability to:

1. Organize experiments by sample conditions, where each condition may have one or more replicate samples. If the database has samples with many different conditions, this allows users to perform pairwise comparisons of all of these condition sets.

2. Reduce the set of all genes on the array to a working set of genes that pass the selected data filters. This process may be thought of as a data

prefilter. There are three main categories of filters: set membership, data ranges, and statistical tests.

3. Cluster the working gene set using various clustering methods, possibly generating additional gene subsets corresponding to particular clusters.

4. Explore, compare, and record these gene sets using direct manipulation graphics and spreadsheet reports to gain different insights into the underlying relationships of the data.

5. Save various gene sets during this analysis for subsequent analysis, documentation, and the accessing of public Internet genomic databases.

Because of the iterative nature of this analysis process, it is important to keep a record of the analysis steps you used or the measurements made during this process. MAExplorer lets you record command history and measurements in log files.

10.2.2   Particular Analysis Methods

The primary way that the user invokes analysis steps is by selecting commands from pull-down menus. Since all of the menus are described in detail in Section 2 of the Reference Manual, we will not repeat that level of detail here but rather summarize their functionality.

10.2.3   Data Conversion

One of the problems in analyzing data from a variety of systems is the need to handle different formats. When the microarray community adopts the final MGED-ML MIAME standard data format, we will adopt it as well. Until then, to use MAExplorer with your array data, you must convert your data files into MAExplorer input format. Although you can do this by manually editing your data files into the required format (see Reference Manual), it is a nontrivial process. Therefore, we developed a Java "wizard" conversion tool called Cvt2Mae to automate data conversion. It handles most commercial and nonstandard array data formats. For user-defined chips or special exceptions to common formats, the wizard Q&A lets you describe or edit the specification for your chip and experiment project. We call this description the *array layout*. Part of this process is the specification of which fields of interest are defined in the GIPO (Gene-In-Plate-Order or Print) file, and which fields contain the quantified intensity data. After you have defined the chip layout, you can save it for use in future conversions or share it with a collaborator. After you have created the array layout and specified the input files to be converted, you run the converter and it generates the proper set of converted data files for MAExplorer.

10.3   Software

We give an overview of MAExplorer functionality by summarizing the menus. Menus contain commands (such as to pop-up a scatterplot) as well as checkboxes and ratio buttons to toggle the status of a command modifier (e.g., use sets of X, Y samples rather than individual samples in computations). Menus can also contain submenus where the organization requires it.

Files Menu—Select Database to Be Analyzed

The File menu includes commands for providing access to database data from disk and Web servers as well as for saving the data-mining session. The user may select the database subset to be loaded from either a Web server or a local file system. When you open an existing database, you do so by specifying a .mae startup file. As was mentioned in the introduction, the initial database includes a .mae startup file created prior to running MAExplorer. If you are on a system where programs are invoked by clicking on a data file (e.g., in MS Windows, MAExplorer is associated with files ending in .mae), then you may start MAExplorer directly by just clicking on the startup file. If not, then you can open the .mae file using the (File | Databases | Open disk DB) menu command.

Samples Menu—Selecting Sets of Data Sample Conditions

The Samples menu includes commands to select the current hybridized sample or samples to be analyzed. There is a pop-up sample chooser that lets you graphically assign samples to the HP-X "set," HP-Y "set," and HP-E "list" of hybridized samples. You may also manipulate these sample sets and lists using menu lists of samples or through a wild-card name-guesser interface. Cy3 and Cy5 channel data may be swapped on an individual sample basis. This menu contains the preference for treating X and Y sample data as individual samples or as sets for computation and graphic display purposes.

Edit Menu—Edit EGL, Gene Sets, Condition Lists, and Preferences

The Edit menu includes operations to modify the Edited Gene List (EGL). You may also define and edit named gene sets and named sample condition lists using Boolean operations to create new sets and lists. Various user preferences, such as the database title, X and Y sample set names, and others may be set in this menu. It is sometimes useful to preset threshold parameters used in data filters and clustering. Therefore, a window of threshold scrollbars may be popped up to adjust these preferences to their desired range. A user-defined named gene set may be used as an additional data filter or normalization gene set or be assigned to the EGL. You may define named gene sets for particular gene set ontologies using the pop-up *gene name guesser* and gene set Boolean operations in the Edit menu.

Analysis Menu—Organized as First Approximation of an Analysis

The Analysis menus is an ordered list of six primary menus that may be used sequentially to perform an initial analysis. In practice, all analyses will be much more complex, but it is a useful way to think about the tasks that would be required. In more complex analyses, the sequence of operations will vary and include commands selected from other menus. The analysis tasks are as follows:

1. *GeneClass*—define the primary subset of genes to be analyzed
2. *Normalization*—select initial normalization method
3. *Filter*—select gene data filters and adjust parameters for these filters
4. *Plot*—view various graphic representations of the data
5. *Cluster*—cluster genes by expression profile data
6. *Report*—generate dynamic or exportable reports

Gene Class Menu—Restrict Database to a Primary Subset of Genes

This menu lets you specify the current gene class from a set of built-in gene sets that are characteristic of your array and are extracted from the GIPO data. The data filter can be enabled to test for gene class membership. The *gene class* is a specific subset of genes (e.g., all genes, all named genes, ESTs, ESTs similar to named genes, all genes and ESTs, and so on). Some of the other built-in gene sets are good genes, replicate genes, housekeeping genes, calibration DNA, your plates, or empty wells, depending on their availability in a specific database.

Normalization Menu—Scale Data so Samples Are Comparable

This menu includes operations to normalize raw gene channel intensity data between hybridized samples by applying the transform to all spots within each sample in the database. Currently, each sample is normalized by itself, not taking other samples into account. The built-in methods do not currently take the possibly nonlinear dye response to intensity or gene sequence into account. These could be handled by creating MAEPlugin normalization methods to take these factors into account.

The current normalization methods that transform raw intensity include Z-score of intensity, median intensity, $\log_{10}$ median intensity, Z-score $\log_{10}$ intensity using either standard deviation or mean absolute deviation in its calcuation, by the sum of calibration DNA genes (if any), by the sum of genes in a user-defined normalization gene set, by scaling to 65K/maximum intensity, and unnormalized. In addition to the normalization method, you may adjust the raw data in other ways prior to performing the normalization. If background data are included in your database, you may subtract the background value.

Filter Menu—Find a Subset of Genes Meeting Data Filter Criteria

The working set of genes computed by the data filter is the intersection of gene sets passing the selected tests. Tests are grouped into three main types with any number of tests being allowed for any of these groups. They include: (1) membership in particular sets of genes, (2) testing whether data are in specified intensity or ratio ranges, and (3) statistical or clustering tests. Many of these data filters require a threshold parameter, and will automatically pop-up parameter sliders for you to adjust if required. You then interactively set the thresholds using visual feedback from the active displays and reports. This process of interactively adjusting threshold parameters is one of the tools used in direct manipulation.

Gene set filters include filtering by membership one or more of the following sets: gene class, user defined gene set, EGL, global *good genes list*, genes with replicates on the same array. If you created ratio or intensity histograms, you can select bins to automatically generate ranges of data to be filtered. Data may be restricted to only positive data (after background subtraction). Several filters let you set channel intensity, sample intensity, and ratio ranges by threshold sliders on a various sublists of samples. The coefficient of variation of intensity may be used in a thresholded filter computed over various sublists of samples. The p-value computed on a t-test between X and Y sets and the absolute difference between mean X and Y samples are other filters. You can also filter by the highest or lowest X/Y ratios of individual samples or sets.

Plot Menus—Display Graphical Representations of the Data

The Plot menu lets you specify the pseudoarray image in the main MAExplorer window. Scatterplots, ratio and intensity histograms, and expression profile plots are generated in separate pop-up windows. Depending on the particular plot, multiple instances may be allowed.

The Show Microarray submenu sets the pseudoarray for the current samples presented as either a grayscale or pseudo-color image. The grayscale image represents spot intensity. Several types of pseudo-color images may be used, including a *sum* of HP-X (red) and HP-Y (green) that generates a red-yellow-green image. A similar display may be generated for Cy3 (green) + Cy5 (red). An alternate display generates a ratio X/Y or Cy3/Cy5 image by scaling values > 1.0 as red, values near 1.0 as black, and values < 1.0 as green. Where appropriate, the p-value for all spots can be displayed as a pseudo-color image when comparing HP-X and HP-Y sets of samples.

Depending on the origin of the array data, a pseudoarray may have the same geometric verisimilitude as the original arrays. If there is no grid, row, or column data associated with array data, they are displayed as a generated pseudoarray image containing grids, rows, and columns computed to fit the window for visualization purposes. It will not have the same geometry as the original array image. However, the pseudoarrays may be useful to get a rough idea of the global changes in the data between arrays and how many genes pass the data filter. When performing gene clustering, clusters are reported as blue circles or squares drawn as overlays on the pseudoarray with the size of the overlay being proportional to similarity. If you are doing K-means clustering, the cluster currently selected is displayed in the scatterplot with the number of genes in the cluster proportional to the circle size, and so forth.

Scatterplots may be generated for X vs. Y data, mean X vs. Y "set" data, or Cy3 vs. Cy5 data. You may compare different permutations of Cy3 and Cy5 channel data between different X and Y samples. Cy3/Cy5 plots are available if the data exist in your particular database. That might be the case with replicate spots or with Cy3/Cy5 data. You may zoom into any area of the scatterplot using the horizontal and vertical scrollbars on the left and right edges of the scatterplot. If there are duplicate spots (F1, F2) for each gene on a sample, you may plot F1 vs. F2 intensity. For each scatterplot, the correlation coefficient for the filtered data is displayed in the plot. Data plotted are the intensity values of channels, sample, or mean samples using the current normalization method. The plot scales are changed when the normalization method changes. Clicking on spots in an array image or points in scatterplots sets the current gene and will bring up data on the gene or (optionally) access corresponding data from a genomic database in a pop-up Web browser.

Intensity histograms may be generated for the current sample, and ratio histograms may be generated for the mean X/Y or X/Y "set" data, or Cy3/Cy5 data. The histogram plots are active, letting you select a histogram bin and then use it to define data filters for all genes as a symmetric function (around a ratio of 1.0) of the range specified by that bin ($=$, $<$, $>$, $<>$, $><$).

Expression profile plots (*EP plots*) may be generated for either an individual gene or a list of genes using the ordered HP-E list of samples. In the latter case, it generates a scrollable list of EP plots that may alternatively be presented as an overlay plot. These plots are active and may be zoomed and interrogated for the expression data of a particular gene and sample.

By clicking on a spot (i.e., gene) in the pseudoarray image or on a point (i.e., gene) in the scatterplot, that gene is defined as the *current gene* that is used in other operations. The current gene is indicated by a yellow circle in the pseudoarray image and a green circle in the scatterplot. Similarly, you may add (remove) genes from the EGL from either the pseudoarray image or the scatterplot by clicking on a gene with the Control (Shift) key press. When viewing is enabled, it overlays those genes in the EGL with magenta squares.

Cluster Menu—Cluster Genes into Similar Groups

Cluster analysis finds a subset or subsets of similar genes based on expression-profile similarity measures across the HP-E list of samples. The three methods perform various types of gene clustering operations on the working gene set. When you invoke a clustering operation, it will pop-up a *cluster report* window and may modify the pseudoarray image using overlay graphics indicating which genes are part of the cluster(s). Direct-manipulation sliders let you specify the number of clusters (K-means or K-median) or a similarity distance. Changing these parameters recomputes the cluster(s) each time they are changed. The current gene is used as the *seed* gene for the similar genes and K-means methods. The center of the clusters is replaced by the gene closest to the centroid and the entire data set is then reclustered. Silhouette plots are generated for similar gene and K-means clustering. When doing K-means clustering, selecting the current gene defines all genes in its cluster as the *current cluster*. The gene clusters may be saved as named gene sets. The clustergram (i.e., heat map) and dendrogram analysis dynamic plots are used with the hierarchical clustering and may be interrogated, and gene subsets may be selected. These graphic plots show genes belonging to particular clusters or genes that cluster well with specified genes. Additional lists of EP plots, mean EP plots, and reports including statistics on clusters may be generated. For all of these methods, you may select the intergene distance metric as either the Euclidean distance or correlation coefficient. Various weighting, scaling, and normalization options are available for the hierarchical clustering method.

Reports Menu—Generate Dynamic or Exportable Summary Reports

Tabular reports summarizing gene or sample data may be generated and appear in pop-up windows. These include: data on all of the hybridized samples in the database, which may include links to related Web databases; hybridized sample calibration; Samples vs. Samples correlation coefficient of filtered genes, and mean and variance tables. You can generate reports of all named genes, genes in the EGL, genes in the current gene class, genes passing the data filter, or the N genes with the highest or lowest ratios (X/Y, Cy3/Cy5, or F1/F2) of the filtered genes. Additional features may be added to the data filtered gene reports, including expression-profile data values, X vs.Y "set" statistics, and correlation coefficient statistics. Report tables are presented as either (a) an active *dynamic* (i.e., clickable) spreadsheet that may contain links to genomic Web sites and that pops up a Web browser window to that site or (b) a scrollable tab-delimited text window that may be cut and then pasted into a Microsoft Excel-type spreadsheet for further analysis. Clicking on a column name in the dynamic report will sort the report by data in the report in ascending and descending order. If using the tab-delimited format, you may save the table into a text file.

View Menu—Adjust Views in Plots, Reports, and Genomic DB Access

The View menu options are used to modify the view of genes visible in the pseudoarray image and other displays. Genes may be displayed with additional properties or capabilities, including access to Web-based genomic database entries for specific genes if those database identifiers are available in your data and you are connected to the Internet.

Plugins Menu—Add New User-Defined Analysis Methods

Users can add new analysis methods to MAExplorer using Java plug-ins we call MAEPlugins. These plug-ins can include those written by us, collaborators, the research community, or commercial groups. If you have created a Java .jar MAEPlugin file, you may load it at run time using a (Plugin | Load Plugin) menu command that adds the plug-in command to the appropriate menu in the MAExplorer menu tree. You may also remove unwanted plug-ins if they are no longer needed to save memory or when debugging a MAEPlugin. When you save a data-mining session, the names of plug-ins you were using are saved and are reloaded when you restart MAExplorer. Figure 10.4 shows the MAEPlugin design used to add new analysis methods to MAExplorer. Additional analysis methods are written as Java MAEPlugins and can be 100% portable Java, Java stubs to connect to other local programs such as the R statistics package, or Java stubs to perform client-server access to genomic Internet databases. For example, a new cluster method plug-in called *Cluster* might be added and then invoked from the Cluster menu. MAEPlugins may be loaded at startup time or during an analysis as needed. Once loaded and started, they access MAExplorer data using the Open Java API (Application Programming Interface described on the Web site under MAEPlugins and Javadocs) to request data be fetched or saved.

Help Menu—Online Help in Pop-Up Web Browsers

Online documentation is available if you are connected to the Internet. This appears in a separate pop-up Web browser window so you may view it while working with MAExplorer. The documents include the Reference Manual, tutorials, menu overview, index, glossary, and other information. If you are setting up a database, you may include links to other Web pages describing key information on your databases.

10.3.1 System Design—Software Implementation

MAExplorer was written in Java to make it available as a Web applet, although its use as an applet has been deprecated. Java as an implemen-

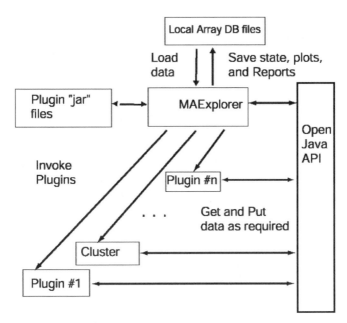

FIGURE 10.4. The MAEPlugin design is enables adding new analysis methods to MAExplorer. Plug-ins access MAExplorer data through an Open Java API (Application Programming Interface). For example, the box marked Cluster might be a new cluster method plug-in. It accesses data through the Open Java API which in turn accesses data in MAExplorer.

tation language has an advantage over many other languages in that it is highly portable between operating systems as well as being reasonably efficient. For this reason, the stand-alone application was also written in Java. Users have the option of including a Java Virtual Machine (JVM) with the downloaded program. Since we know that the JVM is current, we are able to ensure that MAExplorer runs correctly.

The primary design concept is that the base program should contain the minimum set of core functionalies needed to do a basic analysis. Since no software can contain all possible analysis methods, which is not desirable in any case because of the complexity of learning and maintaining such a system, we added Java plug-ins to allow users to intelligently extend MAExplorer's functionality.

MAExplorer was constructed using a number of fundamental data objects, including genes, hybridized samples (arrays), tables, plots, and so on, organized using an object-oriented methodology enforced by Java. Sets of genes are implemented as bit sets for efficiency in both storage and set-theoretic operations. With a set being implemented as 64 bits/word, a set intersection, union or difference can be performed on 64 genes in parallel in one logical (i.e., AND, OR, XOR) computer instruction. This makes the data filter quite efficient when computing the intersection of many gene sets. When ordered gene lists are required, memory and computationally intensive lists are used, but only when needed. Tab-delimited ASCII is used as the basic I/O file type for all types of data. This simplifies I/O and allows data to be prepared with a variety of systems, including Excel, array quantification programs, and relational database systems.

Another major decision was to use multiple pop-up windows for 2D plots, histograms, expression profiles, clustergrams, reports, and dialog boxes rather than sharing a single window. These windows are maintained by a special pop-up registry class that handles many of the bookkeeping chores involved with tracking and updating multiple windows viewing the same underlying data. Whenever an event occurs that may change the set of data-filtered genes, the current gene, or the current cluster set of genes, the registry is notified. Some of the events are: the current gene changed, the data filter was recomputed, the filter parameters changed, the sample labels changed, and the normalization method changed. The pop-up registry in turn invokes methods in all registered active windows (plots, tables, reports, MAEPlugins) to make the actual updates for windows that requested notification for particular events. This object-oriented design greatly simplifies the process of synchronizing the various data presentations with changes in the database and makes it easier to implement a direct manipulation.

We use direct manipulation methodology rather than a strictly command or menu-driven paradigm because of the intuitive way users interact with their data—they want to "grab" it and manipulate it. This meant that users should be able to select genes and samples by clicking on their representations in various ways. Scrollbars are used for adjusting zoom views as well as setting parameters for data filtering. These interactions let users select the data visually rather than having to enter specific values without the visual feedback.

Because data-mining is a complex process, possibly extending over multiple sessions, the use of a *data-mining* state was introduced to help users keep track of intermediate results. We associated the state with the startup file so restarting the database on a startup file restores the state to where it was when it was last saved to that file.

Extendable Analyses via MAEPlugins Modules

More information on MAEPlugins is available on the MAExplorer Web site which includes documentation on the Open Java API, open-source Java code examples to serve as a basis for you to create your own plug-ins, our plug-ins and donated plug-ins, and links to plug-ins at other Web sites. Typical plug-ins may include new methods for normalization, data filters, PCA, clustering, client-server, Web-server functional analysis of cluster results, and more. The MAExplorer Open Java API allows users to access all data structures without having to understand the low-level details of the system. Specialized application interfacing classes called MJA classes were written for the plug-in developer to call. These and a special MaeJavaAPI Java class let plug-in developers access all of the internal MAExplorer data structures in a protected manner. This has the added advantage of allowing us to improve and change the internal data structures without causing future problems with user-written plug-ins using those internal data structures directly. Figure 10.4 shows the top-level plug-in design.

10.3.2   How to Download the Software

The distribution of executable and source files on our MAExplorer home page Web site may be freely downloaded for use by academic or commercial interests and may be redistributed without restriction under the Mozilla1.1 license (available on the Web site). The ready-to-run installation download includes the MAExplorer.jar file, a set of 50 hybridized sample MGAP data for use as a demonstration database, an optional JVM, and other startup support files. You may also download a "slimmed-down" version that does not include the JVM. The latter may be downloaded separately and should be installed in the same directory where you installed MAExplorer. These files are packaged using the commercial InstallAnywhere^{TM} packaging software by ZeroG which provides simple download installations for a wide variety of computers. If you do not have a recent JVM or are having problems with the program on your computer, you might want to initially download the full version with the JVM to install the JVM. You can later download slimmed-down versions for subsequent releases and omit the JVM. Note that the downloaded JVM is used only with MAExplorer and does not overwrite your existing JVM if it exists (except for MacOS 8/9). InstallAnywhere generates a program called MAExplorer.exe (Windows), MAExplorer.bin (MacOS, Linux, Solaris, Unix, and others) that may be used to start MAExplorer. For some systems, it will make files ending with .mae (the startup files) clickable so it starts MAExplorer on that particular file.

Open Source Web Site http://maexplorer.sourceforge.net/

The Open Source Web site gives full access to downloads for MAExplorer, examples of MAEPlugins, and the Cvt2Mae data converter wizard. You may download ready-to-run installers, Java source files, or Java .jar files (for subsequent updates). In addition, many documents are available, including the hyperlinked Reference Manual (both online and downloadable versions as well as Adobe PDF format), tutorials, and PDF and Microsoft PowerPoint versions of training slide shows. A history of revisions reflecting ongoing changes in the status of MAExplorer is constantly updated. To help elucidate the Open Java API for MAExplorer, "javadocs" are available online showing various levels of detail from writing plug-ins to the entire core system. Information on and examples of writing plug-ins are also available. We encourage the research community to write MAEPlugins implementing functionality that is missing that they would like to see and, if possible, make that plug-in available on the Open Source Web site or provide a link to their Web site. The original site (http://www.lecb.ncifcrf.gov/MAExplorer) contains a mirror of most of the materials (but no source code) as well as older documentation and the MGAP database. You can reach all of these data through either the SourceForge Web site or the LECB/NCI Web site.

10.3.3   Strengths and Weaknesses of the Approach

There are a range of approaches for performing data-mining of microarray data over the Internet between 100% servercentric and 100% clientcentric methods. However, both methods assume rapid access to underlying databases and the ability to transform data from one presentation mode to another where differences might be easily observed. One extreme is the servercentric model using CGI, servlets, or lightweight applets in a Web browser (see Vilo et al., Chapter 6, this volume) as an example of a server-based system with minimal browser requirements). This assumes that all data search and analysis is performed on a back-end server and graphic or tabular results from the server are sent back to the researcher over the Internet. The servercentric model has the advantage of keeping all user data up-to-date but the disadvantage of performing all computations and graphics generation on the back-end server. Relying on the server for major computations and graphics generation may result in significant delays if the networks or servers are heavily loaded. The other extreme is the client-centric model, which may include stand-alone programs or data-intensive applets. Here all of the data being analyzed are copied to a user's computer and then computationally expensive analyses are done there. This has the disadvantage for the user of possibly not having the most up-to-date data to analyze unless the copy is kept up-to-date. However, it does distribute the computational load, allowing more effective data-mining with many alternate views and avoiding excessive delays during a data-mining session (see Gentleman and Carey, Chapter 2, this volume) for an example of a client-based system using methods written for the R statistical package in a stand-alone environment). In both the stand-alone application and the Web browser applet, data need to be downloaded to MAExplorer. A major difference between these paradigms is being able to read and write data locally. Such data might include array database caching, state and report saving, and the data-mining state.

   A good intersection of the servercentric and clientcentric methods is to distribute the computation and data to the systems where they can be handled most effectively. Because Java enables computation in a Web browser, PCs currently available have enormous power, large memory and high-

TABLE 10.1. Comparison of clientcentric versus servercentric methods for data mining comparing some of the features of clientcentric and servercentric data-mining analysis methods. The clientcentric approach presented here primarily uses Java with data downloaded to the client's computer. A servercentric approach might use a mix of HTML, CGI programs, servlets, and Java. However, even a clientcentric approach may take advantage of server support for additional functionality (e.g., accessing genomic servers to gain additional information about specific genes or sets of genes). Advantages and disadvantages are indicated with a + and − respectively.

Clientcentric

+ Java runs on all operating systems as either stand-alone or browser applets
+ handles rapid response required for direct manipulation on desktop computers
+ stand-alone version may be restarted quickly from local or cached data
+ size limitations are not a problem with stand-alone Java applications
+ Java plug-ins allow prototyping new analysis methods by any group of users
+ easy to build large, stable stand-alone programs handling very large data sets
− for applet version, slow startup since program and data downloaded when run
− difficult to build large, stable Web applets handling very large datasets
− for stand-alone application, must be installed on client's computer

Servercentric

+ may have better resources for very large datasets but with dependence on server
+ faster startup than full applet since minimal GUI required and little data downloaded
+ easier to prototype and distribute new functionality using centralized CGI or servlets
− susceptible to Internet traffic bandwidth problems for large numbers of users
− susceptible to server-load dependencies for large numbers of users
− difficult to get very rapid response for direct manipulation for data mining

speed Internet connections are readily available, and it is now possible to distribute some of the data and computations to the desktop. If high-speed direct manipulation methodology is to be made available on the Internet for microarray data-mining, then it must be brought to the user's desktop browser or local computer rather than residing solely on the back-end server. This is the approach taken in designing the MAExplorer. Table 10.1 summarizes this comparison of the two approaches.

10.4 Applications

Tutorials for Learning How to Use MAExplorer

The Reference Manual has a short tutorial in Appendix A and a more advanced tutorial in Appendix B. There are also many examples and screen captures illustrating the commands in the Reference Manual and in PDF slide shows on the Web site. The tutorials show the sequences of menu and other commands used to get a rudimentary understanding of MAExplorer in performing various aspects of an analysis. They also suggest examples of different analysis methods to try. The tutorials were written in a generic way to be used with any database. When you install the full version of MAExplorer, it includes a subset of the MGAP database data set that could be used with the tutorials. The MGAP data are also available as a separate download from the MAExplorer Web site. If you have access to other data, you can use those data. As with all tutorials, they are only starting points for getting you started—in this case into understanding the MAExplorer data-mining analysis environment.

A MAExplorer database is started using a .mae startup file generated when the database is created using the Cvt2Mae converter or other method. If your operating system lets you start programs by clicking on a particular type of file (e.g., a .doc file in Windows will start Microsoft Word), then clicking on a .mae file will start MAExplorer on that database. Alternatively, you can start MAExplorer with no data and then specify the .mae startup file from File menu. Note that the MGAP database array data distributed with MAExplorer includes a number of .mae startup files described in Appendix D.6 of the Reference Manual. In the following example, we show how to start this for MS Windows. For other operating systems, see the Reference Manual.

1. To start MAExplorer after you have installed it, go to the Windows Start Menu and click on MAExplorer or click on the MAExplorer startup icon. If it is not in your Start Menu, you can go to where you installed it (typically C:\Program Files\MAExplorer) and click on MAExplorer.exe.
2. Then, after it starts, go to the File menu, select the Databases submenu, and finally select the Open disk DB command. It will pop-up a file browser to let you select the startup file you want. Alternatively, you may launch the file directly by going to the list of startup files in the C:\Program Files\MAExplorer\MAE folder and selecting a startup file.

When MAExplorer starts, the main pseudoarray window will pop up. When it is ready for you to begin interaction, the menu bar becomes active and it displays a green *Ready—click on a gene to query database* message. You might print the tutorial and then read the instructions from the printout rather than trying to keep this window visible.

Tutorial Notation

The following example is from the self-guided tutorial (you issue the commands). The notation go to A:B:C means go pull-down menu A, then submenu B, and then make selection C. Selecting a gene from the microarray image or scatterplot means clicking on a spot in the pseudoarray image or a point in any of the plots. We show a few examples from the tutorial to illustrate this notation.

A.3.1.5 Filter by expression ratio between two conditions X and Y

Step 1: Go to Analysis : Plot : Histograms : HP-X/HP-Y.

The histogram shows the ratios of the data passing the data filter.

Step 2: Move the pop-up plot so you can see it and the array simultaneously.

Step 3: Choose (click on) a ratio bin. Genes filtered by the ratio range of the bin will have white circles as they passed the data filter.

Step 4: Click on a different bin in the histogram to select another bin.

Step 5: Click on the word "Freq" in the histogram to remove the histogram bin filter.

Note: if the signal is close to background the X/Y ratio is probably incorrect. You can filter out low intensity genes by filtering by spot intensity.

A.3.1.6 Filter by spot intensity range

Step 1: Go to Analysis: Filter : Filter by spot intensity [SI1: SI2] sliders : Use spot intensity [SI1: SI2] sliders.

Step 2: Adjust intensity lower bound (SI1) to remove low-ratio genes.

Step 3: When done, remove the "Filter by intensity sliders" by toggling it off (redo step 1 to toggle it off).

Step 4: Repeat steps 1-3, but this time use Filter : Filter by [I1:I2] sliders : Use spot intensity (or Cy3/Cy5) [I1:I2] sliders.

A.3.1.7 Multiple conditions—expression profile plots of HP-E data

Step 1: Go to Analysis : Plot: Expression profile : Display a gene's expression profile for HP-E.

Step 2: After the expression profile window pops up, click on a gene in the array to see its profile.

Step 3: Click on a line in the profile plot to see its intensity.

Step 4: Click on a different gene in the array to see its profile.

Step 5: Press "Show HPs" button to see the list of samples used.

Step 6: Press "Close" button to remove pop-up windows.

10.5   Discussion

MAExplorer is a flexible microarray data-mining tool running on the user's computer. It uses direct manipulation, data filtering, built-in graphics, statistics, clustering, and gene and sample set operations, with results presented as reports. It has tools for managing multiple samples, replicates, gene sets, and expression profile lists. The data-mining exploration state may be saved any time during a session and restored at a subsequent session. Genomic Internet databases may be accessed for genes of interest in pop-up Web browsers. The Cvt2Mae "wizard" tool may be used to convert commercial and academic chip array data for use with MAExplorer. Extensibility for new analytic methods may be provided using MAEPlugins extensions.

Acknowledgments. We wish to thank the members of Lothar Hennighausen's Laboratory of Genetics and Physiology (NIDDK), who inspired the initial development of MAExplorer and its continued development. Thanks also to many others for useful discussions and suggestions that have helped improve the MAExplorer's capabilities and usability. We also want to thank Tom Schneider, Eric Shen, and Ellen Burchill for useful comments on this chapter.

References

Alizadeh A, Eisen MB, Davis RE, Ma C, Lossos IS, Rosenwald A, Boldrick JC, Sabet H, Tran T, Yu X, Powell JI, Yang L, Marti GE, Moore T, Hudson Jr J, Lu L, Lewis DB, Tibshirani R, Sherlock G, Chan WC, Greiner TC, Weissenburger DD, Armitage JO, Warnke R, Levy R, Wilson W, Grever MR, Byrd JHC, Botstein D, Brown PO, Staudt LM (2000). Distinct types of diffuse large B-cell lymphoma identified by gene expression profiling. *Nature*, 403:503–511.

Cleveland WS (1985). *The Elements of Graphing Data*. Wadsworth Press: Monterey, CA.

Cooper CS (2001). Applications of microarray technology in breast cancer research—Review. *Breast Cancer Research*, 3:158–175.

DeRisi J, Penland L, Brown PO, Bittner ML, Meltzer PS, Ray M, Chen Y, Su YA, Trent JM. (1996). Use of a cDNA microarray to analyze gene expression patterns in human cancer. *Nature Genetics* 14:457–460.

Dudoit S, Yang YH, Callow MJ, Speed TP (2000). Statistical methods for identifying differentially expressed genes in replicated microarray experiments. Technical Report No. TR-578 (August 2000). Department of Statistics, Stanford University: Stanford, CA.

Eisen MB, Spellman PT, Brown, PO, Botstein D (1998). Cluster analysis and display of genome-wide expression patterns. *Proceedings of the National Academy of Science USA*, 95:14863–14868.

Ermolaeva O, Rastogi M, Pruitt KD, Schuler GD, Bittner, Chen Y, Simon R, Meltzer P, Trent JM, Boguski MS (1998). Data management and analysis for gene expression arrays. *Nature Genetics*, 20:19–23.

Hughes TR, Marton MJ, Jones AR, Roberts CJ, Stoughton R, Armour CD, Bennett HA, Coffey E, Dai H, He YD, Kidd MJ, King AM, Meyer MR, Slade D, Lum PY (2000). Functional discovery via a compendium of expression profiles. *Cell*, 102:109–126.

Ideker T, Thorsson V, Ranish JA, Christmas R, Buhler J, Eng JK, Bumgarner R, Goodlet DR, Aebersold R, Hood L (2001). Integrated genomic and proteomic analyses of a systematically perturbed metabolic network. *Science*, 292:929–934.

Jagota A (2001). *Microarray Data Analysis and Visualization*. Bioinformatics By The Bay Press, Santa Cruz, CA.

Kerr MK, Churchill GA (2001). Statistical design and the analysis of gene expression microarray data. *Genetical Research*, 77:123–128.

Lemkin PF, Thornwall GC, Walton K, Hennighausen L (2000). The Microarray Explorer tool for data-mining of cDNA microarrays—application for the mammary gland. *Nucleic Acids Research*, 20:4452–4459.

Lemkin PF, Thornwall GC, Hennighausen L (2001). MicroArray Explorer—A Java-based tool for data-mining microarrays. AMS-IMS-SIAM Summer Conference on Statistics in Functional Genomics, June 10-14, 2001. (http://www.lecb.ncifcrf.gov/MAExplorer/PDF/AMS-IMS-SIAM-Web-9-28-2001.pdf).

Schena M, Shalon D, Davis RW, Brown PO (1995). Quantitative monitoring of gene expression patterns with a complementary DNA microarray. *Science*, 270:467–470.

Schneiderman B (1997). *Designing the Human Interface*, 3rd ed. Addison-Wesley: New York.

Schulze A, Downward J (2001). Navigating gene expression using microarrays—A technology review. *Nature Cell Biology*, 3: E190–E195.

Simon R, Radmacher MD, Dobbin K (2001). Design of Studies Using DNA Microarrays. Technical Report #4.NCI, BRB: Rockville, MD.

Sneath PHA, Sokol RR (1973). *Numerical Taxonomy*. W.H. Freeman and Co: San Francisco.

Sorlie T, Perou CM, Tibshirani R, Aas T, Geisler S, Johnsen H, Hastie T, Eisen MB,van de Rijn M, Jeffrey SS, Thorsen T, Quist H, Matesse JC, Brown PO, Botstein D, Lonning PE, Borresen-Dale A-L (2001). Gene expression patterns of breast carcinomas distinguish tumor subclasses with clinical implications. *Proceedings of the National Academy of Science USA*, 98:10869–10874.

Strausberg RL, Austin MJF (1997). Functional genomics: Technological challenges and opportunities. *Physiological Genomics*, 1:25–32.

Tufte E (1997). *Visual Explanations. Images and Quantities, Evidence and Narrative*. Graphics Press: Cheshire, CT.

Tukey J (1977). *Exploratory Data Analysis*. Addison-Wesley: Reading, MA.

Weinstein JN, Myers TG , O'Conner PM, Friend SH, Fornace AJ, Kohn KW, Fojo T, Bates SE, Rubinstein LV, Anderson NL, Buolamwini JK, van Osdol WW, Monks AP, Scudiero DA, Sausville EA, Zaharevitz DW, Bunow B, Viswanadhan VN, Johynson GS, Wittes RE, Paull KD (1997). An information-intensive approach to the molecular pharmacology of cancer. *Science*, 275:343–349.

11

Parametric Empirical Bayes Methods for Microarrays

Michael A. Newton
Christina Kendziorski

Abstract

We have developed an empirical Bayes methodology for gene expression data to account for replicate arrays, multiple conditions, and a range of modeling assumptions. The methodology is implemented in an R library called EBARRAYS. Functions in the library calculate posterior probabilities of patterns of differential expression across multiple conditions. This chapter provides an overview of the methodology and its implementation in EBARRAYS.

11.1 Introduction

Empirical Bayes (EB) methods are well-suited to high-dimensional inference problems and thus provide a natural approach to microarray data analysis. In contrast to methods that apply classical statistical inferences separately for different genes (e.g., t-tests, ANOVA), there is a kind of information sharing among genes in an EB analysis. This can be beneficial because experiments often involve tens of thousands of genes but only tens of microarrays, so the amount of information per gene can be relatively low. It may seem paradoxical that an inference about differential expression of a certain gene between two conditions, say, should be influenced by the expression levels of other genes, but for some time it has been recognized that related inference problems can be combined to gain advantage (e.g., Efron and Morris, 1973, 1977; Carlin and Louis, 1996). Roughly speaking, the data from other genes provide some information about the typical variability in the system. The general framework provided by EB analysis is quite flexible; probability distributions are specified in several layers and account for multiple sources of variation. Our calculations explore two particular specifications that may be effective for typical microarray studies.

We present here an R library called EBARRAYS that is designed to implement several EB methods for gene expression data analysis. The methodology is focused on specific questions concerning differential expression; it presumes that data have been sufficiently preprocessed so that they represent bona fide approximations to some relative gene expression in the cells being sampled. These may be the probe-level intensity signals from an Affymetrix oligonucleotide system or intensity measurements from a spotted cDNA array. The methods that we discuss provide estimates of differential expression, hypothesis tests of differential expression between two conditions, and the assessment of patterns of differential expression among multiple conditions; in each case, we allow replicated expression profiles within each condition.

The first calculation is to estimate the magnitude of differential expression between two conditions. Although the common practice is to report a fold change on a raw or log scale, the EB framework leads to various shrinkage estimates of fold change that differ from the standard estimators, especially for genes of relatively low total abundance. When there is interest in rank ordering genes by the extent of differential expression, the EB ranking can be different from a ranking based on apparent fold change; the present methods may provide better estimates of the ranking because they implicitly account for differential variability in apparent fold-change estimates.

When comparing expression profiles from two groups of samples, an effective gene-specific inference summary is the posterior probability of differential expression. This transforms evidence from the scale of expression score or fold change to the familiar scale of probability. Underlying the calculation is a simple parametric mixture model. Some genes are differentially expressed and some are not, and we view the question as stochastic. Measured expression data have different distributions depending on this outcome, so the posterior probability of differential expression is readily computed once the mixture model is estimated. Posterior probability calculations carry over naturally to comparisons among multiple conditions. In an example that we consider from a study of rat mammary epithelium cells, there are four cellular conditions and four distinct patterns of differential expression of particular interest.

The work presented in Newton et al. (2001) may have been the first application of EB methodology for gene expression data analysis. There we considered both shrinkage estimation of fold change and the posterior probability of differential expression between two conditions. Our results were restricted to so-called "single-slide" data (rather than replicated profiles in each condition) and to one particular hierarchical model. These analyses are extended in Kendziorski et al. (in press); we review these extensions and describe a software implementation in the present chapter.

11.2   EB Methods

11.2.1   Canonical EB Example

Although empirical Bayes methods have been widely studied, they may not be familiar to many investigators analyzing microarray data, so we consider a canonical EB example in the present context. The expression profile of some cell type in some growth condition is a numerical quantification of the abundances (or relative abundances) of all mRNAs. Biological as well as other sources of variation cause measured expression levels to fluctuate around the hypothetical target profile $(\mu_1, \mu_2, \ldots, \mu_J)$, where μ_j is the true expression level of gene j and J is the number of genes under study. To simplify matters, suppose that our measured profile is the single vector $(x_1, x_2, \ldots, x_J)$ (single array) and that a scale of measurement has been chosen so that all the x_j's have the same variance. It would seem that the measured profile $(x_1, \ldots, x_J)$ is the obvious estimate of $(\mu_1, \ldots, \mu_J)$ in the absence of other information since each x_j approximates its respective μ_j. Indeed, under mild conditions, $(x_1, \ldots, x_J)$ is the maximum likelihood estimate and has the least variation among all unbiased estimates of the whole profile.

The obvious estimate has some problems when J is relatively large, as was first shown by Stein (1956). He observed that $\sum_j x_j^2$ (the squared Euclidean length of the vector $(x_1, \ldots, x_J)$) tends to be much greater than the squared length of the target profile $(\mu_1, \ldots, \mu_J)$. This fact implies that better estimates of the whole profile can be found. Interestingly, improved estimates can be found using an empirical Bayes argument. We treat the μ_j's themselves as random, work out a summary feature of their conditional distribution given the data, and then plug in estimates of remaining parameters using the whole dataset. For example, an improved estimator might have component j equal to $w x_j + (1 - w)\bar{x}$, where $\bar{x}$ is the average of expression measurements and w is some data-dependent weight in the unit interval.

Roughly speaking, optimal estimates of the whole profile involve sharing information among genes and are not obtained by simply combining optimal estimates of the individual components. The empirical Bayes argument provides an effective approach to statistical inference in high-dimensional problems. It can be used not only to improve the estimation of profiles but also to enable inference about differential gene expression, as we now show.

11.2.2   General Model Structure: Two Conditions

Our models attempt to characterize the probability distribution of expression measurements $\mathbf{x}_j = (x_{j1}, x_{j2}, \ldots, x_{jI})$ taken on a gene j. (As we clarify below, the parametric specifications that we adopt allow either that these x_{ji} are recorded on the original measurement scale or that they have been log-transformed.) A baseline hypothesis might be that the I samples are exchangeable (i.e., that potentially distinguishing factors, such as cell-growth conditions, have no bearing on the distribution of measured expression levels). We would thus view measurements x_{ji} as independent random deviations from a gene-specific mean value μ_j and, more specifically, as arising from an observation distribution $f_{obs}(\cdot|\mu_j)$. For example, a gene with a large μ_j typically exhibits high expression measurements and high variability.

When comparing expression samples between two groups (e.g., cell types), the sample set $\{1, 2, \ldots, I\}$ is partitioned into two subsets, say s_1 and s_2; s_k contains the indices for samples in group k. The distribution of measured expression may not be affected by this grouping, in which case our baseline hypothesis above holds and we say that there is equivalent expression, EE_j, for gene j. Alternatively, there is differential expression, DE_j; our formulation requires that there now be two different means, say μ_{j1} and μ_{j2}, corresponding to measurements in s_1 and s_2, respectively. Arguably, the fold change of interest is $\rho_j = \mu_{j1}/\mu_{j2}$ rather than the apparent fold change computed either by averaging separate fold-change estimates from pairs of arrays or by taking the ratio of average expression from each group. Empirical Bayes estimates of ρ_j are obtained by specifying a probability distribution on the gene effects μ_{jk} themselves and then computing some measure $\hat{\rho}_j$ of the center of the posterior distribution of each ρ_j given the data. In our calculations, we take the simplest assumption that the gene effects arise independently and identically from a system-specific distribution $\pi(\mu)$. The precise form of $\hat{\rho}_j$ depends on details of the model specification, but in the two cases that we have worked out the value $\hat{\rho}_j$ is *shrunk* toward unity compared to the apparent fold-change estimate. This tends to provide more conservative inferences than the naive approach and has been shown to reduce estimation error (Newton et al., 2001). Were we instead to treat the μ_j's as fixed effects, we would not derive any such advantage and there would be no information sharing across genes.

In addition to reporting fold-change estimates, it may also be useful to gauge the significance of differential expression. Here, the empirical Bayes formulation has an appealing structure compared to the classical one-gene-at-a-time tests, which treat the gene effects as fixed and separate. Some unknown fraction p of the genes are differentially expressed (DE_j) and the remainder are equivalently expressed, and the state for each gene is considered to be a matter of chance. Putting this together with the framework above, we have that a gene j respecting equivalent expression (EE_j) presents data $\mathbf{x}_j = (x_{j1}, \ldots, x_{jI})$ according to a distribution

$$f_0(\mathbf{x}_j) = \int \left(\prod_{i=1}^{I} f_{obs}(x_{ji}|\mu) \right) \pi(\mu) \, d\mu. \tag{11.1}$$

This predictive distribution is an average (i.e., a mixture) over the possible gene effects μ_j. Among other things, the averaging induces a positive dependence among measurements on the same gene, a property that we regularly observe in data. Alternatively, if gene j is differentially expressed, the data $\mathbf{x}_j = (\mathbf{x}_{j1}, \mathbf{x}_{j2})$ are governed by the distribution

$$f_1(\mathbf{x}_j) = f_0(\mathbf{x}_{j1})f_0(\mathbf{x}_{j2}) \tag{11.2}$$

owing to the fact that different mean values govern the different subsets $\mathbf{x}_{j1}$ and $\mathbf{x}_{j2}$ of samples. We have both continuous mixing over the unknown values of μ_j and discrete mixing over the two patterns (DE_j and EE_j) for each gene. The marginal distribution of the data becomes

$$pf_1(\mathbf{x}_j) + (1-p)f_0(\mathbf{x}_j). \tag{11.3}$$

With estimates of p, f_0, and thus f_1, the posterior probability of differential expression is calculated by Bayes' rule as

$$\frac{p\, f_1(\mathbf{x}_j)}{p\, f_1(\mathbf{x}_j) + (1-p)\, f_0(\mathbf{x}_j)}. \tag{11.4}$$

To review, the distribution of data involves an observation component, a component describing variation of mean expression μ_j, and a discrete mixing parameter p governing the pattern of expression between conditions. The first two pieces combine to form a key predictive distribution $f_0(\cdot)$ (see (11.1)), which enters both the marginal distribution of data (11.3) and the posterior probability of differential expression (11.4).

11.2.3 Multiple Conditions

Many studies take measurements from more than two cellular conditions, and this leads us to consider more patterns of mean expression than simply DE_j and EE_j. For example, with three conditions, there are five possible patterns among the means, including equivalent expression across the three conditions (1), altered expression in just one condition (3), and distinct expression in each condition (1). We view a pattern of expression for a gene j as a grouping or clustering of conditions so that the mean level μ_j is the same for all conditions grouped together. With microarrays from four cell conditions, there are 15 different patterns, in principle, but with extra information we might reduce the number of patterns to be considered. We discuss an application in Section 11.4 in which ten array sets are measured across four cell conditions, but the context tells us to look only at a particular subset of four patterns.

We always entertain the null pattern of equivalent expression among all conditions. Consider m additional patterns so that $m+1$ distinct patterns of expression are possible for a data vector $\mathbf{x}_j = (x_{j1}, \ldots, x_{jI})$ on some gene j. Generalizing (11.3), $\mathbf{x}_j$ is governed by a mixture of the form

$$\sum_{k=0}^{m} p_k f_k(\mathbf{x}_j),\qquad(11.5)$$

where $\{p_k\}$ are mixing proportions and component densities $\{f_k\}$ give the predictive distribution of measurements for each pattern of expression. Consequently, the posterior probability of expression pattern k is

$$P(k|\mathbf{x}_j) \propto p_k f_k(\mathbf{x}_j).\qquad(11.6)$$

The pattern-specific predictive density $f_k(\mathbf{x}_j)$ may be reduced to a product of $f_0(\cdot)$ contributions from the different groups of conditions, just as in (11.2), and this suggests that the multiple-condition problem is really no more difficult computationally than the two-condition problem except that there are more unknown mixing proportions p_k. (See Kendziorski et al. (in press) for details.) The posterior probabilities (11.6) summarize our inference about expression patterns at each gene. They can be used to identify genes with altered expression in at least one condition, to order genes within conditions, or to classify genes into distinct expression patterns.

We emphasize that our use of the term pattern refers to a grouping or clustering of conditions according to equal mean expression of a given gene. More generic concepts, such as a pattern of increasing means in a time-course experiment, are not so easily represented by groups of conditions having the same mean. Still, as we demonstrate in Section 11.3, there is a great flexibility provided by the present framework.

11.2.4 The Gamma–Gamma and Lognormal–Normal Models

Following Kendziorski et al. (in press), we consider two particular specifications of the general mixture model described above. Each is determined by the choice of observation component and mean component, and each depends on a few additional parameters θ to be estimated from the data.

In the Gamma–Gamma (GG) model, the observation component is a Gamma distribution having shape parameter $\alpha > 0$ and a mean value μ_j; thus, with scale parameter $\lambda = \alpha/\mu_j$,

$$f_{obs}(x|\mu_j) = \frac{\lambda^\alpha x^{\alpha-1} \exp\{-\lambda x\}}{\Gamma(\alpha)}$$

for measurements $x > 0$. Note that the coefficient of variation in this distribution is $1/\sqrt{\alpha}$, taken to be constant across genes j. Matched to this observation component is a marginal distribution $\pi(\mu_j)$, which we take to be an inverse Gamma. More specifically, fixing α, the quantity $\lambda = \alpha/\mu_j$ has a Gamma distribution with shape parameter α_0 and scale parameter ν.

Thus, three parameters are involved, $\theta = (\alpha, \alpha_0, \nu)$, and, upon integration, the key density $f_0(\cdot)$ has the form

$$f_0(x_1, x_2, \ldots, x_I) = K \frac{\left(\prod_{i=1}^{I} x_i\right)^{\alpha-1}}{\left(\nu + \sum_{i=1}^{I} x_i\right)^{I\alpha + \alpha_0}}, \tag{11.7}$$

where

$$K = \frac{\nu^{\alpha_0} \Gamma(I\alpha + \alpha_0)}{\Gamma^I(\alpha)\Gamma(\alpha_0)}.$$

In the lognormal normal (LNN) model, the gene-specific mean μ_j is a mean for the log-transformed measurements, which are presumed to have a normal distribution with common variance σ^2. Like the GG model, LNN also demonstrates a constant coefficient of variation: $\sqrt{\exp(\sigma^2) - 1}$ on the raw scale. A conjugate prior for the μ_j is normal with some underlying mean μ_0 and variance τ_0^2. Integrating as in (11.1), the density $f_0(\cdot)$ for an n-dimensional input becomes Gaussian with mean vector $\underline{\mu}_0 = (\mu_0, \mu_0, \ldots, \mu_0)^t$ and exchangeable covariance matrix

$$\boldsymbol{\Sigma}_n = \left(\sigma^2\right)\mathbf{I_n} + \left(\tau_0^2\right)\mathbf{M_n},$$

where $\mathbf{I}_n$ is an $n \times n$ identity matrix and $\mathbf{M_n}$ is an $n \times n$ matrix of ones.

The GG and LNN models characterize fluctuations in array data using a small number of parameters, and both involve the assumption of a constant coefficient of variation (CV). This property is often observed in microarray data, at least approximately (see Section 11.4). The models may provide useful results even if this assumption does not hold. We recommend that assumptions be checked on a case-by-case basis.

11.2.5 Model Fitting

For both the GG and LNN models, we use the method of maximum (marginal) likelihood to obtain estimates of the unknown parameters θ. In the GG model, $\theta = (\alpha, \alpha_0, \nu)$ and in LNN, $\theta = (\mu_0, \sigma^2, \tau_0^2)$. We estimate parameters using the EM algorithm. (Any gene with at least one negative intensity value in one condition is omitted from this step of the calculation.) With data $\mathbf{x}_j$ governed by a mixture of the form (11.5), we introduce pattern indicators ϕ_{jl} defined as 1 if the expression pattern of gene j is pattern l and 0 otherwise. The *complete* data log likelihood is

$$l_c(\theta) = \sum_j \left\{ \sum_{k=0}^{m} \phi_{jk}[\log f_k(\mathbf{x}_j) + \log(p_k)] \right\}. \tag{11.8}$$

Using a current estimate θ_0, the expectation given the observed data amounts to replacing ϕ_{jl} with $\hat{\phi}_{jl}$, the posterior probability of expression pattern l for gene j (see 11.6) (i.e., the E-step). For the M-step, we use the arithmetic mean of $\hat{\phi}_{.,k}$ to estimate p_k; then optim in R provides updated estimates of θ. The two steps are repeated until there is convergence in the estimates. Results should be checked from various starting configurations.

Certain initial values can be determined from simple summary statistics. Recall from subsection 11.2.4 that the coefficient of variation (CV) for the GG model is $1/\sqrt{\alpha}$, which gives $\alpha = \left(\frac{1}{CV}\right)^2$. Using the average coefficient of variation across genes as an estimate of CV gives an initial estimate of α. Similar reasoning applies to estimating σ^2 in the LNN model since there CV is completely determined by σ^2.

11.3   Software

The empirical Bayes hierarchical modeling approach is implemented in EBARRAYS, an R library available from

http://www.biostat.wisc.edu/~kendzior

and written for R version 1.4.0. EBARRAYS can be used to calculate posterior probabilities of a predefined set of expression patterns under either GG or LNN assumptions. The main functions available in version 1.0 are:

cleanup	reads in data and records data characteristics
em	EM algorithm to fit the EB model
postprob	posterior probabilities for expression patterns

These functions are supplemented by other functions that are not called directly by the user, including complete.loglik, which calculates the complete data log likelihood (11.8), and functions to evaluate the key predictive distribution f_0. The current package contains two possible forms for f_0, f0gg for the GG model and f0lnn for the LNN model, and we note that other families could be added by the user.

Each analysis requires input files that contain the normalized intensities (*datafile*), identify the replicate samples (*repfile*), and specify the patterns to be considered (*patternfile*). EBARRAYS assumes that input files are tab or single-space delimited ASCII text files. The *datafile* contains intensities in J rows and I columns; row and column names should be provided. The *repfile* specifies which of the I samples are considered replicates. It contains one row with I columns $(r_1, r_2, \ldots, r_I)$, where $r_1 = 1$ and for $k = 2, 3, \ldots, I$, $r_k = r_{k-1}$ if samples k and $k-1$ are considered replicates and $r_k = r_{k-1}+1$ otherwise. The *patternfile* specifies the patterns to be considered in the analysis. The kth row identifies which samples are assumed to have the same mean level of expression in pattern k. Typically, the $(k, 1)$ element is 1 for every k and for $l = 2, 3, \ldots, I$, $(k, l) = (k, l-1)$ if samples l and $l-1$

are assumed to have the same mean level of expression in pattern k and $(k, l) = (k, l-1) + 1$ otherwise.

As an illustration, consider a dataset with $I = 10$ arrays taken from two conditions (five arrays in each condition ordered so that the first five columns contain data from the first condition). In this case, there are two, possibly distinct, levels of expression for each gene and two potential patterns ($\mu_{j1} = \mu_{j2}$ and $\mu_{j1} \neq \mu_{j2}$). The replicate and pattern files are, respectively,

```
1 1 1 1 1 2 2 2 2 2
```

and

```
1 1 1 1 1 1 1 1 1 1
1 1 1 1 1 2 2 2 2 2
```

The exception to this convention is that zero columns can be used to identify arrays that are not used in model fitting or analysis. An example of this is given in Section 11.4. The initial data files (*datafile*, *repfile*, and *patternfile*) are loaded into R and processed with the function `cleanup`:

```
cleanup(path,"datafile","repfile","patternfile")
```

path is a character string specifying the directory containing the data. All output is written to this directory. `cleanup` returns a list. A few of the key components are:

> `int.data`: a data frame containing the intensity measurements.

> `gene.names`, `sample.names`, `gene.num`, `sample.num`: the names and total number of genes and samples.

> `all.pos`: a T/F vector indicating which genes have positive intensity measurements in every condition.

After initial data organization with `cleanup`, exploratory analyses can be carried out using standard R functions to ensure data quality and identify any potential problems. The function `em` can then be used to obtain parameter estimates under either GG or LNN model assumptions:

```
em(theta.init,p.init,data,family,num.iter,write.output)
```

Here, *family* is a character string indicating whether the calculations are for the GG or LNN model, the two `.init` arguments are starting values for the EM iterations, `num.iter` indicates how many iterations of EM to run, and `write.output` allows the results to be written to a file. The output of `em` is a list including:

> `params.p.mat`: a matrix containing estimates of the mixing proportions at each step of the EM algorithm.

`params.theta.mat`: a matrix containing estimates of theta at each step of the EM algorithm.

`params.p.est`: a vector containing final mixing proportion estimates.

`params.theta.est`: a vector containing final estimates of θ.

If `write.output` is set to TRUE, the matrices and final estimates are written to files (named as above) in the directory *pathname* as specified in `cleanup`.

After fitting the mixture model with `em`, the function `postprob` can be used to calculate the posterior probability for each gene in each expression pattern. `postprob` requires output from `cleanup`, output from `em`, and a flag (TRUE or FALSE) indicating whether output should be written to a file.

```
postprob(data,em.out,write.output)
```

Output is a list with components `postprob` and `table`. Both are matrices with rows corresponding to genes. The (j,k)th entry of `postprob` is the posterior probability that gene j is in pattern k. The `table` component includes this information and also appends the sample mean intensities averaged across conditions. If `write.output` is set to TRUE, these matrices are written to files `postprobs` and `summary.postprobs` in the directory *pathname* as specified in `cleanup`.

11.4   Application

In collaboration with Dr. M.N. Gould's laboratory in Madison, we have been investigating gene expression patterns of mammary epithelial cells in a rat model of breast cancer. We use a subset of the data from this study to illustrate the mixture model calculations and the EBARRAYS library. For a more extensive analysis, see Kendziorski et al. (in press).

In summary, the experiment that we consider used Affymetrix U34 chip sets and thus measured the expression of 26,379 rat genes. Each of $I = 10$ mRNA samples was probed using one chip set. The mRNA was derived from rat mammary epithelium cells of same-aged females living in a controlled environment. Four inbred strains of rats contributed mRNA, including two parental strains and two derived *congenic* lines. The parental strains differ greatly in their resistance to carcinogen-induced mammary cancer, with the Copenhagen (COP) strain almost completely resistant to certain carcinogens, while the inbred Wistar Furth (WF) strain was highly susceptible (Gould et al., 1989). By careful breeding, two intermediate inbred lines (CI and CII) were produced that carry the homozygous WF/WF genotype throughout the genome except on a relatively small region of interest where the animals are homozygous COP/COP (Figure 11.1). Un-

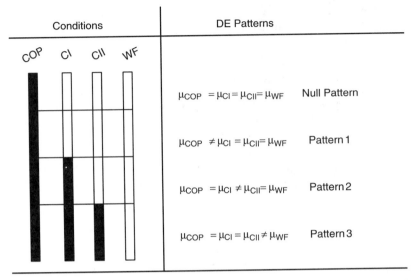

FIGURE 11.1. Schematic diagram showing animal lines (conditions) from which mRNAs were obtained (left) along with differential expression patterns (right). Genotypes shown in black (COP/COP) and white (WF/WF) are not drawn to scale (the homozygous COP/COP region is approximately 30 cM in congenic line CI and 1.5 cM in CII). True expression intensities for each group are denoted by μ. Note that differences in genotype do not imply differences in expression. Genes classified into the null pattern show equivalent expression across groups but differ in genotype.

derstanding expression patterns among these strains may be helpful in the study of breast cancer resistance.

As an initial processing step, we ran all the data through DNA-Chip Analyzer (Li and Wong, 2001) which uses a statistical model for probe-level data to account for artifacts such as probe-specific biases. Corrected and normalized model-based estimates of gene expression were obtained for $J = 25{,}248$ genes (1131 were identified as outliers) and were stored in a *datafile* called `data.txt`. EBARRAYS uses the *repfile* to characterize the organization of replicates among the $I = 10$ samples across the four conditions. We note that, in column order, there is one COP sample, five CI samples, two CII samples, and two WF samples. To reflect this arrangement, the *repfile* (called `reps.txt`) is

    1 2 2 2 2 2 3 3 4 4

The last bit of input is the *patternfile*, which tells EBARRAYS what patterns of mean expression will be considered in the analysis. Let us first ignore the derived congenic lines and perform an analysis on the parental strains only. There are two possible expression patterns ($\mu_{COP} = \mu_{WF}$ and $\mu_{COP} \neq$

μ_{WF}). These are represented in the *patternfile* (called `patterns.txt`)

```
1 0 0 0 0 0 0 0 0 1 1
1 0 0 0 0 0 0 0 0 2 2
```

Recall that a zero column indicates that the data in that condition are not considered in the analysis. An alternative approach would be to define a new data matrix containing intensities from the parental strains only and define the associated *patternfile* as

```
1 1 1
1 2 2
```

The corresponding *repfile* would be

```
1 2 2
```

This may be useful in some cases, but in general we recommend importing the full data matrix and defining the pattern matrix as a 2×10 matrix with intermediate columns set to zero. Doing so facilitates comparisons of results among different analyses since attributes of the data, such as the number of genes that are positive across each condition, remain constant.

With EBARRAYS installed, we use the `library` function in R to connect it to the current session (see Parmigiani et al., Chapter 1, this volume). The illustration below assumes that the input files `data.txt`, `reps.txt`, and `patterns.txt` are in the local directory. First, we organize the data:

```
> data <- cleanup(NULL,"data.txt","reps.txt","patterns.txt")
[1] "There are 25248 genes and 10 samples in this data set.
    The samples are arranged across 4 conditions."
[1] "Out of the 25248 genes, 24452 have positive intensities
    across each sample."
[1] "3 samples and 2 patterns will be considered in this
    analysis."
```

Preliminary data analysis can be done using standard R functions. For example, we can check to see if there is any relationship between the mean expression level and the coefficient of variation; recall that both GG and LNN models assume a constant CV. Figure 11.2 shows that the assumption is remarkably good for the three parental chip sets, with just a slight violation at very low mean expression. (Similar results were obtained using all ten samples).

Figure 11.3 shows a second diagnostic plot for nine subsets of 100 genes spanning the range of mean expressions. Shown are *qq* plots against the best-fitting Gamma distribution. Again, the goodness of fit is striking. (Note that we only expect these *qq* plots to hold for equivalently expressed genes, so some violation is expected in general.)

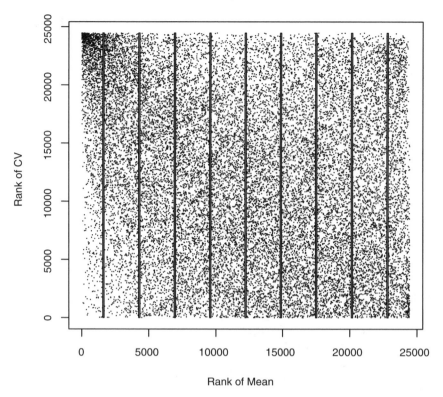

FIGURE 11.2. Coefficient of variation (CV) (ranked) as a function of the mean (ranked) for the WF and COP data. The CV averaged across all genes is 0.215. Shaded vertical bars indicate nine subsets of 100 genes, each examined more closely in Figure 11.3.

Using **em**, we can fit either the GG or the LNN model. Here we show how to do the GG fit, but we note that the LNN fit is done by the same approach. We recommend fitting both for the sake of comparison.

```
> emgg.out <- em(theta.init, p.init,data,"GG",20,TRUE)
[1] "GG model parameter estimates:
    12.480, 0.920, and 35.955"
[1] "Estimates of mixing proportions:
    0.99778, 0.00222"
```

Posterior probabilities can then be obtained; highlighted below are a few genes of interest.

```
> ppgg.out <- postprob(data,emgg.out,TRUE)
> attributes(ppgg2.out)
    $names
```

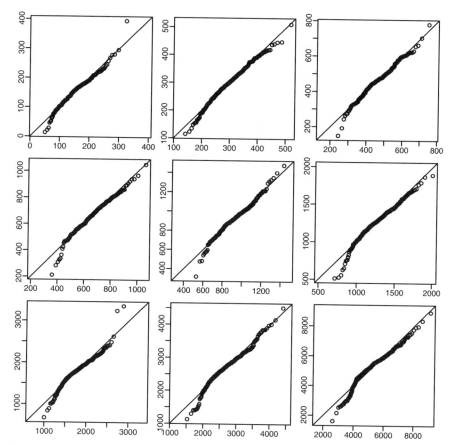

FIGURE 11.3. Gamma *qq* plots for the nine subsets shown in Figure 11.2. Each subset contains 100 genes.

```
    [1] "postprob" "table"
> ppgg2.out$table[c(2008, 6516, 1987),]
                  AVE_C1      AVE_C2       P0       P1
    J00801.at   3066.300   9082.8800   0.99286  0.00714
    L08100.at   4367.490  14162.3000   0.98984  0.01016
    J00772.s.at  391.968    678.8935   0.99897  0.00103
> ppgg2.out$postprob[c(2008, 6516, 1987),]
                  P0       P1
    J00801.at   0.99286  0.00714
    L08100.at   0.98984  0.01016
    J00772.s.at 0.99897  0.00103
```

We identify 50 genes that are probably differentially expressed.

```
> sum(ppgg2.out$postprob[,2]>0.5)
  [1] 50
```

Of these, 49 are also identified by the LNN model.

A nice feature of EBARRAYS is that comparisons among more than two groups can be carried out simply by changing the pattern matrix. For the four conditions, there are 15 possible expression patterns; however, if latent expression in each congenic matches one of the parentals, only four expression patterns are possible (see Figure 11.1). This is the case that we will consider (thus, $m + 1 = 4$). The null pattern consists of equivalent expression across the four conditions. The three other patterns allow for differential expression between the parental strains, with the congenic lines exhibiting the same mean expression as one of the parentals. Specifically, differential expression of the COP parent only is specified in pattern 1, between the congenics in pattern 2, and of the WF parent only in pattern 3.

The *repfile* is the same as before, but now the pattern matrix (`patterns .txt`) for the four group analysis is given by

```
1 1 1 1 1 1 1 1 1 1
1 2 2 2 2 2 2 2 2 2
1 1 1 1 1 1 2 2 2 2
1 1 1 1 1 1 1 1 2 2
```

The data are imported as in the comparison between two groups:

```
> data <- cleanup(NULL,"data.txt","reps.txt", "patterns.txt")
[1] "There are 25248 genes and 10 samples in this data set.
     The samples are arranged across 4 conditions."
[1] "Out of the 25248 genes, 24452 have positive intensities
     across each sample."
[1] "10 samples and 4 patterns will be considered in this
     analysis."
```

em and `postprob` are also called as before.

```
> emgg.out <- em(theta.init, p.init,data,"GG",20,TRUE)
[1] "GG model parameter estimates:
     16.720, 0.885, and 24.517"
[1] "Estimates of mixing proportions:
     0.98492, 0.01228, 0.00186, and 0.00094"
> ppgg.out <- postprob(data,emgg.out,TRUE)
```

Under the GG model, 24,797 genes had posterior probability greater than 0.5 of being in the null pattern; 250, 84, and 111 genes were classified into patterns 1, 2, and 3, respectively. For six genes, no pattern had posterior probability greater than 0.5. The LNN model identified slightly more genes as differentially expressed. Specifically, 24,168 were classified into the null pattern; 447, 343, and 280 were classified into patterns 1, 2, and 3, respectively; there were ten genes with posterior probabilities less than 0.5 for each pattern.

Of much interest in this analysis were genes showing differential expression between the first three conditions (COP, CI, and CII) and the WF condition. This corresponds to pattern 3 (see Figure 11.1). Genes with high posterior probability of being in pattern 3 are easily identified, and their probability distributions can be obtained:

```
> ppgg.out$postprob[c(2008, 6516, 1987),]
             P0      P1      P2       P3
J00801.at    0.04646 0.00021 0e+00    0.95332
L08100.at    0.00000 0.00000 0e+00    1.00000
J00772.s.at 0.04459 0.00013 2e-05    0.95527
```

The intensity values averaged within conditions show differences between the average intensities in the first three conditions and that in the WF condition C4. (Columns 5 through 8 contain the posterior probabilities as shown above.)

```
> ppgg.out$table[c(2008, 6516, 1987), 1:4]
             AVE_C1      AVE_C2      AVE_C3     AVE_C4
J00801.at    3066.300    9082.8800   4777.0160  995.295
L08100.at    4367.490   14162.3000   4002.6490 1278.285
J00772.s.at  391.968     678.8935    325.8258  121.734
```

11.5   Discussion

The proposed empirical Bayes methodology accounts for replicate arrays, multiple conditions, and a range of modeling assumptions. The methodology is implemented in an R library called EBARRAYS. Functions in the library calculate posterior probabilities of patterns of differential expression across multiple conditions.

The hierarchical modeling approach accounts for patterns of differential expression for a given gene among cell types. It also considers differences among genes in their average expression levels and measurement fluctuations. Since properties of individual experiments affect these features, the distributions governing them are to some extent experiment-dependent. However, there are features of microarray data that are generally observed and well-described using certain parametric models, such as a constant coefficient of variation. Two parametric formulations have been implemented, and model checks indicate that the assumptions are reasonable for the data considered here (see Figures 11.2 and 11.3). We have observed similar results for other datasets as well, and we expect that simple parametric analyses will often be suitable for array data. Of course, they must be accompanied by diagnostic model checks. EBARRAYS is flexible enough to incorporate a user-supplied predictive distribution in place of GG or LNN if necessary.

The methodology proposed here is not the only mixture-model-based approach to expression data analysis. Other approaches have been developed to address the question of comparisons between two conditions. Efron et al. (2001) describe calculations in the context of a specific experimental design. After a long series of preprocessing steps, each gene yields a one-dimensional test statistic whose marginal distribution turns out to be known and whose null distribution (i.e., on equivalent expression) can be nonparametrically estimated. Lee et al. (2000) also use the idea of a two-group mixture model for expression analysis; their calculations were in a slightly different context and were applied to parameter estimates from a first-stage analysis. The method of Lonnstedt and Speed (2002) has features in common with ours.

Another general approach that applies to expression data from two conditions is to conduct a hypothesis test at each gene and then correct for multiple comparisons. Most of the test statistics currently used are t (or t-like) and differ primarily in the estimation of the variance. (Dudoit et al., 2002) use a t-statistic with variance estimated by the within-gene sample variance and go on to address the multiple comparisons problem extensively using permutation analysis (see Dudoit and Yang, Chapter 3, this volume). Tusher et al. (2001) also use the within-gene sample variance, but adjust the denominator of their test statistic by adding a constant to account for the dependence between the relative difference in expression and absolute intensity (see Storey and Tibshirani, Chapter 12, this volume).

Methods that treat genes as separate fixed effects may have reduced efficiency compared to methods that treat the genes as arising from some population since they do not take full advantage of the level of information sharing among genes. Furthermore, classifying genes into expression patterns by the posterior probability is an optimal procedure in the context of the mixture model: it minimizes the expected number of errors. Interestingly, this goal is different from the goal in classical testing, which is to bound the type I error rate and then aim to maximize the power. Further investigation is required to more formally assess the advantages and disadvantages of these approaches.

References

Carlin BP, Louis TA (1996). *Bayes and Empirical Bayes Methods for Data Analysis*, Chapman & Hall: New York.

Dudoit S, Yang YH, Speed TP, Callow MJ (2002). Statistical methods for identifying differentially expressed genes in replicated cDNA microarray experiments. *Statistica Sinica*, 12(1):111-139.

Efron B, Tibshirani R, Storey JD, Tusher V (2001). Empirical Bayes analysis of a microarray experiment. *Journal of the American Statistical Association*, 96:1151-1160.

Efron B, Morris C (1973). Combining possibly related estimation problems (with discussion). *Journal of the Royal Statistical Society, Series B*, 35:379-421.

Efron B, Morris C (1977). Stein's paradox in statistics. *Scientific American*, 236:119-127.

Gould MN, Wang B, Moore CJ (1989). Modulation of mammary carcinogenesis by enhancer and suppressor genes. In: Colburn, NH (ed) *Genes and Signal Transduction in Multistage Carcinogenesis*, 19–38. Marcel Dekker: New York.

Kendziorski CM, Newton MA, Lan H, Gould MN (in press). On parametric empirical Bayes methods for comparing multiple groups using replicated gene expression profiles. *Statistics in Medicine*.

Lee MLT, Kuo FC, Whitmore GA, Sklar J (2000). Importance of replication in microarray gene expression studies: Statistical methods and evidence from repetitive cDNA hybridizations. *Proceedings of the National Academy of Sciences USA*, 97(18):9834-9839.

Li C, Wong W (2001). Model-based analysis of oligonucleotide arrays: Expression index computation and outlier detection. *Proceedings of the National Academy of Sciences USA*, 98(1):31-36.

Lonnstedt I, Speed TP (2002). Replicated microarray data. *Statistica Sinica*, 12(1):31-46.

Newton MA, Kendziorski CM, Richmond CS, Blattner FR (2001). On differential variability of expression ratios: Improving statistical inference about gene expression changes from microarray data. *Journal of Computational Biology*, 8:37-52.

Stein CM (1956). Inadmissibility of the usual estimator of the mean of a multivariate normal distribution. *Proceedings of the Third Berkeley Symposium on Mathematical Statistics and Probability*, 1:197-206.

Tusher V, Tibshirani R, Chu G (2001). Significance analysis of microarrays applied to the ionizing radiation response. *Proceedings of the National Academy of Sciences USA*, 98:5116-5121.

12

SAM Thresholding and False Discovery Rates for Detecting Differential Gene Expression in DNA Microarrays

JOHN D. STOREY
ROBERT TIBSHIRANI

Abstract

SAM is a computer package for correlating gene expression with an outcome parameter such as treatment, survival time, or diagnostic class. It thresholds an appropriate test statistic and reports the q-value of each test based on a set of sample permutations. SAM works as a Microsoft Excel add-in and has additional features for fold-change thresholding and block permutations. Here, we explain how the SAM methodology works in the context of a general approach to detecting differential gene expression in DNA microarrays. Some recently developed methodology for estimating false discovery rates and q-values has been included in the SAM software, which we summarize here.

12.1 Introduction

In this chapter, we discuss the problem of identifying differentially expressed genes from a set of microarray experiments. Statistically speaking, this task falls under the heading of "multiple hypothesis testing." In other words, we must perform hypothesis tests on all genes simultaneously to determine whether each one is differentially expressed or not. Recall that in statistical hypothesis testing, we test a null hypothesis versus an alternative hypothesis. In this example, the null hypothesis is that there is no change in expression levels between experimental conditions. The alternative hypothesis is that there is some change. We reject the null hypothesis if there is enough evidence in favor of the alternative. This amounts to rejecting the null hypothesis if its corresponding statistic falls into some predetermined rejection region. Hypothesis testing is also concerned with measuring the probability of rejecting the null hypothesis when it is really true (called a false positive) and the probability of rejecting the null hypothesis when the alternative hypothesis is really true (called power).

There are four important steps one must take in testing for differential gene expression. The first is that a statistic must be formed for each gene. The choice of this statistic is important in that one wants to make sure that no relevant information is lost with respect to the test of interest, yet all measurements on the gene are condensed into one number. The second step is to calculate the null distribution(s) for the statistics. One can assume that each gene has a different null distribution or one can calculate a null distribution for each gene. The third step is choosing the rejection regions. One can take a priori symmetric or one-sided rejection regions, or one can choose them adaptively. This involves comparing the original statistics to null versions of the statistics. Asymmetric rejection regions are most appropriate because we do not know beforehand what proportion of differentially expressed genes are in the positive or negative direction. If either of these two steps is not performed well, a loss in statistical power can result. The fourth step is to assess or control the number of false positives in some fashion. If one were simply to perform uncorrected hypothesis tests at the traditional 5% level, then $6000 \times 0.05 = 300$ false positives would be expected under the full null hypothesis if we were testing 6000 genes. This is clearly unacceptable, so some procedure must be performed to control the false positive rate in a reasonable manner.

In the remainder of this work, we discuss some simple methods for achieving these three steps in testing for differential gene expression. The methodology presented in this chapter is implemented in the SAM software.

12.2   Methods and Applications

For simplicity, we limit our discussion of the methodology to the case where there are two experimental conditions and each sample is independent. We begin by discussing multiple hypothesis testing.

12.2.1   Multiple Hypothesis Testing

We are interested in determining which genes show a statistically significant difference in gene expression between two or more conditions. Therefore, the null hypothesis for each gene is that the data we observe have some common distributional parameter among the conditions (usually the mean or average expression level). For each gene we form a *statistic* that is a function of the data (e.g., a t-statistic). A significance region is then defined so that if the statistic lies in the region, then the gene is said to exhibit differential gene expression; otherwise, there is no significant differential gene expression.

There are two kinds of errors that can be committed *at each particular gene* when testing for differential gene expression. The first is a *Type I error* (false positive), which is when the statistic is significant yet the gene is not truly differentially expressed. The second is a *Type II error* (false negative), and this occurs when the statistic is not significant yet the gene is truly differentially expressed. Power is defined to be $1 - $ Type II error rate. See Rice (1995) for an introduction to hypothesis testing.

When testing multiple hypotheses, the situation becomes much more complicated. Now each gene has possible Type I and Type II errors, and it becomes unclear how one should measure the overall error rate. Specifically, consider Table 12.1 which lists the possible outcomes when testing J hypotheses simultaneously.

For example, V is the number of Type I errors (false positives), T is the number of Type II errors (false negatives), and $R = V + S$ is the total number of significant hypotheses. In order to measure the errors incurred in multiple hypothesis testing, it is convenient to define a compound error measure. Four that have been considered in the multiple hypothesis testing literature are defined as follows:

$$PCER = \mathbf{E}[V]/J \qquad \text{per comparison error rate,}$$
$$FWER = \mathbf{Pr}(V \geq 1) \qquad \text{familywise error rate,}$$
$$FDR = \mathbf{E}[V/R|R > 0] \cdot \mathbf{Pr}(R > 0) \qquad \text{false discovery rate,}$$
$$pFDR = \mathbf{E}[V/R|R > 0] \qquad \text{positive false discovery rate.}$$

The PCER and FWER have been used for many years, but the FDR was more recently proposed by Benjamini and Hochberg (1995) and the pFDR was proposed by Storey (2001). In multiple hypothesis testing, the familywise error rate (FWER) has traditionally been the main quantity of interest. Also, the goal has been to *control* the error rate at a particular prechosen level. We choose a level α at which to control the error rate (e.g., we want $FDR \leq \alpha$). Then an algorithm is applied to the statistics and the rejection regions to estimate the rejection region to use. If the algorithm guarantees that *on average* we will have $FDR \leq \alpha$ regardless of J_0, then the algorithm is said to provide *strong control*. If this only occurs when $J = J_0$ (i.e., when all null hypotheses are true), then the algorithm only provides *weak control*.

TABLE 12.1. Outcomes from m hypothesis tests.

	Accept	Reject	Total
Null True	U	V	J_0
Alternative True	T	S	J_1
Total	W	R	J

The FWER is the probability of one or more false positives occurring among all significant hypotheses. However, in microarray studies, the number of tests (genes) is large, and the nature of the analysis is exploratory; that is, we want to identify many differentially expressed genes without too many false positives resulting. Hence, we expect more than one false positive, but we do not want too many in proportion to true positives. If we estimate the FDR or pFDR to be 5% among a list of 100 significant genes, then we roughly expect that 5 of those 100 are false positives. This kind of information is very useful for choosing genes to examine more carefully, and it is much more statistically rigorous than taking the top k genes for some arbitrarily chosen k.

Therefore, the FWER is not a relevant quantity in this setting. Instead, we focus here on the FDR and pFDR. In Storey (2002), Storey (2001), Storey and Tibshirani (2001), and Storey et al. (2002) we study false discovery rates from a unified point estimation approach, making connections to Bayesian procedures, introducing the pFDR, and developing more powerful methodology than that in Benjamini and Hochberg (1995). Moreover, we show that the methodology is robust against the kind of dependence encountered in DNA microarrays.

12.2.2 An Application

Here is a simple example of the methodology to detect differential gene expression. The exact methods are a bit more complicated than we present here, but this example serves as a useful reference.

We obtained expression data from Stanford collaborators on $J = 3000$ genes, comparing normal untreated samples to a set of treated samples. (Since the data are unpublished, we must omit the biological details, but many similar datasets are publicly available.) There are 15 samples in group 1 (normal) and 13 in group 2 (treated). Figure 12.1 is a histogram of the 3000 two-sample t-statistics from the genes.

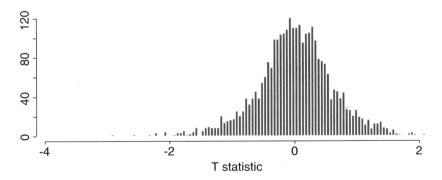

FIGURE 12.1. Histogram of 3000 t-statistics from the example.

They range from -4.54 to 3.72. Suppose that we decide to reject all genes whose t-statistic is greater than 2 in absolute value; there are 146 such genes. What is the FDR among these 146 genes? To assess this, we do a random permutation of the sample labels

$$(1, 1, 1, 1, 1, 1, 1, 1, 1, 1, 1, 1, 1, 1, 1, 2, 2, 2, 2, 2, 2, 2, 2, 2, 2, 2, 2, 2),$$

recompute the t-statistics and count how many exceed ± 2. (A "random permutation" is a random scrambling of the labels so that each label configuration is equally likely.) Doing this for 100 permutations, we find that the average number is 12.3. Thus a simple estimate of the FDR is $12.3/146 = 8.42\%$.

Now it turns out that this simple estimate tends to be biased upward. The reason is clear from Table 12.1. The permutations make all J genes null, but in our data only a proportion $\pi_0 = J_0/J$ are null. Hence to improve our estimate of the FDR, we multiply it by an estimate of π_0. To obtain the latter, we look at small values of the t-statistic (in absolute value), where null statistics are much more abundant than alternatives. Looking, for example, at t-statistics below 0.15 in absolute value, we find that 668 of the observed t-statistics fall in that range, while on the average 750 of the t-statistics from the permutations fall in that range. (The 0.15 cutoff is arbitrary in this example, but it can be automatically chosen taking bias and variance into account as in Storey and Tibshirani (2001).) Hence our estimate of π_0 is $\hat{\pi}_0 = 668/750 \approx 0.890$. Finally, our revised estimate of the FDR is $0.890 \cdot 8.42 = 7.49\%$.

12.2.3 Forming the Test Statistics

Suppose that we have J genes measured on I arrays under two different experimental conditions. Let $\bar{x}_{j1}$ and $\bar{x}_{j2}$ be the average gene expression for gene j under conditions 1 and 2, and let s_j be the pooled standard deviation for gene j:

$$s_j = \sqrt{\left(\frac{1}{I_1} + \frac{1}{I_2}\right) \cdot \frac{\sum_1 (x_{ji} - \bar{x}_{j1})^2 + \sum_2 (x_{ji} - \bar{x}_{j2})^2}{I - 2}}.$$

Here, I_k is the number of arrays in condition k, and each summation is taken over its respective group. Then, a reasonable test statistic for assessing differential gene expression is the standard (unpaired) t-statistic:

$$t_j = \frac{\bar{x}_{j2} - \bar{x}_{j1}}{s_j}.$$

Another statistic that could be formed is the rank-sum statistic. Let r_{ji} for $j = 1, \ldots, J$ be the rank of the ith expression value within gene j. Then, the rank-sum statistic for gene j is

$$r_j = \sum_1 r_{ji},$$

where the summation is taken over the genes in group 1. An extreme r_j value in either direction would indicate a difference in gene expression. The statistic t_j tests for a difference in means, whereas r_j tests for a more general difference in distributions. Usually, one is more concerned with a difference in mean gene expression, so t_j is a more powerful statistic to use for this test.

Each of these statistics is formed using only information from the gene itself. It is possible to model the data in such a way that one can "borrow strength" across the genes. For example, if we view the s_j as coming from some overall random process, then they can be jointly modeled. This can lead to reducing the overall variance of the s_j, giving the tests more power on average. Tusher et al. (2001) take a nonparametric approach to this and shrink the s_j toward an adaptively chosen s_0. The modified t-statistic is then

$$d_j = \frac{\bar{x}_{j2} - \bar{x}_{j1}}{s_j + s_0}.$$

Specifically, s_0 is chosen as the percentile of the s_j values that makes the coefficient of variation of d_j approximately constant as a function of s_j. This has the added effect of dampening large values of d_j that arise from genes whose expression is near zero.

Sometimes one wants to test for a difference between three or more conditions or for a trend in time-course data. Statistics that can be used for these and other situations are described in Tusher et al. (2001) and Section 12.3.

12.2.4   Calculating the Null Distribution

Concentrating on the two-sample case, the null distribution can most easily be calculated by permuting the group labels as we did in the example, or one can use the bootstrap. For the latter method, one can use the standard methodology described in Efron and Tibshirani (1993). The permutation method has a strength in that if the null hypothesis is true, then we are able to calculate the null distribution exactly. Westfall and Young (1993) discuss the differences between these two approaches more thoroughly, and Romano (1989) shows that they are asymptotically equivalent (in I, the number of arrays) in the two-sample case as well as many other cases.

As in forming the statistic, one can decide to treat each gene individually or borrow strength across the genes. If we calculate the null distribution individually for each gene, then this has two drawbacks. The first is that one suffers from a granularity problem: the null distribution has a resolution on the order of the number of permutations. For example, if we perform 1000 permutations, then the genes' resulting p-values will be on a resolution of $1/1000$. If we test 30,000 genes with 1000 permutations, then we will reject about 30 at a time. If we assume that each gene has the same null distribution, and combine them for each permutation, then the resolution of the null distribution is much finer. With J genes and B permutations, the resolution is on the order of $1/B$ for individual null distributions, but $1/(JB)$ for a pooled null distribution.

The second drawback of calculating a separate null distribution for each gene is that one is not then able to estimate better rejection regions. With individual null distributions, one essentially treats each gene as a different experiment. Therefore, for each "experiment," we have B observations from the null distribution and one observation from the original data. One cannot compare the null distribution to the observed statistic to derive more powerful, asymmetric rejection regions. This can lead to a loss of power and a waste of information in the data. On the other hand, if we pool the statistics and consider them to follow a mixture distribution as in Efron et al. (2001), then we now have many observations from a mixture of the null and affected distributions, as well as from the pure null distribution. We can use these observations jointly to derive more powerful rejection regions as in SAM and Efron et al. (2001).

The drawback of pooling the null statistics is that one can end up using different distributions to estimate an overall error rate. This is actually not a problem for using false discovery rates if the conditions discussed in Storey et al. (2002) hold. The null statistics just have to converge to some overall "null" distribution in J, the number of genes. This is likely met because J is typically very large, and one can easily see that upon repeats of the experiments, the null distribution is similar. Therefore, in the methodology that we describe here, the null statistics are pooled across the genes, making for more accurate and powerful procedures as well as cutting down on the computation time.

12.2.5 The SAM Thresholding Procedure

The SAM (significance analysis of microarrays) method (Tusher et al., 2001) and software forms cut points based on the data. Rather than using a standard rule of the form $|d_j| > t$ to call genes significant (i.e., having symmetric cut points $\pm t$), SAM derives cut points t_1 and t_2 and uses the rejection rule $d_j < t_1$ or $d_j > t_2$. This can lead to a more powerful test in situations where more genes are overexpressed than underexpressed, or vice-versa. Given its cut points, the SAM method estimates the false discovery rate quantities described below in subsection 12.2.6. The derivation of the cut points is illustrated in Figure 12.2 and detailed below:

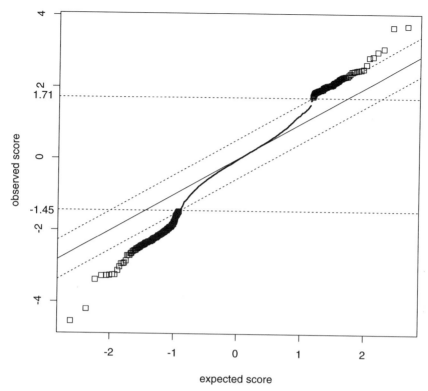

FIGURE 12.2. SAM plot for microarray example. We draw a band of two parallel (broken) lines a distance $\Delta = 0.531$ from the 45° (solid) line. Moving up and to the right, we find the first time that the points go outside the band. All genes to the right of that point are called significant (even if they fall inside the band). We do the same thing in the bottom left corner. The upper and lower excursion values imply upper and lower cut points—here 1.71 and -1.45.

The SAM Procedure

1. Compute the ordered statistics $d_{(1)} \leq d_{(2)} \cdots \leq d_{(J)}$.
2. Take B sets of permutations of the group labels. For each permutation b compute statistics d_j^{*b} and corresponding order statistics $d_{(1)}^{*b} \leq d_{(2)}^{*b} \cdots \leq d_{(J)}^{*b}$. From the set of B permutations, estimate the expected order statistics by $\bar{d}_{(j)} = (1/B) \sum_{b=1}^{B} d_{(j)}^{*b}$ for $j = 1, 2, \ldots J$.
3. Plot the $d_{(j)}$ values versus the $\bar{d}_{(j)}$. For a fixed threshold Δ, starting at the origin, and moving up to the right, find the first $j = j_2$ such that $d_{(j)} - \bar{d}_{(j)} \geq \Delta$. All genes past j_2 are called "significant positive." Similarly, start at the origin, move down to the left and find the first $j = j_1$ such that $\bar{d}_{(j)} - d_{(j)} \leq \Delta$. All genes past j_1 are called "significant negative." For each Δ, define the upper cut point $t_2(\Delta)$ as the smallest

d_j among the significant positive genes, and similarly define the lower cut point $t_1(\Delta)$.

4. If $t_1(\Delta) > t_2(\Delta)$, then we set $t_2(\Delta) = t_1(\Delta) = 0$.

In Figure 12.2, we used $\Delta = 0.531$ to yield the same number of genes (252) that are obtained using the cutoffs ± 1.5. As the figure shows, the resulting cut points from SAM are asymmetric; they are -1.45 and 1.71. As a result, the number of genes called significant at the negative and positive ends are quite unequal—they are 167 and 85 (and they are 157 and 95 for the cut points ± 1.5). As mentioned earlier, the actual SAM procedure adds (the same) constant s_0 to the denominator of each statistic d_j. Here we set s_0 equal to the median value of the s_j, which is 17.7.

Note that under this procedure, if $\Delta' \geq \Delta$, then $t_2(\Delta') \geq t_2(\Delta)$ and $t_1(\Delta') \leq t_1(\Delta)$, giving a nested set of rejection regions as is most useful in hypothesis testing.

12.2.6 Estimating False-Discovery Rates

For a fixed rejection region (fixed Δ), the FDR and pFDR are

$$FDR(\Delta) = \mathbf{E}\left[\frac{V(\Delta)}{R(\Delta)}\middle| R(\Delta) > 0\right] \mathbf{Pr}(R(\Delta) > 0),$$

$$pFDR(\Delta) = \mathbf{E}\left[\frac{V(\Delta)}{R(\Delta)}\middle| R(\Delta) > 0\right],$$

where

$$V(\Delta) = \#\{d_j : \text{ gene } j \text{ unchanged and } d_j \leq t_1(\Delta) \text{ or } d_j \geq t_2(\Delta)\},$$

$$R(\Delta) = \#\{d_j : d_j \leq t_1(\Delta) \text{ or } d_j \geq t_2(\Delta)\}.$$

Therefore, $V(\Delta)$ is the number of false positive genes and $R(\Delta)$ is the number of significant genes for the SAM threshold indexed by Δ.

In Storey (2002) we develop the following estimates of the FDR and pFDR for a given Δ:

$$\widehat{FDR}_{\Delta'}(\Delta) = \widehat{\pi}_0(\Delta') \cdot \frac{R^0(\Delta)}{R(\Delta) \vee 1},$$

$$\widehat{pFDR}_{\Delta'}(\Delta) = \widehat{\pi}_0(\Delta') \cdot \frac{R^0(\Delta)}{\mathbf{Pr}(R^0(\Delta) > 0) \cdot [R(\Delta) \vee 1]}.$$

$R(\Delta)$ is defined as before, but

$$R^0(\Delta) = \frac{\sum_{b=1}^{B} \#\{d_j^{b*} : d_j^{b*} \leq t_1(\Delta) \text{ or } d_j^{b*} \geq t_2(\Delta)\}}{B},$$

$$\mathbf{Pr}(R^0(\Delta) > 0) = \frac{\#\left\{b : \#\{d_j^{b*} : d_j^{b*} \le t_1(\Delta) \text{ or } d_j^{b*} \ge t_2(\Delta)\} > 0\right\}}{B}.$$

Lastly, $\widehat{\pi}_0(\Delta')$ is an estimate of the overall proportion of true null hypotheses (unchanged genes). This estimate depends on our choosing another Δ'. The SAM methodology takes Δ' such that $R^0(\Delta') = J/2$ (i.e., half the null statistics fall in the rejection region defined by Δ'), but Δ' can be chosen more adaptively to minimize the bias and the variance of the estimates (Storey and Tibshirani, 2001). Nevertheless, the estimate is

$$\widehat{\pi}_0(\Delta') = \frac{J - R(\Delta')}{J - R^0(\Delta')}.$$

The denominator represents the number of genes not significant at Δ' when all null hypotheses are true. The numerator is the number of observed genes that are not significant at Δ'. For a reasonably chosen Δ', this is going to consist mostly of unchanged genes, so $\widehat{\pi}_0(\Delta') \approx J_0/J$. It is easy to show that $\mathbf{E}[\widehat{\pi}_0(\Delta')] \ge J_0/J$.

A natural question to ask is how to pick the rejection region (i.e., the Δ value). We can use these point estimates in one of the following three options:

1. Fix Δ and calculate $\widehat{FDR}_{\Delta'}(\Delta)$ or $\widehat{pFDR}_{\Delta'}(\Delta)$. Then, for large J and weak dependence, or if the unchanged genes are independent, we have

$$\mathbf{E}[\widehat{FDR}_{\Delta'}(\Delta)] \ge FDR(\Delta)$$

and

$$\mathbf{E}[\widehat{pFDR}_{\Delta'}(\Delta)] \ge pFDR(\Delta).$$

2. Choose a level α beforehand at which we wish to control the FDR. Then, take the smallest Δ such that $\widehat{FDR}_{\Delta'}(\Delta) \le \alpha$. For large J and weak dependence, this guarantees that $FDR(\Delta) \le \alpha$. If the unchanged genes are independent and we limit the Δ that we consider to $\Delta \ge \Delta'$, then regardless of the size of J, we have $FDR(\Delta) \le \alpha$.

3. Calculate $\widehat{FDR}_{\Delta'}(\Delta)$ and $\widehat{pFDR}_{\Delta'}(\Delta)$ over all $\Delta > 0$ simultaneously. Then, for large J and weak dependence, we have

$$\min_{\Delta}[\widehat{FDR}_{\Delta'}(\Delta) - FDR(\Delta)] \ge 0$$

and

$$\min_{\Delta}[\widehat{pFDR}_{\Delta'}(\Delta) - pFDR(\Delta)] \ge 0;$$

that is, the estimates are asymptotically simultaneously conservative.

4. For each gene j, $j = 1, \ldots J$, estimate its q-value as

$$\widehat{q}\text{-value(gene } j) = \min_{\Delta:\text{gene } j \text{ significant}} p\widehat{FDR}_{\Delta'}(\Delta).$$

The true q-value(gene j) is calculated analogously with the true $pFDR$. Then, for large J and weak dependence, we have

$$\min_{\text{all gene } j} [\widehat{q}\text{-value(gene } j) - q\text{-value(gene } j)] \geq 0;$$

that is, the estimated q-values are asymptotically simultaneously conservative.

In an attempt to keep the mathematical details within the scope of this book, we have not stated these options in precise detail. See Storey et al. (2002) for a rigorous mathematical justification of the above claims as well as the precise conditions of "weak dependence." (For example, we must bound $\Delta \geq \delta$ for some prechosen arbitrarily small δ.) These results are valid for fixed rejection regions; that is, those chosen independently of the data. Since SAM chooses the rejection regions and does false-discovery-rate calculations from the same data, some bias may be incurred. This is yet to be explored. See Storey et al. (2002) and Storey and Tibshirani (2001) for arguments as to why weak dependence holds for the problem of detecting differential gene expression in DNA microarrays, including simulation results that show that $J \geq 3000$ usually suffices. Notice that option 2 does not include the pFDR estimate. This is because the pFDR is 1 when all genes are unchanged, and therefore it cannot be controlled in the sense described in that option.

The SAM software allows one to interactively change Δ, so option 3 is potentially the most useful. It is argued in Storey (2001) that the pFDR has a more sound interpretation, so it is perhaps the better quantity to use. In single hypothesis testing, one can assign a p-value to a statistic, which is usually defined to be the probability of a statistic more extreme than the one observed given that the null hypothesis is true. The more technical definition of the p-value for an observed statistic is the minimum Type I error rate that can be attained over all rejection regions containing that observed statistic. In Storey (2001), a natural false-discovery-rate analog of the p-value is introduced, which is called the q-value. The q-value of a statistic is the minimum pFDR that can be attained over all rejection regions containing that observed statistic. Its name is derived from its special relationship to the p-value when considering the Bayesian posterior probability form of the pFDR. Whereas the p-value is the probability of observing a statistic as extreme or more extreme given that the null hypothesis is true, the q-value is the probability that the null hypothesis is true given a statistic as extreme or more extreme than the one observed. It is easy to see how the q-value can be interpreted among many tests, whereas the p-value cannot.

The q-value of a particular gene can be estimated by taking the minimum $\widehat{pFDR}_{\Delta'}(\Delta)$ over all Δ for which the gene is found to be significant. The q-value estimate is conservatively consistent under the conditions that we assumed above (Storey et al., 2002). In testing for differential gene expression, we suggest estimating the q-value for each gene. It gives a measure of the strength of evidence for differential gene expression in terms of the pFDR. This is an individual measure for each gene that simultaneously takes into account the multiple comparisons. Note that by using the q-value, it is not necessary to pick the rejection region or the desired error rate beforehand. The q-value estimate is included in the SAM software.

12.3   Software

SAM is a package for implementing the significance analysis of microarrays technique of Tusher et al. (2001). The false-discovery-rate methodology of Storey (2002) was recently included. The software was written by Balasubramanian Narasimhan.

The input to SAM is gene expression measurements from a set of microarray experiments, as well as a response variable from each experiment. The response variable may be a grouping such as untreated/treated (either unpaired or paired), a multiclass grouping (e.g., breast cancer, lymphoma, colon cancer, etc.), a quantitative variable (e.g., blood pressure) or a possibly censored survival time. SAM computes a statistic d_j for each gene j, measuring the strength of the relationship between gene expression and the response variable. It uses repeated permutations of the data to determine if the expression of any genes is significantly related to the response. The cutoff for significance is determined by a tuning parameter Δ, chosen by the user based on the false positive rate. One can also choose a fold-change parameter to ensure that called genes change by at least a prespecified amount.

12.3.1   Obtaining the Software

SAM can be downloaded at `http://www-stat.stanford.edu/~tibs/SAM`. QVALUE, a program for calculating q-values from a list of p-values, can be downloaded at `http://www.stat.berkeley.edu/~storey/qvalue/`.

12.3.2   Data Formats

The data should be put in a Microsoft Excel spreadsheet. The first row of the spreadsheet has information about the response measurement; all remaining rows have gene expression data, one row per gene. The columns represent the different experimental samples. The first line of the file contains the response measurements, one per column, starting at column 3.

This is further described below in subsection 12.3.3. The remaining lines contain gene expression measurements, one line per gene. We describe the format below.

Column 1 This should contain the gene name. It is for the user's reference.

Column 2 This should contain the gene ID for the user's reference. Note that the gene ID column is the column that is linked to the SOURCE Web site by SAM. Hence, a unique identifier (e.g., clone ID, accession number or gene name/symbol) should be used in this column if SOURCE Web site gene lookup is desired.

Remaining columns These should contain the expression measurements as numbers. Missing expression measurements should be coded as NA. This is done easily in good editor or in Excel. In Excel, to change blank fields to NA, choose all columns, pull down the Edit menu, choose Replace and then **nothing (Blank)** with NA.

12.3.3 Response Format

Table 12.2 shows the formats of the response for various data types. A look at the example files is also informative. A *quantitative* response is real-valued, such as blood pressure. *Two class (unpaired)* groups are two sets of measurements, in which the experiment units are all different in the two groups (for example, control and treatment groups, with samples from different patients). With a *multiclass* response there are more than two groups, each containing different experimental units. This is a generalization of the *unpaired* setup to more than two groups. *Paired* groups are two sets of measurements in which the same experimental unit is measured in

TABLE 12.2. Response formats.

Response type	Coding
Quantitative	Real number (e.g., 27.4 or -45.34)
Two class (unpaired)	Integer 1, 2
Multiclass	Integer $1, 2, 3, \ldots$
Paired	Integer $-1, 1, -2, 2$, etc.
	(e.g., $-$ means before treatment, $+$ means after treatment)
	-1 is paired with 1, -2 is paired with 2, etc.
Survival data	(Time, status) pair such as (50,1) or (120,0).
	First number is survival time, second is status
	(1 = died, 0 = censored).
One class	Integer, every entry equal to 1

each group (for example, samples from the same patient, measured before and after a treatment). *Survival data* consists of a time until an event (such as death or relapse), possibly censored. In the *one class* problem we are testing whether the mean gene expression differs from zero. (For example, each measurement might be the log(red/green) ratio from two labeled samples hybridized to a cDNA chip, with green denoting before treatment and red after treatment). Here the response measurement is redundant and is set equal to all 1s.

Sometimes, it is difficult to enter blocking information (see subsection 12.3.5) without confusing Excel. Excel thinks such entries are formulas. Therefore, SAM allows any response to be enclosed within quotes (not apostrophes) and strips the quotes off before doing any computation.

12.3.4   Example Input Data File for an Unpaired Problem

The response variable is 1 = *untreated*, 2 = *treated*. The columns are gene name, gene ID, and then by the expression values. The first row contains the response values. Note that there are two blank cells at the beginning of line 1. The gene expression measurements can have an arbitrary number of decimal places. See Table 12.3.

12.3.5   Block Permutations

Response labels can be specified to be in blocks by adding the suffix *blockN*, where N is an integer, to the response labels. Suppose, for example, that in the two-class data of Section 12.3, samples 1,3,5, and 7 came from one batch of microarrays, and samples 2,4,6, and 8 came from another batch. We call these batches "blocks." Then, we might not want to mix up the batches in our permutations of the data, in order to control for the array differences; that is, we would like to allow permutations of the samples within the set 1,3,5, and 7 and within the set 2,4,6, and 8, but not across the two sets. We indicate the blocks (batches) as follows in Table 12.4.

For example, "1block1" means treatment 1, block (or batch) 1. "1block2" means treatment 1, block (or batch) 2. In this example, there are $4! = 24$ permutations within block 1 and $4! = 24$ permutations within block 2. Hence the total number of possible permutations is $24 \cdot 24 = 196$. If the block information is not indicated in line 1, all permutations of the eight samples would be allowed. There are $8! = 40,320$ such permutations. Please note that block permutations cannot be specified with paired response as there is an implicit blocking already in force.

12.3.6   Normalization of Experiments

Different experimental platforms require different normalizations. Therefore, *the user is required to normalize the data from the different exper-*

TABLE 12.3. Example dataset for an unpaired problem.

		1	1	2	2	1	1	2	2
GENE1	101	7.64	−0.50	−1.95	10.12	−10.77	−4.47	−7.65	7.58
GENE2	102	38.10	4.86	7.87	−13.59	−9.79	−13.46	−8.91	−5.07
GENE3	103	21.15	5.96	3.20	−4.74	−3.70	−12.35	−10.17	0.63
GENE4	104	187.21	−23.81	16.76	14.10	−99.76	−89.11	−10.92	5.52

TABLE 12.4. Example dataset for a blocked unpaired problem.

		1block1	1block2	2block1	2block2	1block1	1block2	2block1	2block2
GENE1	101	7.64	−0.50	−1.95	10.12	−10.77	−4.47	−7.65	7.58
GENE2	102	38.10	4.86	7.87	−13.59	−9.79	−13.46	−8.91	−5.07
GENE3	103	21.15	5.96	3.20	−4.74	−3.70	−12.35	−10.17	0.63
GENE4	104	187.21	−23.81	16.76	14.10	−99.76	−89.11	−10.92	5.52

iments (columns) before running SAM. For cDNA data, centering the columns of the expression matrix (that is, making the column means equal to zero) is often sufficient. For oligonucleotide data, a stronger calibration may be necessary: for example, a linear normalization of the data for each experiment versus the rowwise average for all experiments.

12.3.7 Handling Missing Data

There are currently two options for imputing missing values in SAM.

Row average Each value is imputed with the average of nonmissing values for that gene.

K nearest neighbor In the other (default) option, missing values are imputed using a k-nearest neighbor average in gene space (default $k = 10$):

1. For each gene j having at least one missing value:
   (a) Let S_j be the samples for which gene j has no missing values.
   (b) Identify all other genes G_j having no missing values for samples S_j.
   (c) Find the k nearest neighbors to gene j among genes G_j, using only samples S_j to compute the Euclidean distance.
   (d) Impute the missing values in gene j, using the averages of the nonmissing from the k nearest neighbors for that sample.
2. If a gene still has missing values after the steps above, impute them with the average available expression values for that gene.

12.3.8 Running SAM

To begin, you highlight an area of the spreadsheet that represents the data by first clicking on the top left-hand corner and then shift-clicking on the bottom right-hand corner of the rectangle. Then, click on the SAM button in the toolbar. A dialog form shown in Figure 12.3 (see color insert) now pops up. You have to select the type of response variable and, if desired, change any values of the default parameters. For two-class and paired data, one has to specify whether the data are in the logged (base 2) scale or not. Click the OK button to do the analysis.

If you had any missing data in your spreadsheet, a new worksheet named "SAM imputed dataset" containing the imputed dataset is added to the workbook. This data can be used in subsequent analyses to save time. If there are no missing data, this worksheet is not added. The software adds three more worksheets to the workbook. There is one that is hidden called "SAM plot data" and should be left alone. The sheet named "SAM plot" contains the plot with which the user can interact. The sheet named "SAM output," is used for writing any output.

Initially, a slider pops up along with the plot shown in Figure 12.4 (see color insert) that allows one to change the Δ parameter and examine the effect on the false discovery rate. It you want a more stringent criterion, try setting a nonzero fold-change parameter. Positive significant genes are labeled in red on the SAM plot, negative significant genes are green. When you have settled on a value for Δ, click on the "List significant genes" button for a list of significant genes. The "List delta table" button lists the number of significant genes and the false positive rate for a number of values of Δ. Please note that all output tables are sent to the worksheet named "SAM output" erasing whatever was previously present in the worksheet. While the slider is present, all interaction with the workbook is only possible via the slider. It can be killed anytime and recreated by clicking on the SAM plot control button.

12.3.9   Format of the Significant Gene List

For reference, SAM numbers the original genes, in their original order, as 1, 2, 3, and so on. In the output, this is the row number. The output for the list of significant genes has the following format:

Row Number The row in the selected data rectangle.

Gene Name The gene name specified in the first column selected data rectangle. This is for the user's reference.

Gene ID The gene ID specified in the second column selected data rectangle. This is for the user's reference, but is also linked to the SOURCE Web site for gene information.

SAM score(d) The t-statistic value.

Numerator The numerator of the t-statistic.

Denominator($s + s_0$) The denominator of the t-statistic.

Bayesian probability This is an univariate measure of the evidence that a gene is significant, computed using a Bayesian mixture model and ignoring the multiple comparisons. See Efron et al. (2001) for details.

q-**value** This is the lowest pFDR at which the gene is called significant. It is analogous to the familiar p-value, but interpreted as the probability that a gene is a false positive given its statistic is as or more extreme than the observed statistic. (The p-value has the opposite interpretation.) The q-value measures how significant the gene is: as $d_j > 0$ increases, the corresponding q-value decreases. See Storey (2002), Storey (2001), and subsection 12.2.6 for details.

The numerator, denominator and q-value were explained in Section 12.2. The list is divided into positive and negative genes, having positive or negative score d_j. A positive score means positive correlation with the response variable: e.g., for group responses 1 then 2, a positive score means expression is higher for group 2 than group 1. For a survival-time response, a positive score means people with higher expression have longer survival times. All statements are reversed for negative scores.

12.4   Discussion

There have been several other methods suggested for detecting differential gene expression, including those discussed in Dudoit and Yang (Chapter 3, this volume) and Newton and Kendziorski (Chapter 11, this volume). A Bayesian method was developed in Newton et al. (2001), although the multiple hypothesis testing was not taken into account. In fact, only posterior probabilities are reported according to this methodology. They do suggest taking into account both fold change and the variance of the fold change (analogous to a t-statistic). An empirical Bayes approach is suggested in Efron et al. (2001). Posterior probabilities are calculated as well, and no distributional assumptions are made. In both of these methods, it is important to realize that a posterior probability of differential gene expression given the value of the t-statistic for only the gene at hand is not sufficient for looking at several genes at once. In Dudoit et al. (2002), symmetric rejection regions and the FWER as the multiple hypothesis testing measure are used. They use the adjusted p-value methodology of Westfall and Young (1993).

A strength of the approach we have presented is that the rejection regions are chosen to accommodate the proportion and degree to which the affected genes are overexpressed or underexpressed. Also, the multiple comparisons are taken into account via a multiple hypothesis testing error rate that is appropriate for finding many significant genes, while limiting the proportion of false positives among those called significant. Of course other methods may exist for choosing rejection regions or for assessing the false positives that provide a more powerful, but just as efficient detection of differential gene expression. This is an active area of research, and some additional, interesting statistical methodology is bound to emerge.

References

Benjamini Y, Hochberg Y (1995). Controlling the false discovery rate: A practical and powerful approach to multiple testing, *Journal of the Royal Statistical Society, Series B* 85: 289–300.

Dudoit S, Yang Y, Callow M, Speed T (2002). Statistical methods for identifying differentially expressed genes in replicated cDNA microarray experiments, *Statistica Sinica* 12:111–139.

Efron B, Tibshirani RJ (1993). *An Introduction to the Bootstrap*, Chapman & Hall.

Efron B, Tibshirani R, Storey JD, Tusher V (2001). Empirical Bayes analysis of a microarray experiment, *Journal of the American Statistical Association* 96:1151–1160.

Newton M, Kendziorski C, Richmond C, Blatter F, Tsui K (2001). On differential variability of expression ratios: Improving statistical inference about gene expression changes from microarray data, *Journal of Computational Biology* 8:37–52.

Rice JA (1995). *Mathematical Statistics and Data Analysis*, 2nd ed., Duxbury Press, Belmont, CA.

Romano JP (1989). Bootstrap and randomization tests of some nonparametric hypotheses, *Annals of Statistics* 17:141–159.

Storey JD (2001). The positive false discovery rate: A Bayesian interpretation and the q-value. Submitted. Available at `http://www.stat.berkeley.edu/~storey/`.

Storey JD (2002). A direct approach to false discovery rates, *Journal of the Royal Statistical Society, Series B* 64:479–498.

Storey JD, Taylor JE, Siegmund D (2002). A unified estimation approach to false discovery rates. Submitted. Available at `http://www.stat.berkeley.edu/~storey/`.

Storey JD, Tibshirani R (2001). Estimating false discovery rates under dependence, with applications to DNA microarrays. Submitted. Available at `http://www.stat.berkeley.edu/~storey/`.

Tusher V, Tibshirani R, Chu C (2001). Significance analysis of microarrays applied to transcriptional responses to ionizing radiation, *Proceedings of the National Academy of Sciences* 98:5116–5121.

Westfall PH, Young SS (1993). *Resampling-based Multiple Testing: Examples and Methods for p-value Adjustment*, Wiley.

13

Adaptive Gene Picking with Microarray Data: Detecting Important Low Abundance Signals

Yi Lin
Samuel T. Nadler
Hong Lan
Alan D. Attie
Brian S. Yandell

Abstract

DNA microarrays to evaluate gene expression present tremendous opportunities for understanding complex biological processes. However, important genes, such as transcription factors and receptors, are expressed at low levels, potentially leading to negative values after adjusting for background. These low-abundance transcripts have previously been ignored or handled in an ad hoc way. We describe a method that analyzes genes with low expression using normal scores and robustly adapts to changing variability across average expression levels. This approach can be the basis for clustering and other exploratory methods. Our algorithm also assigns p-values that are sensitive to changes in variability with gene expression. Together, these two features expand the repertoire of genes that can be analyzed with DNA arrays.

13.1 Introduction

Microarray technology to measure gene expression is now widespread. The application of microarray analysis to such diverse biological processes as aging Lee et al. (1999), cancer (Golub et al., 1999; Perou et al., 1999), diabetes (Nadler et al., 2000), and obesity (Nadler et al., 2000; Soukas et al., 2000) have provided important insights. The power of microarrays to simultaneously evaluate the level of expression of thousands of genes creates the challenge of identifying those few genes that demonstrate significant changes in expression from among numerous genes that show little or no change.

Several approaches have been proposed to interpret microarray data. Clustering methods (Eisen et al., 1998; Tamayo et al., 1999) to search for genes showing similar changes in expression across experimental conditions require extensive prefiltering to eliminate genes with low intensity or modest fold changes. Furthermore, it has become apparent that at different gene expression levels, different thresholds for significant changes are needed (Wittes and Friedman, 1999; Hughes et al., 2000; Roberts et al., 2000). More recent methods model the variability across average expression levels to establish thresholds but still rely on ad hoc methods for genes expressed at very low abundance (Newton et al., 2001).

We present a robust statistical approach to pick genes showing significant differential expression across abundance levels from microarray experiments. The application of this method to mouse experiments studying diabetes and obesity uncovered changes in gene expression at low abundance that were missed by other methods. Details of the method are provided, including information on how to obtain public domain software.

13.2   Methods

Our gene array analysis algorithm uses rank order to normalize data for each experimental condition and estimates the variability at each level of gene expression to set varying significance thresholds for differential expression across levels of mRNA abundance. This procedure can be used to prefilter data in detecting patterns of differential gene expression, for instance using clustering methods. We propose assigning Bonferroni-corrected p-values, which requires only minimal assumptions. Although expression data may be acquired from a variety of technologies, we focus attention on the oligonucleotide arrays in Affymetrix chips used in a mouse experiment on diabetes and obesity.

Our approach was motivated by a series of experiments on diabetes and obesity. Nadler et al. (2000) used Affymetrix MGU74AV2 chips with over 13,000 probes representing about 12,000 genes on mRNA from adipose tissue to examine the relationship between obesity and mouse genotype (B6, BTBR, or F1). Further experiments have grown out of this collaboration using replicates and will be reported elsewhere. The primary goal was to find patterns of differential gene expression in mouse tissue between strains. Thus, we have a two-factor experiment with possible replication for each chip mRNA.

13.2.1   Background Subtraction

Raw microarray measurements are typically normalized to account for systematic bias and noise to attempt to restore expression levels from raw data (Lockhart et al., 1996). One important source of bias is background fluorescence. Other factors that require attention include variations in array, dye, thickness of sample, and measurement noise. Background fluorescence may be measured in several ways, depending on chip technology, and is typically removed by subtraction (see Lockhart et al., 1996; Li and Wong, 2001; Schadt et al., 2001; Irizarry et al., Chapter 4, this volume; Li and Wong, Chapter 5, this volume). Affymetrix chips handle background by comparing perfect match (PM) with mismatch (MM) intensity. We use weighted averages PM and MM across oligo probe pairs using recent "low-level" analysis (Li and Wong, 2001; Schadt et al., 2001; Li and Wong, Chapter 5, this volume) to reduce measurement variability.

The following model motivates the simple subtraction for background, although it is not required for our methodology; see Hughes et al. (2000) and Li and Wong (2001) for other approaches. The background intensity b for a gene may be attenuated at some level $\alpha \leq 1$ that could depend on the array. The gene signal g may be affected in a relative way by the degree of hybridization h, blurred by intrinsic noise ϵ (with variance depending on g) and the abundance (β) of material administered to an array. Measurements for gene j are further subject to reading error ω:

$$MM_j = \alpha b_j + \omega_{Mj},$$

$$PM_j = \alpha[b_j + \beta \exp(g_j + h_j + \epsilon_j)] + \omega_{Pj}.$$

Notice that the gene signal g is confounded with h unless hybridization efficiency is independent of the experimental condition. Subtracting the background intensity MM from PM yields the adjusted measurements

$$\Delta_j = PM_j - MM_j = \alpha\beta \exp(g_j + h_j + \epsilon_j) + \delta_j = \alpha\beta G_j + \delta_j,$$

where the measurement error $\delta = \omega_P - \omega_M$ is symmetric around 0 and $\log(G) = g + h + \epsilon$ is the log expression level. Thus G would be observed if there were no measurement error and no array attenuation. Hence, it is natural under this model to consider the log transformation $x_j = \log(\Delta_j)$. This model forms the basis for simulations presented later.

13.2.2 Transformation to Approximate Normality

Background-adjusted intensities are typically log-transformed to reduce the dynamic range and achieve normality. Various authors have noted that comparisons based on such log-transformed gene expression levels appear to be approximately normal (see Kerr and Churchill, 2001). However, negative adjusted values can arise from low expression levels swamped by background noise (Figure 13.1). Some authors have proposed adding a small value before taking the log to recover some of these data (Kerr and Churchill, 2001). Our alternative normalization method leverages this idea

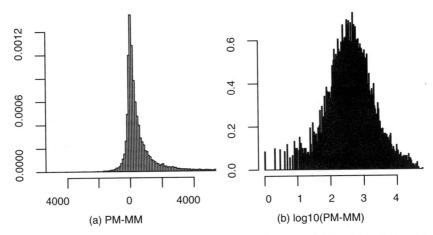

FIGURE 13.1. Log transform. Expression data from one chip: (a) relative histogram of raw values $\Delta = PM - MM$, deleting 5% of values beyond ± 5000 for display; (b) relative histogram of $\log(\Delta)$ with 23% of values being negative and hence dropped.

while providing comparisons that are more robust to difficulties with the lognormal assumption. For further discussion on normalization, see Dudoit and Yang (Chapter 3, this volume) and Colantuoni et al. (Chapter 9, this volume).

Our procedure converts the background-adjusted expression values into normal-scores without discarding negative values. This normal-scores transformation has been employed for microarray data using a different approach (Efron et al., 2001). If expression data are really lognormal, then this normal-scores transformation is indistinguishable from a log transformation after rescaling. We have found that log-transformed data (Figure 13.1b) appear roughly normal in the middle of the distribution, while the normal scores (Figure 13.2) are normal throughout.

Our procedure depends on the existence of some unknown monotone transformation of the data to near multivariate normal. There is always such a transformation in one dimension: let F be the cumulative distribution of adjusted values Δ and Φ be the cumulative normal distribution. Then $\Phi^{-1}(F(\Delta))$ transforms Δ to normal. If F is lognormal, then $\Phi^{-1}(F(\Delta)) = \log(\Delta)$, but we prefer not to make this assumption up-front. Instead, we approximate the transformation by $\Phi^{-1}(F_J(\Delta))$, where F_J is the empirical distribution of the J adjusted values $\Delta_1, \ldots, \Delta_J$. The difference between this approximate transformation and the ideal one is small (on the order of $1/\sqrt{J}$). This is known as the normal-scores transformation, and is readily computed as

$$x = \Phi^{-1}(F_J(\Delta)) = \text{qnorm}(\text{rank}(\Delta)/(J+1)),$$

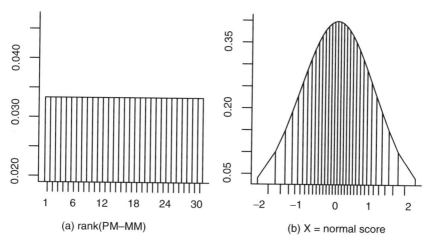

FIGURE 13.2. Normal scores transform. Sample of 30 from one chip for illustration: (a) relative histogram of ranks with equal mass; (b) normal-scores transformation with equal probability mass.

where rank(Δ) is the rank order of adjusted gene measurements $\Delta = PM - MM$ among all J genes under the same condition. The normal quantiles, qnorm(), transform the ranks to be essentially a sample from standard normal: a histogram of these x is bell-shaped and centered about zero (Figure 13.2b), with normal scores equally spaced in terms of probability mass (Figure 13.2a). Thus, these normal scores are close to a transformation that would make the data appear normal (Efron et al., 2001). If done separately by condition, this normalization automatically standardizes the center to 0 and the scale (standard deviation) to 1. Alternatively, if the experimental conditions are viewed as a random sample of a broader set of possible conditions, data across all conditions could be transformed together by normal scores. Normal scores are unaffected by monotone transformations of adjusted intensities or by global factors such as array, dye, and thickness of chip sample. Ranks may be disturbed by local noise, but that effect is unavoidable in any analysis of such an experiment.

13.2.3 Differential Expression Across Conditions

Differential expression across conditions of interest can be computed by comparing their transformed expression levels. Information on comparison of two conditions, 1 and 2, is summarized in pairs of normal scores, x_1 and x_2, across the genes; plotting x_1 against x_2 yields points dispersing from the diagonal. However, differential gene expression between experimental conditions may depend on the average level of gene expression, with genes of different average expression having intrinsically different variability. Thus, we recommend plotting the average intensity $a = (x_1 + x_2)/2$ against the difference $d = x_1 - x_2$, which involves just a 45 degree rotation (Roberts et al., 2000; Dudoit and Yang, Chapter 3, this volume; Irizarry et al., Chapter 4, this volume; Lee and O'Connell, Chapter 7, this volume; Colantuoni et al., Chapter 9, this volume; Wu et al., Chapter 14, this volume). Since our normal scores may be considered a forgiving approximation to the log transform, we prefer to represent the plotting axes as if the data were log-transformed; that is, use an antilog or exp scale. Thus, the a axis is centered on 1 and suggests a fold change in intensity, while the d axis suggests a fold change in differential expression.

This method can be extended to experiments with multiple conditions, multiple readings (e.g., dyes) per gene on a chip, and replication of chips (Kerr et al., 2001b; Wu et al., Chapter 14, this volume). Consider an ANOVA model

$$x_{ijk} = \mu + c_i + g_j + (cg)_{ij} + \epsilon_{ijk}$$

with $i = 1, \dots, I$ conditions, $j = 1, \dots, J$ genes, $k = 1, \dots, K$ replicate chips per condition, $\epsilon_{ijk} \sim \Phi(0, \sigma_j^2)$ being the measurement error for the kth replicate, and $c_i = 0$ if there is separate normalization by condition. Both the gene effect g_j and the condition by gene interaction $(cg)_{ij}$ are random effects. In general, all variance components may depend on the gene effect g_j. Adding multiple readings per chip introduces a nested structure to the experimental design that we do not develop further here (see Lee et al., 2000).

The major biological research focus is on differential gene expression, the condition i by gene j interaction. We assume that most genes show no differential expression; thus with some small probability π_1, a particular interaction $(cg)_{ij}$ is nonzero, say from $\Phi(0, \delta_j^2)$. Let $z_j = 1$ indicate differential expression, Prob$\{z_j = 1\} = \pi_1$. The variance of the expression score is

$$\text{Var}(x_{ijk}) = \gamma_j^2 + \delta_j^2 + \sigma_j^2 \quad \text{if } z_j = 1 \text{ (differential expression)},$$
$$\text{Var}(x_{ijk}) = \gamma_j^2 + \sigma_j^2 \qquad \text{if } z_j = 0 \text{ (no differential expression)},$$

for $i = 1, \dots, I, k = 1, \dots, K$, with γ_j^2 the variance for the gene j random effect. This differential expression indicator has been effectively used for microarray analysis (Lee et al., 2000; Kerr et al., 2001b; Newton et al., 2001). This ANOVA framework allows isolation of the $(cg)_{ij}$ differential expression from the g_j gene effect by contrasting conditions. Suppose that w_i are condition contrasts such that $\sum_i w_i = 0$ and $\sum_i w_i^2 = 1$. The standardized contrast $d_j = (\bar{x}_{1j\cdot} - \bar{x}_{2j\cdot})\sqrt{K/2}$ with $\bar{x}_{ij\cdot} = \sum_k x_{ijk}/K$ compares condition 1 with condition 2. More generally, the contrast

$$d_{jk} = \sum_i w_i \bar{x}_{ij\cdot} \cdot \sqrt{K} = \sum_i w_i \sqrt{K} [c_i + (cg)_{ij} + \bar{\epsilon}_{ij\cdot}]$$

with $\bar{\epsilon}_{ij.} = \sum_k \epsilon_{ijk}/K$ has $E(d_j) = \sum_i w_i c_i \sqrt{K}$ and

$$\mathrm{Var}(d_j) = \delta_j^2 + \sigma_j^2 \quad \text{if } z_j = 1 \text{ (differential expression)},$$
$$\mathrm{Var}(d_j) = \sigma_j^2 \qquad \text{if } z_j = 0 \text{ (no differential expression)}.$$

Again, condition effects c_i drop out and $E(d_j) = 0$ if each chip is standardized separately, but in general they remain part of the contrast.

Although microarray experiments began by contrasting two conditions, this approach adapts naturally to contrasts capturing key features of differential gene expression across design factors. Time or other progressions over multiple levels, such as a linear series of glucose concentrations, might be examined for linear or quadratic trends using orthogonal contrasts (Lentner and Bishop, 1993). For instance, with five conditions the linear and quadratic contrasts are, respectively (dropping subscripts except for condition),

$$d_{\text{linear}} = (2x_5 + x_4 - x_2 - 2x_1)\sqrt{K/8},$$
$$d_{\text{quadratic}} = (2x_5 - x_4 - 2x_3 - x_2 + 2x_1)\sqrt{K/14}.$$

With conditions resolved as multiple factors, such as obesity and genotype in our situation, separate contrasts can be considered for main effects and interactions. Each contrast can be analyzed in a fashion similar to that above. Alternatively, one can examine factors with multiple levels, say three genotypes, by an appropriate ANOVA evaluation (Lee et al., 2000).

13.2.4   Robust Center and Spread

For the majority of genes that are not changing, the difference d_j reflects only the intrinsic noise. Thus, genes that do change can be detected by assessing their differential expression relative to the intrinsic noise found in the nonchanging genes. Although it is natural to use replicates when possible to assess the significance of contrasts for each gene, microarray experiments have typically had few replicates K, leading to unreliable tests. Some authors have considered shrinkage approaches that combine variance information across genes (Efron et al., 2001; ?).

Measurement error seems to depend on the gene expression level $a_j = \sum_{ik} x_{ijk}/IK$, and it may be more efficient to combine variance estimates across genes with similar average expression levels (Hughes et al., 2000; Roberts et al., 2000; Baldi and Long, 2001; Kerr et al., 2001b; Long et al., 2001; Newton et al., 2001). Further, if there were no replicates, as in early microarray data, then it would be important to combine across genes in some fashion. There may in addition be systematic biases that depend on the average expression level (Dudoit et al., 2000; Yang et al., 2000). We noticed that empirically the variance across nonchanging genes seems to depend approximately on expression level in some smooth way, decreasing as a increases, due in part to the mechanics of hybridization and reading spot measurements. Here, we consider smooth estimates of abundance-based variance to account for these concerns. In a later paper, we will investigate shrinking the gene-specific variance estimate using our abundance-based estimate and an empirical Bayes argument similar to that of Lönnstedt and Speed (2001).

Our approach involves estimating the center and spread of differential expression as it varies across average gene expression a_j to standardize the differential expression. Specifically, we use smoothed medians and smoothed median absolute deviations, respectively, to estimate the center and spread. Smoothing splines (Wahba, 1990) are combined with standardized local median absolute deviation (MAD) to provide a data-adapted, robust estimate of spread $s(a)$. A smooth, robust estimate of center $m(a)$ can be computed in a similar fashion by smoothing the medians across the slices. We use these robust estimates of center and scale to construct standardized values

$$T_j = (d_j - m(a_j))/s(a_j)$$

and base further analysis on these standardized differences.

For convenience, we illustrate with two conditions and drop explicit reference to gene j. Revisiting the motivating model helps explain our specification for spread. Consider again $\log(G) = g + h + \epsilon$ and suppose that hybridization error is negligible or at least the same across conditions. The intrinsic noise ϵ may depend on the true expression level g: for two conditions 1 and 2, the difference d is approximately

$$d \approx \log(G_1) - \log(G_2) = g_1 - g_2 + \epsilon_1 - \epsilon_2.$$

If there is no differential expression, $g_1 = g_2 = g$, then $\mathrm{Var}(d|g) = s^2(g)$, and the gene signal g may be approximated by a. However, the true formula for $\mathrm{Var}(d|a)$ is not exactly $s^2(a)$ and cannot be determined without further assumptions.

Thus, differential contrasts standardized by estimated center and spread that depend on a should have approximately the standard normal distribution for genes that have no differential expression across the experimental conditions. Comparison of gene expressions between two conditions involves finding genes with strong differential expression. Typically, most genes show no real difference, only chance measurement variation. Therefore, a robust method that ignores genes showing large differential expression should capture the properties of the vast majority of unchanging genes.

The genes are sorted and partitioned based on a into many (say 400) slices containing roughly the same number of genes and summarized by the median and the MAD for each slice. For example, with 12,000 genes, the 30 contrasts d for each slice are sorted; the average of ordered values 15 and 16 is the median, while the MAD is the median of absolute deviations from that central value. These 400 medians and MADs should have roughly the same distribution up to a constant. To estimate the scale, it is natural to regress the 400 values of log(MAD) on a with smoothing splines (Wahba, 1990), but other nonparametric smoothing methods would work as well. The smoothing parameter is tuned automatically by generalized cross-validation (Wahba, 1990). The antilog of the smoothed curve, globally rescaled, provides an estimate of $s(a)$, which can be forced to be decreasing if appropriate. The 400 medians are smoothed via regression on a to estimate $m(a)$.

Replicates are averaged over in the robust smoothing approach, that is, contrasts $d_j = \sum w_i \bar{x}_{ij} \cdot \sqrt{K}$ factor out replicates. We are currently investigating shrinkage variance estimates of the form

$$s_j^2 = \frac{\nu_0 s^2(a_j) + \nu_1 \hat{\sigma}_j^2}{\nu_0 + \nu_1}$$

with $\hat{\sigma}_j^2 = \sum_k (x_{ijk} - \bar{x}_{ij.})^2/\nu_1$, $\nu_1 = I(K-1)$, and ν_0 is the empirical Bayes estimate (see Lönnstedt and Speed, 2002) of the degrees of freedom for $\hat{\sigma}_j^2/s^2(a_j)$.

It should be possible to combine estimates of spread across multiple contrasts; say, by using the absolute deviations $|x_{ijk} - a_j|$ for all genes with average intensity a_j within the range of a particular slice to estimate the slice MAD. This is sensible since these absolute deviations estimate the measurement error for most genes and most conditions. Those few genes with large differential effects across conditions would have large absolute deviations that are effectively ignored by using the robust median absolute deviation.

It may be reasonable in some cases to use "housekeeping genes" that are generally believed not to change over different conditions (see Chen et al., 1997; Baldi and Long, 2001). However, this may not capture the finer details of the center and scale as average intensity changes over the microarray. We use a robust estimation procedure to guard against the influence of the small proportion of changing genes that "contaminate" microarray data when we estimate the intrinsic noise level. Notice, however, that this contamination is of primary interest in a fashion similar to the problem of outlier detection.

13.2.5   Formal Evaluation of Significant Differential Expression

Formal evaluation of differential expression may be approached as a collection of tests for each gene of the "null hypothesis" of no difference or alternatively as estimating the probability that a gene shows differential expression (Kerr et al., 2001b; Newton et al., 2001). Testing raises the need to account for multiple comparisons, here we use p-values derived using a

Bonferroni-style genome-wide correction (Dudoit et al., 2000). Genes with significant differential expression are reported in order of increasing p-value. Further details of this procedure and the software can be found below.

We can use the standardized differences T to rank the genes. The conditional distribution of these T given a is assumed to be standard normal across all genes whose expressions do not change between conditions. Hypothesis testing here amounts to comparing the standardized differences with the intrinsic noise level. Since we are conducting multiple tests, we should adjust the test level of each gene to have a suitable overall level of significance. We prefer the conservative Zidak version of the Bonferroni correction: the overall p-value is bounded by $1 - (1 - p)^J$, where p is the single-test p-value. For example, for 13,000 genes with an overall level of significance of 0.05, each gene should be tested at level 1.95×10^{-6}, which corresponds to 4.62 score units. Testing for a million genes would correspond to identifying significant differential expression at more than 5.45 score units. Guarding against overall type I error may seem conservative. However, a larger overall level does not substantially change the normal critical value (from 4.62 to 4.31 with 13,000 genes for a 0.05 to 0.20 change in p-value). This test can be made one-sided if preferred. We report gene-by-gene results in the obesity data analysis below in terms of the overall significance level rather than the single-test level in order to avoid confusion.

Apparently less conservative multiple-comparison adjustments to p-values are proposed in Yang et al. (2000). However, the results are essentially the same with all such methods, except when more than 5–10% of the genes show differential expression across conditions. Figure 13.3 shows p-values from a typical mouse experiment with about $J = 13{,}000$ genes. The Bonferroni (p/J) and Zidak ($1 - (1 - p)^J$) significance levels are virtually identical (3.64×10^{-6} and 3.74×10^{-6}, respectively). Further, the Holms method discussed in Dudoit et al. (2000), which adjusts for the number of genes remaining to be tested, agrees with these two for the most significant 5–10% of the genes, picking the top 124 while Bonferroni picks only 123. The Westfall–Young method recommended by Dudoit et al. (2000) is not shown but should have properties similar to that of Holms. Thus, we are quite comfortable using the Zidak method in situations where only a small portion of the genes show differential expression.

It may be appropriate to examine a histogram of standardized differences T using these critical values as guidelines rather than as strict rules. The density f of all the scores is a mixture of the densities for nonchanging f_0 and changing f_1 genes,

$$f(T) = (1 - \pi_1)f_0(T) + \pi_1 f_1(T).$$

By our construction, f_0 is approximately standard normal. Following Efron et al. (2001), set π_1 just large enough so that the estimate

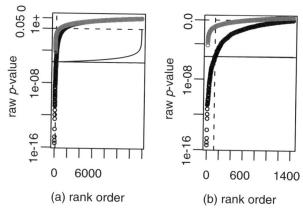

(a) rank order   (b) rank order

FIGURE 13.3. Multiple comparisons criteria. Typical mouse chip evaluation with roughly 13,000 genes. Circles are raw p-values on a semilog plot in rank order. The gray curved line at the top is for the uniform p-value ideal under the null hypothesis, dotted line at the nominal 5% level, dashed lines at about 3.7×10^{-6}, where Bonferroni, Zidak, and Holms methods meet. The curved solid line is Holms. (a) all p-values; (b) top 5% of p-values.

$$f_1(T) = [f(T) - (1 - \pi_1)f_0(T)]/\pi_1$$

is positive. This in some sense provides a 'liberal' estimate of the distribution of differentially expressed genes. It lends support to examination of a wider set of genes, with standardized scores that are above 3 or below –3. We suggest using this set as the basis for hierarchical clustering. Notice also that this provides an estimate of the posterior probability of differential expression ($z_j = 1$) for each mRNA,

$$\text{Prob}\{z_j = 1 | T_j\} = \pi_1 f_1(T_j)/f(T_j).$$

Gross errors on microarrays can be confused with changing genes. Replicates can be used to detect outliers in a fashion similar to the approach for differential gene expression. Residual deviations of each replicate from the condition × gene mean, $x_{ijk} - \bar{x}_{ij.}$, could be plotted against the average intensity, a_j. Robust estimates of center and scale could be used as above in formal Bonferroni-style tests for outliers. Separate smooth robust estimates of center and scale are needed for each contrast. Perhaps an additional Bonferroni correction may be used to adjust for multiple contrasts.

13.2.6 Simulation Studies

Three simulation studies were conducted to examine properties of our procedure. The first study shows how well the smoothed median absolute deviations can estimate the variability among unchanging genes. The second study verifies that the normal-scores procedure can essentially extract the "true differential expression" that would be observed if there were no measurement error. Simulated data from the second study were used to compare our procedure with other procedures that have been previously proposed.

The following simulation demonstrates the effectiveness of the robust standardization. We generated 9500 (a, d) pairs, with a from standard normal and d normally distributed with mean 0 and standard deviation $\sigma(a) = 1/[a/3+2.5]$. Then, we generated another 500 pairs by adding independent standard-normal random numbers to each d value. Thus, given the same X, the standard deviation of the contaminated d is $[1+(a/3+2.5)^2]^{1/2}$ times that of the uncontaminated d (1.8 to 3.64 as a goes from -3 to 3). We applied our robust scaling function to the combined data of 10,000 pairs. Figure 13.4 shows scatterplots of the simulated data before and after the addition of contamination. Figure 13.5a shows the closeness of the true (solid line) and estimated (dotted line) scales. Although there is always some bias with nonparametric estimation, the key bias problem arises in estimating spread in the presence of differentially expressed genes. The robust procedure reduces the influence of this contamination. The normal quantile plot of $Y/s(X)$ in Figure 13.5b shows the middle portion to be almost straight, as expected with normal data, while the tails diverge due to the "contamination" by differentially expressed genes.

We tested the normal-scores procedure on simulated data with two conditions and constant intrinsic variance across average expression levels. We generated samples with 10,000 genes and 5% differential expression and increasing amounts of measurement error. First, we randomly generated 9500 normal variates with mean 4 and variance 2. Next, we generated 500 random numbers from the same distribution and added normal "contami-

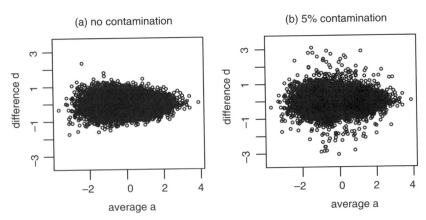

FIGURE 13.4. Simulation of expression data. Scatterplots of difference d against average intensity a (a) before and (b) after adding 5% contamination.

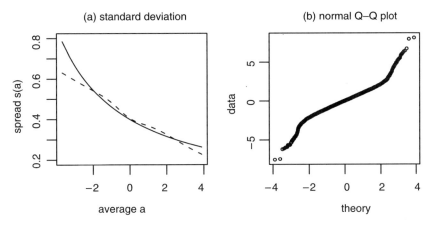

FIGURE 13.5. Simulation study of spread. Simulated data have 5% contamination shown in Figure 13.4b. (a) True (solid line) and estimated (dotted line) spread $s(a)$; (b) $Q - Q$ plot reveals "contamination" by differentially expressed genes.

nation" which was either up-regulated or down-regulated with probability $1/2$. This contamination had variance $1/2$ and mean tending from 3 to 2 as average expression level ranged from low to high abundance. The intrinsic noise ϵ was generated with variance 0.5, attenuations β were set at 1. We considered a range of measurement error variances from none to high ($\delta = 0, 1, 2, 5, 10, 20$). In the ideal situation of no measurement error, the "best" ranking would be based on the differential expression between the two conditions. We use the performance of this ideal "best" ranking as the benchmark against which to test our procedure. Figure 13.6 compares the top 500 "best" ranks when there is no measurement error with the ranks produced by our procedure under different level of measurement error. In the absence of measurement error, our ranks essentially matches the "best" ranking (line 0). When a typical level of noise is applied to the simulation, the ranks produced by our procedure still comes close to the "best" ranking.

In practice, analysis of low-abundance mRNAs leads to negative adjusted values, which are ignored or set to an arbitrary value by most other procedures. In the absence of measurement error, previously proposed methods perform well when they are first rank-ordered as done in our algorithm (Figure 13.7a; see color insert). In practice, measurement error becomes high with genes of low abundance and therefore background correction masks changes in gene expression. Despite a high level of noise, our method successfully detected numerous differentially expressed low-abundance mRNAs (Figure 13.7b). None of the nonchanging genes were identified in our simulations. In contrast, an early analytical method assuming a constant coefficient of variation (Chen et al., 1997) yielded conservative, flat thresh-

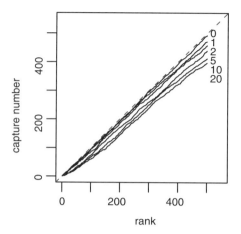

FIGURE 13.6. Capture efficiency. The top 500 "best" ranks when there is no measurement error with the ranks produced by our procedure under different levels of measurement error ($\delta = 0$, 1, 2, 5, 10, 20).

olds (Figure 13.7b, dashed line). The Bayesian approach (Newton et al., 2001) missed the pattern of changing variation with average gene intensity and missed most of the differentially expressed genes (Figure 13.7b, dotted line).

13.2.7 Comparison of Methods with E. coli Data

We reexamined some of the *E. coli* data reported in Newton et al. (2001). Figure 13.8 (see color insert) shows (a) log-transformed and (b) normal-scores-transformed data, with decisions based on Newton et al. (2001) as red dotted lines and our method as purple solid lines. Both methods and both figures agree in choosing the three extreme genes associated with IPTG-b. However, the normal-scores method (b) handles these in a natural way, while the log-transform (a) uses an ad hoc approach, leaving negative values out for computation and then including them at the margin for inference. Note in addition that our line based on a nonparametric estimate of variance better reflects the variability in the data, particularly at both extremes of mean intensity.

13.3 Software

The analysis procedures are written as an R language module. The R system is publicly available from the R Project, and our code is available from the corresponding author as the R **pickgene** library. The function `pickgene()` plots d against a, after backtransforming to show fold changes, and picks the genes with significant differences in expression. Examples include the simulations and graphics presented here. This library can be found at `www.stat.wisc.edu/~yandell/statgen`.

In its simplest form, `pickgene()` takes a data frame (or matrix) of microarray data, one column per array. We assume that housekeeping genes have already been removed. Columns are automatically contrasted using the prevailing form of orthonormal contrast (default is polynomial, `contrasts = "contr.poly"`).

```
library( pickgene )
result <- pickgene( data )
```

This produces a scatterplot with average intensity a along the horizontal axis and contrasts d along the vertical, with one plot for each contrast (typically one fewer than the number of columns of `data`).

With two columns, we are usually interested in something analogous to the log ratio, which can be achieved by renormalizing the contrast. If desired, the log transform can be specified by setting `rankbased = F`. Gene ideas can be preserved in the results as well.

```
result <- pickgene( data, geneID = probes,
                    renorm = sqrt( 2 ), rankbased = F )
print( result$pick[[1]])
```

The `pick` object is a list with one entry for each contrast, including the probe names, average intensity a, fold change ($\exp(d)$, as if $\Phi^{-1}(F(\Delta)) = \log(\Delta)$), and Bonferroni-adjusted p-value. The result also contains a score object with the average intensity a, score T, lower and upper Bonferroni limits, and probe names.

The `pickgene()` function relies on two other functions. The function `model.pickgene()` generates the contrasts, although this can be bypassed. More importantly, the function `robustscale` slices the pairs (a, d) into 400 equal-sized sets based on a, finds medians and $\log$(MAD)s for each slice, and then smoothes them using splines (Wahba, 1990) to estimate the center, $m(a)$, and spread, $s(a)$, respectively.

Estimates of density are based on the `density()` function, packaged in our `pickedhist()` routine.

```
pickedhist(result, p1 = .05, bw = NULL)
```

We pick a bandwidth `bw` that provides smooth curves and then adjust π_1 = `p1` so that f_1 is positive.

The standard deviation $s(a)$ is not returned directly in result. However, it is easily calculated as $\log(\text{upper}/\text{lower})/2$.

13.4   Application

13.4.1   Diabetes and Obesity Studies

Our approach was motivated by a series of experiments on diabetes and obesity (Nadler et al., 2000) using Affymetrix mouse MGU74AV2 chips with over 13,000 RNA probes representing roughly 12,000 genes or ESTs. Here obesity was controlled by the leptin gene. Each chip was treated with RNA extracted from adipose tissue combined over sets of four mice. Three chips were assigned to lean mouse sets from strains of B6, BTBR, or F1 mice; the three other chips were for obese mice from the same strains. Thus, we had a 3×2 factorial design with genotype and obesity as the main effects. The primary goal was to find patterns of differential gene expression in adipose tissue between obese and lean mice.

The majority of individuals with Type 2 diabetes mellitus are obese. Adipose tissue is thought to influence whole-body fuel partitioning and might do so in an aberrant fashion in obese and/or diabetic subjects. Almost half of these genes had at least one negative adjusted value in the dataset at low expression (Figure 13.9a, green dots; see color insert) and were missed by other methods. Our earlier study using clustering methods found interesting genes at high expression levels, including many at smaller fold changes (purple dots). Clearly, fold change is not the whole story.

However, this earlier analysis ignored genes at low expression, which we detected here. Roughly 100 genes were determined to have significant ($p < 0.05$) fold changes in gene expression using the robust normal-scores procedure (blue circles). A handful of significant genes with low expression (green dots in blue circles) were evaluated using RT-PCR, with a false-positive rate of about 50%, reflecting high noise in Affymetrix data at low intensity. Nevertheless, true positives detected by this method corresponding to transcription factors or receptors have potential to shed important light on adipocyte signaling pathways.

We also examined genotype effects (Figure 13.9b), finding numerous genes that seemed to have either additive or dominance effects, but not both. Red lines in Figure 13.9 correspond to Bonferroni limits for one contrast; the blue circle in Figure 13.9b is a simultaneous limit for the two genotype contrasts. The 5% criterion is based on the Zidak adjustment for multiple comparisons, which essentially agrees with both the Bonferroni and Holms methods when only a small fraction of genes show differential expression. We employed a permutation check of our method to verify the size of the test, that is, we took two of the chips (B6 and BTBR), randomly permuted the data for one chip relative to the other, and applied our method on the permuted data. This procedure was repeated 100 times. Our method picked out one significant gene in four permutations, and a fifth permutation had one gene right on the boundary; this agrees well with the expected 5% error rate.

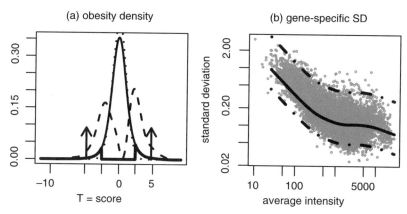

FIGURE 13.10. Investigation of differential density and gene-specific variance. (a) Density f of obesity T scores: all values (solid) overlaid on standard normal f_0 (dotted), with good agreement to 2 SD; dashed for f_1 differentially expressed genes; arrows at Bonferroni critical value. Thick lines near ± 2.5 show where $f = 2f_0$. (b) Gene-specific SDs for experiment with five replicates per two conditions: solid line is our smoothed MAD estimate of $s(a)$; dashed lines are upper and lower limits based on χ_8^2.

Figure 13.10a shows the density f of standardized differences T for obesity overlaid on the standard normal density f_0, with good agreement for standardized differences between –2 and 2. The long-dashed line shows an estimate of the density f_1 for differentially expressed genes as described above, picking the proportion of changing genes just large enough to ensure that the density is positive (see Efron et al., 2001). This illustrates just how conservative the Bonferroni approach is. We have since begun examining the genes with differential expression exceeding 3 SDs using hierarchical clustering after removing the mean expression level. Initial results suggest important rearrangement that might imply functional association among genes in clusters.

Whereas this early experiment had no replication, subsequent experiments have employed more chips to increase power. One study, to be reported elsewhere, had ten mice, five from each of two strains, that were separately applied to ten chips. Using our procedure on the mean differences, we computed the spread estimate $s(a)$ described above. Figure 13.10b shows $s(a)$ overlaid on gene-specific estimates of standard deviation, that is, for each gene, an SD was estimated using 8 df (10 chips – 2 groups). The abundance-based $s(a)$ appears to capture the central tendency in spread over the range of average intensities for this experiment. Notice that the SD confidence band based on χ_8^2 variability of gene-specific SDs covers most of the SDs, suggesting that our abundance-based approximation may not be too bad. However, since this band spans an almost tenfold difference in SD, a shrinkage-based estimator that combines gene-specific and average intensity based estimates of spread is advisable.

In conclusion, this novel method adapts to the dynamic range of expression data while handling low-intensity signals, including negative adjusted values. No data need be ignored, as the method finds a transformation to identify differentially expressed genes from large microarray datasets. Further, we have demonstrated the feasibility of putting p-values on differential gene expression without making many of the assumptions that other methods require.

This method can be extended to general experimental designs (Lee et al., 2000) by adjusting for variability in expression across all conditions relative to the average gene expression. The utility of clustering (Eisen et al., 1998; Tamayo et al., 1999) and classification (Golub et al., 1999) methods can be extended by relying on the standardized normal scores rather than log-transformed values. This can uncover novel relationships, particularly involving low-abundance transcripts. The p-values proposed here can further refine relationships uncovered by these omnibus methods.

Transcriptional regulation plays a particularly important role in the biology of low-abundance mRNA transcripts. This new algorithm now extends the powerful techniques of DNA array analysis to the world of low-abundance mRNAs.

13.4.2 Software Example

Data analysis was based on a data frame from Affymetrix chip processing, which after some manipulation has six columns of Affymetrix "Log.Avg" values plus the probe-set name. These data have the "housekeeping genes" and other Affymetrix references already removed. Suppose that the data frame has the probe names in column 1 and the next six columns contain the three lean chips, followed by the three obese chips, with B6, F1, and BTBR genotypes within each obesity class. The `pickgene` routine will show the three main-effects plots and return a data analysis list with the following command:

```
> library( pickgene )
> Leanob.pick <- pickgene( data[,-1], data[,1],
    faclevel = c(2,3), facnames = c("Obese","Genotype"),
    marginal = T, mfrow = c(2,2),
    renorm = c( sqrt(2)/3, sqrt(2)/2, sqrt(6)/4 ) )
```

Contrasts for the experimental design are automatically created with `faclevel` and `facnames` through a call to `model.pickgene`, that is, there are two levels of factor `Obese` and three levels of factor `Genotype`. The default normalization is for the sum of squared weights $\sum_i w_i$ to be K times the product of levels of other model factors. Often, it is useful to change this using the `renorm` option.

```
> apply(model.pickgene(faclevel = c(2,3),
+     facnames = c("Obese","Genotype")),
+     2, function(x) sum( x^2 ) )
(Intercept)     Obese.L  Genotype.L  Genotype.Q
     6            3          2           2
## renormalize so contrasts have natural meaning
> model.pickgene( faclevel = c(2,3),
+     facnames = c("Obese","Genotype"),
+     renorm = c( sqrt(2)/3, sqrt(2)/2, sqrt(6)/4 ) )
   (Intercept)   Obese.L Genotype.L Genotype.Q
1       1       -0.333333    -0.5      0.25
2       1       -0.333333     0.0     -0.50
3       1       -0.333333     0.5      0.25
4       1        0.333333    -0.5      0.25
5       1        0.333333     0.0     -0.50
6       1        0.333333     0.5      0.25
attr(,"assign")
[1] 0 1 2 2
attr(,"contrasts")
attr(,"contrasts")$Genotype
[1] "contr.poly"
```

It is also possible to explictly enter a design matrix for conditions using the model.matrix option to pickgene.

The overlay of green dots in Figure 13.4a is found by simple use of apply to identify genes with any(x<=0). In order to plot these values, one has to explictly recompute the normal scores and find the average and contrast:

```
> datanorm <- apply(ddata[,-1], 2, function( x )
+     qnorm( rank( x )/( 1 + length( x ))))
> datamean <- apply( datanorm, 1, mean )
> dataleanob <- apply( datanorm, 1, function( x )
+     mean( x[1:3] ) - mean( x[4:6] ) )
```

Figure 13.4b comes directly from the pickgene results. Note that all three sets of scores are ordered by the same average intensity per gene.

```
> names( Leanob.pick$score )
[1] "Obese.L"     "Genotype.L" "Genotype.Q"
> plot( Leanob.pick$score[[2]]$score,
+     Leanob.pick$score[[3]]$score )
```

The red lines come from the Zidak adjusted value,

```
> zidak <- qnorm( 1 - exp(( log( 1 - .05 ))/nrow( data ))/2,
+     lower.tail = F)
```

The density plot in Figure 13.4c was constructed with the `pickedhist` command. This can be applied to a single contrast using the `show` option or to all simultaneously. Since this is an ad hoc procedure (see Efron et al., 2001), a bit of trial and error is required. First, it is advisable to increase the smoothing by raising the bandwidth (`bw`). Then, adjust the prior proportion of changed genes (π_1 = `p1`) until density estimates do not cross negative.

```
## automatic bandwidth selection with p1 = .05
pickedhist( picked )
## refined selection by hand
pickedhist( picked, bw=.5, p1 = .1145)
```

Figure 13.4d came from another experiment. Gene-specific variances were computed by using `apply` on rows of the data frame and then overlaying the abundance-based variance. Details are left to the reader.

References

Baldi P, Long AD (2001) A Bayesian framework for the analysis of microarray expression data: Regularized t-test and statistical inferences of gene changes. *Bioinformatics*, 17:509–519.

Chen Y, Dougherty ER, Bittner ML (1997). Ratio-based decisions and the quantitative analysis of cDNA microarray images. *Journal of Biomedical Optics*, 2:364–374.

Dudoit S, Yang YH, Callow MJ, Speed TP (2000). Statistical methods for identifying differentially expressed genes in replicated cDNA microarray experiments. Technical Report 578, Dept. Biochem., Stanford University: Stanford, CA.

Efron B, Tibshirani R, Goss V, Chu G (2001). Microarrays and their use in a comparative experiment. *Journal of the American Statistical Association*, 96:1151–1160.

Eisen MB, Spellman PT, Brown PO, Botstein D (1998). Cluster analysis and display of genome-wide expression patterns. *Proceedings of the National Academy of Sciences USA*, 95:14863–14868.

Golub TR, Slonim DK, Tamayo P, Huard C, Gaasenbeek M, Mesirov JP, Coller H, Loh ML, Downing JR, Caligiuri MA, Bloomfield CD, Lander ES (1999). Molecular classification of cancer: Class discovery and class prediction by gene expression monitoring. *Science*, 286:531–537.

Hughes TR, Marton MJ, Jones AR, Roberts CJ, Stoughton R, Armour CD, Bennett HA, Coffey E, Dai HY, He YDD, Kidd MJ, King AM, Meyer MR, Slade D, Lum PY, Stepaniants SB, Shoemaker DD, Gachotte D, Chakraburtty K, Simon J, Bard M, Friend SH (2000). Functional discovery via a compendium of expression profiles. *Cell*, 102:109–126.

Kerr MK, Churchill GA (2001). Statistical design and the analysis of gene expression microarrays. *Genetical Research*, 77:123–128.

Kerr MK, Martin M, Churchill GA (2001). Analysis of variance for gene expression microarray data. *Journal of Computational Biology*, 7:819–837.

Lee CK, Klopp RG, Weindruch R, Prolla TA (1999). Gene expression profile of aging and its retardation by caloric restriction. *Science*, 285:1390–1394.

Lee MLT, Kuo FC, Whitmore GA, Sklar J (2000). Importance of replication in microarray gene expression studies: Statistical methods and evidence from repetitive cDNA hybridizations. *Proceedings of the National Academy of Sciences USA*, 97:9834–9839.

Lentner M, Bishop T (1993). *Experimental Design and Analysis*. Valley Book Company, Blacksburg, VA, 2nd edition.

Li C, Wong WH (2001). Model-based analysis of oligonucleotide arrays: Expression index computation and outlier detection. *Proceedings of the National Academy of Sciences USA*, 98:31–36.

Lockhart DJ, Dong H, Byrne MC, Follettie MT, Gallo MV, Chee MS, Mittmann M, Wang C, Kobayashi M, Horton H, Brown EL (1996). Expression monitoring by hybridization to high-density oligonucleotide arrays. *Nature Biotechnology*, 14:1675–1680.

Long AD, Mangalam HJ, Chan BYP, Tolleri L, Hatfield GW, Baldi P (2001). Gene expression profiling in *escherichia coli* K12: Improved statistical inference from DNA microarray data using analysis of variance and a Bayesian statistical framework. *Journal of Biological Chemistry*, 276:19937–19944.

Lönnstedt I, Speed T (2002). Replicated microarray data. *Statistica Sinica*, 12:31–46.

Nadler ST, Stoehr JP, Schueler KL, Tanimoto G, Yandell BS, Attie AD (2000). The expression of adipogenic genes is decreased in obesity and *diabetes mellitus*. *Proceedings of the National Academy of Sciences USA*, 97:11371–11376.

Newton MA, Kendziorski CM, Richmond CS, Blattner FR, Tsui KW (2001). On differential variability of expression ratios: Improving statistical inference about gene expression changes from microarray data. *Journal of Computational Biology*, 8:37–52.

Perou CM, Jeffrey SS, van de Rijn M, Rees CA, Eisen MB, Ross DT, Pergamenschikov A, Williams CF, Zhu SX, Lee JCF, Lashkari D, Shalon D, Brown PO, Botstein D (1999). Distinctive gene expression patterns in human mammary epithelial cells and breast cancers. *Proceedings of the National Academy of Sciences USA*, 96:9212–9217.

Roberts CJ, Nelson B, Marton MJ, Stoughton R, Meyer MR, Bennett HA, He YDD, Dai HY, Walker WL, Hughes TR, Tyers M, Boone C, Friend SH (2000). Signaling and circuitry of multiple mapk pathways revealed by a matrix of global gene expression profiles. *Science*, 287:873–880.

Schadt EE, Li C, Su C, Wong WH (2001). Analyzing high-density oligonucleotide gene expression array data. *Journal of Cellular Biochemistry*, 80:192–202.

Soukas A, Cohen P, Socci ND, Friedman JM (2000). Leptin-specific patterns of gene expression in white adipose tissue. *Genes and Development*, 14:963–980.

Tamayo P, Slonim D, Mesirov J, Zhu Q, Kitareewan S, Dmitrovsky E, Lander ES, Golub TR (1999). Interpreting patterns of gene expression with self-organizing maps: Methods and application to hematopoietic differentiation. *Proceedings of the National Academy of Sciences USA*, 96:2907–2912.

Wahba G (1990). *Spline Models for Observational Data*. Society of Industrial and Applied Mathematics, Philadelphia, PA.

Wittes J, Friedman HP (1999). Searching for evidence of altered gene expression: A comment on statistical analysis of microarray data. *Journal of the National Cancer Institute*, 91:400–401.

Yang YH, Dudoit S, Luu P, Speed TP (2001). Normalization for cDNA microarray data. Technical Report, SPIE BiOS 2001, San Jose, CA.

14

MAANOVA: A Software Package for the Analysis of Spotted cDNA Microarray Experiments

HAO WU
M. KATHLEEN KERR
XIANGQIN CUI
GARY A. CHURCHILL

Abstract

We describe a software package called MAANOVA (MicroArray ANalysis Of VAriance). MAANOVA is a collection of functions for statistical analysis of gene expression data from two-color cDNA microarray experiments. It is available in both the Matlab and R programming environments and can be run on any platform that supports these packages. MAANOVA allows the user to assess data quality, apply data transformations, estimate relative gene expression from designed experiments with ANOVA models, evaluate and interpret ANOVA models, formally test for differential expression of genes and estimate false-discovery rates, produce graphical summaries of expression patterns, and perform cluster analysis with bootstrapping. The development of MAANOVA was motivated by the need to analyze microarray data that arise from sophisticated designed experiments. MAANOVA provides specialized functions for microarray analysis in an open-ended format within flexible computing environments. MAANOVA functions can be used alone or in combination with other functions for the rigorous statistical analysis of microarray data.

14.1 Introduction

This chapter describes the philosophy behind, and the function of, a software package called MAANOVA (MicroArray ANalysis Of VAriance). MAANOVA is implemented in both Matlab and R programming environments. We focus our discussion on the Matlab implementation and address minor differences in the two implementations below. The package is a collection of functions that can be employed by a data analyst for the purpose of investigating gene expression data from two-color cDNA microarray experiments. Our goal is to provide a computing environment for microarray data analysis that is open-ended and flexible. Expert users of the system should be able to write their own functions and scripted analyses to achieve any result desired. At the same time, we provide a core set of functions that can be applied in a routine and prescripted fashion by a novice user to carry out a rigorous statistical analysis of microarray data.

MAANOVA functions were developed for the purpose of analyzing both small- and large-scale microarray experiments with arbitrary design structures ranging from simple two-array dye-swap experiments to elaborate loops. Basic concepts of microarray experimental designs are discussed by Kerr and Churchill (2001a). Although we encourage the use of efficient, balanced designs, the MAANOVA software can be applied to analyze data from any microarray experiment that uses more than one microarray to assay a set of samples. Microarray experiments may be implemented for many different purposes, and we do not wish to prescribe or in any way limit the possibilities for their application. We believe that scientists should design experiments based on the specific goals of their investigations and statistical principles. Design choices should not be determined by the availability of analysis software.

14.2   Methods

In this section, we introduce the statistical model that is at the core of our approach to microarray analysis. We digress in order to describe examples of experimental designs that will be used to illustrate points below. The remainder of the section follows the sequence of steps in a typical microarray analysis session. The process begins with diagnostics for data quality and transformations of data to remove intensity-dependent and spatial effects on relative expression. The next step is fitting the MAANOVA model. We describe the fitting algorithm and some diagnostics that are useful for assessing the quality of the fit. We discuss randomization methods that are available in MAANOVA to assess the significance of statistical tests and to compute interval estimates of relative expression. Although we emphasize randomization techniques, the standard normal theory results are available as well. Lastly, we discuss some clustering techniques that are useful for organizing (long) lists of differentially expressed genes or for investigating relationships among the RNA samples. The application of randomization methods to cluster analysis is a unique and powerful feature of the MAANOVA software package.

14.2.1   Data Acquisition

The technology underlying cDNA microarray experiments is described in Schena (2000). The raw data from a single microarray consist of a pair of images representing the fluorescent intensities detected by a photomultiplier tube when the microarray is scanned with each of two lasers. Images are typically stored in a 16-bit TIFF format, and data are extracted by segmenting the image and quantifying the intensity associated with each spot. We assume that the experimenter is satisfied with the quality of the images and that the data have been extracted with a software package such as SPOT (Yang et al., 2002) or GenePix (Axon Instruments, 1999). Decisions regarding background subtraction and other adjustments to the raw data are left to the user. Data from a set of microarray slides that constitute an experiment can be assembled into a single flat file for analysis. The input file for MAANOVA requires two columns of intensity values for each array and may include additional columns and header lines as needed. For a four-array experiment, the eight columns of intensity data are arranged as $(R_1, G_1, R_2, G_2, R_3, G_3, R_4, G_4)$. We refer to these as the raw intensity data.

14.2.2   ANOVA Models for Microarray Data

An analysis of variance (ANOVA) model for microarray data was proposed by Kerr et al. (2000) and is discussed further by Kerr and Churchill (2001b). Similar models have been described by Pritchard et al. (2001) and by Wolfinger et al. (2001). The ANOVA model is applied to transformed intensity data (for example, a logarithmic transform of raw intensity data). It allows one to account for sources of variation in the data that are attributable to factors other than differential expression of genes; thus, it effectively normalizes the data. There is no loss of information as is the case when raw intensity data are converted to ratios. Furthermore, it allows one to combine the information from many different arrays in a single analysis.

In a cDNA microarray experiment, two differentially labeled samples are applied to each array. Thus, an experiment with k arrays provides $2k$ measurements for each gene. The key to combining information is to assign each sample a label, which we call the *variety*. All samples from a common source are given a common label, and the information about these samples will be combined in the ANOVA analysis.

Denoting the transformed intensity data by $\mathbf{y} = \{y_{ijkgr}\}$, we can express the microarray analysis of variance model as

$$y_{ijkgr} = \mu + A_i + D_j + AD_{ij} + G_g + VG_{kg}$$
$$+ DG_{jg} + AG_{ig} + S_{r(ig)} + \epsilon_{ijkgr}. \tag{14.1}$$

The first four terms in the model (μ, A, D, and AD) capture the overall average intensity and variations due to arrays, dyes, and labeling reactions (i.e., array-by-dye interactions) as average effects across all the genes. The term gene G captures the average intensity associated with a particular gene. Although it is tempting to interpret G as representing the average level of expression of a gene, the effect is confounded with specific labeling and hybridization properties of the gene and should be interpreted cautiously. The variety-by-gene terms VG capture variations in the expression levels of a gene across varieties. The VG terms are the quantities of primary interest in our analysis. When the RNA samples (varieties) in an experiment have a complex design of their own (e.g., a factorial design), the VG terms may be structured to reflect the relationships among the samples (Jin et al., 2001). The dye-by-gene terms DG are included to account for a commonly observed technical artifact in which dyes show gene-specific effects (Kerr et al., 2002). The array-by-gene term AG captures the variation of each gene across spots on different arrays. Any gene-specific perturbations in the labeling reactions will be captured here. The spot term S captures the differences among the duplicated spots within an array. If there are no duplicated spots, this term is automatically dropped from the model.

The subscripting in equation (14.1) requires some comment. Arrays are indexed by i, dyes by j, and genes by g. The index r is nested within i and g to identify individual spots on each array and is only needed when there are multiple spots of the same gene on the arrays. Varieties are indexed by k, but the triplet of indices (i, j, k) for array, dye, and variety is redundant. For any given i and j, there is only one k; that is, if we know the array and dye, we know which sample it is and thus know the variety label. We take advantage of the redundant indexing to represent the experimental design in MAANOVA software. The raw intensity data are arranged in columns with a prescribed order, such as $(R_1, G_1, R_2, G_2, R_3, G_3, R_4, G_4)$. To specify the design (arrangement of varieties among the arrays), we simply have to indicate which variety labels are associated with the samples in each column. This vector of variety labels, called the *varid*, is used to achieve this specification in MAANOVA.

Estimated Relative Expression

When the ANOVA model is fit to data, we obtain estimates for each of the individual terms. Of particular interest in this context are the estimated values of VG. We refer to these as *relative expression values* and denote the estimated values by $\widehat{VG}$. The relative expression value represents the expression level of a gene g in a sample k relative to the weighted average expression of that gene over all samples in the experiment. The matrix of $\widehat{VG}$ values has dimensions number of genes $\times$ number of varieties. These derived data combine information across multiple samples of the same variety (and multiple spots of the same gene). MAANOVA may be run solely for the purpose of obtaining these derived data. The estimated relative expression values may then be analyzed by other methods, including a second-stage analysis of variance or a cluster analysis. The relative expression values are *normalized* data in the sense that effects due to the array, gene, spot, and so forth, have been removed.

Although the use of relative expression values represents a departure from the customary analysis of ratios, differences in normalized expression values are in fact estimates of the log ratio of the relative expression between two samples (assuming that the raw data have been log-transformed). Thus, it is not a radical departure from the norm. Rather, it is a means to achieve a more general interpretation of microarray data. The power of the ANOVA formulation is that it allows investigators to consider experiments that involve more than two samples and to combine information across multiple arrays that are hybridized with experimental samples in (almost) any arrangement. To achieve these goals, a concept of relative expression that is more general than pairwise ratios was required.

Mixed Model ANOVA

In the ANOVA model described above, all terms are viewed as fixed but unknown quantities. In an alternative formulation of the ANOVA model, the mixed-model, some of the terms are considered to be realizations of a random process. A mixed-model formulation for microarray ANOVA in which the spots are treated as random effects has been described by Wolfinger et al. (2001). An application of mixed-model ANOVA to a complex microarray experiment is described by Jin et al. (2001). In its current implementation, MAANOVA will carry out computations for the fixed-effects model. Under the mixed-model, the decomposition of variances is unchanged. However, construction of statistical tests and estimators can be different. Future releases of MAANOVA will include functions for mixed-model analysis.

14.2.3   Experimental Design for Microarrays

Microarray experiments are carried out to compare the relative abundance of specific RNA species in two or more biological samples. There may be many samples involved in an experiment, and they may have been derived from sources with their own experimental design structure. Jin et al. (2001) describe an experiment that involves eight samples in a fully factorial $2 \times 2 \times 2$ arrangement. Other examples include time series experiments (Chu et al., 1998) or treatment versus control comparisons (Callow et al., 2000). It is important that the investigator (and the analyst) understand the structure of the microarray experiment at this level. The distinction between independent biological replicates and technical replicates (obtained from the same biological source) is particularly important for proper construction of test statistics and interpretation of results.

Once the important choices of design at the biological sample level have been made, there is a second layer of design decisions imposed by the paired sample structure of two-color microarrays. A single microarray can only be used to make direct comparisons between two samples. This effectively imposes an incomplete blocking structure on the design (i.e., the samples are paired together on arrays that constitute blocks of size 2). Ideally, one might use a balanced incomplete block design in which all possible pairwise comparisons are made directly. However, due to the expense of the microarrays or limitations of the available sample, this may not be practical. Some solutions to the problem of finding a good partial incomplete block design are discussed by Kerr and Churchill (2001a).

Dye-Swap Design

A simple and effective design for the direct comparison of two samples is the dye-swap experiment (Figure 14.1a). This design uses two arrays to compare two samples. On array 1, the control sample is assigned to the red dye and the treatment sample is assigned to the green dye. On array 2, the dye assignments are reversed. This arrangement can be repeated by using four (or six or more) arrays to compare the same two biological samples (Figure 14.1b). The repeated dye-swap experiment is useful for reducing technical variation in the measurement but should not be confused with the replicated dye-swap experiment in which independent biological samples are compared (Figure 14.1c). The latter experiment accounts for both technical and biological variation in the assay. It may be more difficult

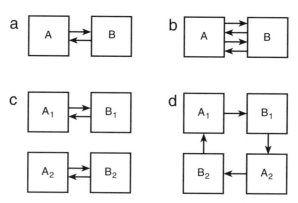

FIGURE 14.1. Experimental designs for the direct comparison of two samples. Boxes, representing RNA samples, are labeled as varieties A or B. Subscripts indicate independent biological replicates. Arrows represent microarrays. The sample at the tail of the arrow is assayed using the red dye, and the sample at the head of the arrow is assayed using the green dye. This figure shows a simple dye swap (a), a repeated dye swap (b), a replicated dye swap (c), and a loop design (d).

to achieve statistical significance using the replicated dye-swap experiment when biological variation is substantial. However, inference in the replicated experiment applies to the biological population from which the samples were obtained. Conclusions from the repeated dye-swap experiment are limited to the samples that were assayed.

Reference Design

The classical microarray experiment (e.g., Chu et al., 1998) employs a special RNA sample, called the reference sample, and all comparisons are made between the test samples and a reference with the same direction of dye labeling (Figure 14.2a). In this unbalanced design, the dye-by-gene (DG) and variety-by-gene (VG) effects are confounded (Kerr and Churchill, 2001a). The MAANOVA software will detect confounding and will drop DG terms from the model. A variation on the classical reference design uses two arrays in a dye-swap configuration to compare each test sample to the reference (Figure 14.2b). This design provides additional technical replicates and eliminates the confounding of relative expression with the gene-specific dye effects.

Fully half of the measurements in a reference experiment are made on the reference sample, which is presumably of little or no direct interest. The consequence is that the number of (technical) replicates available for inference is half of what could be achieved using alternative strategies. Despite this inefficiency, reference designs can have a number of advantages. The path connecting any two samples is never longer (or shorter) than two steps. Thus, all comparisons are made with equal efficiency. Reference designs can be extended easily (as long as the reference sample is available) and can be used to assay large numbers of samples that may have been collected in a (more or less) unplanned fashion.

Loop Designs and More

The loop design (Figure 14.1d), in which samples are compared with one another in a daisy chain fashion, was proposed as an efficient alternative to

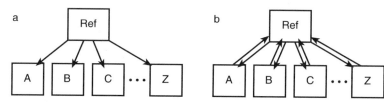

FIGURE 14.2. Experimental designs employing a reference RNA sample. Boxes represent RNA samples and arrows represent microarrays, as in Figure 14.1. The standard reference design (a) uses a single array to compare each test sample to the reference RNA. A variation (b) utilizes a dye swap for each comparison.

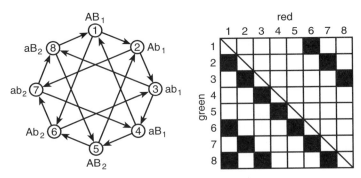

FIGURE 14.3. A woven loop experimental design. In this experiment there are eight samples. The diagram on the left is analogous to those in Figures 14.1 and 14.2. The labels are used to indicate that the experiment has a 2×2 factorial structure with two replicates. The experimental factors are A and B, and subscripts denote independent biological replicates. On the right is an alternative representation of the same experiment. Each box in the grid represents a possible ordered pairing of samples. Boxes corresponding to pairings used in the experiment are highlighted.

the reference design by Kerr and Churchill (2001b). In general, small loops provide good average precision. However, depending on the goals of the experiment, large loops may be inefficient. For example, if an investigator wants to compare every pair of samples, loops become inefficient when there are more than ten samples. In addition, the estimation efficiency of a loop is greatly reduced by the loss of just a single array, so loops are not a robust design. Variations on loop designs can be achieved by interweaving multiple loops together or by combining loops with reference designs (Figure 14.3).

The possibilities for the design of microarray experiments at the level of arranging pairs of samples onto arrays are perhaps bewildering. However, following a few simple guidelines will ensure that a design is effective for the purposes of a given investigation. Potential biases can be minimized by balancing dyes and samples. Create an even number of technical replicates from each biological sample and assign equal numbers of them to each dye label. It is most efficient to make the comparisons of greatest interest directly on the same array. Contrasts between samples that are never directly compared in an experiment are possible provided that there is a path of comparisons linking the two samples. The reference design with dye swapping is a good design for large experiments because it is simple, robust, and the distance between samples is always 2. If conclusions of the analysis will be applied to a biological population, be sure to include independent biological replicates.

Flexibility in the choice of designs has motivated our development of the MAANOVA software. With the confidence of knowing that the complexities of two-color expression assays can be addressed, an investigator can focus on the more important aspects of selecting the relevant samples, with adequate biological replication, to address the scientific question of interest. When planning a microarray experiment, it may be helpful to forget for a moment that measurements will be obtained on thousands of genes and to design the experiment as if only a single measurement would be obtained on each sample. This perspective can reveal flaws, such as inadequate replication, in a potential design.

14.2.4   Data Transformations

Before fitting an ANOVA model, the raw data should be transformed to a scale on which the various effects are additive. An argument can be made for the logarithmic scale; however, it is commonly observed that the mean and variance of log ratios computed from a single microarray will display systematic features that should be removed prior to analysis.

A standard diagnostic tool for assessing the intensity-dependent effects of dyes is the scatterplot of $\log(R/G)$ by $\log(R*G)$. One plot is generated for each array in an experiment. We refer to these as RI plots (for ratio $\times$ intensity), although there is a precedent for calling them MA plots (Yang et al., 2002). The characteristic curvature seen in RI plots can result from background differences in the two dye channels and/or from a differential in the response of the two dyes to laser activation (Cui et al., submitted for publication).

Kerr et al. (2002) proposed a shift-log transform to correct the curvature. This method identifies a single constant c that minimizes curvature in the RI plot when values of R and G are replaced by $R + c$ and $G - c$ prior to taking logarithms. The original motivation for this transformation was to shift the raw data to a scale on which a proportional relationship between $\log(R)$ and $\log(G)$ holds (i.e., the symmetric regression line should pass through the origin; Tanner, 1949). In practice, shift-log is simple and effective. Yang et al. (2001) proposed fitting a smooth curve to the RI plot using the local regression method (lowess) and recentering the log ratio data around the fitted curve. Despite concerns about overfitting, we find that the lowess method has advantages and may be applied in cases where the shift-log transform fails to correct the curvature.

Clones on a microarray are printed on regularly spaced grids, but the arrangement of clones on the array surface is usually arbitrary. Thus, we would not expect to see spatial patterns in the log ratios. We have implemented a version of lowess that simultaneously corrects for spatial and intensity-dependent effects. Again, overfitting may be a concern, but the spatial lowess function has enabled us to recover some otherwise troublesome data. The assumption underlying all of these data transformations is that the bulk of genes are not differentially expressed. MAANOVA implementations of lowess use robust fitting routines to exclude the influence of outliers that may represent differentially expressed genes.

Another commonly observed feature of the RI plot is the excess variability of log ratios at low intensity. Background subtraction often exaggerates this effect. It has been noted that raw microarray data have both additive and multiplicative error components (Rocke and Durbin, 2001). At high intensity, the multiplicative error dominates and a log transform is appropriate. However, at low intensity, the signals are small and the additive component of error dominates. A log transform can inflate the variance here. We have proposed a monotone transformation of the raw intensity data, called *linlog*, that behaves like a logarithm for high intensity and is linear for low intensity signals to stabilize the variance (Cui et al., submitted for publication). In our experience, setting a transition point that includes about 30% (default) of data points in the linear range tends to stabilize the variance. Application of linlog in combination with either shift-log, lowess, or local lowess transformations produces flat RI plots with stable variance across their entire range. This variance-stabilizing transformation should be considered when inference techniques that assume a common error variance across genes will be applied.

Corrections for spatial and intensity-dependent effects on the log ratio are essential to avoid being misled by common artifacts in microarray data. In general, it is best to correct biases at the technical level or through clever design (such as dye swap) rather than rely on post-hoc data adjustments. Simple precautions such as balancing the photomultiplier tube settings when scanning the arrays can be very effective. Correcting biases at the analysis stage is undesirable because the corrections applied can never be perfectly accurate. It is possible in some cases for attempted corrections to introduce biases greater than the ones they remove. Nonetheless, we have found that a small arsenal of data transformation tools is essential for reliable microarray data analysis. Keep in mind that the ANOVA model is an approximation and that transformations are used to improve the quality of this approximation. Our advice is to apply the most gentle transformation that corrects the observed problem.

14.2.5 Algorithms for Computing ANOVA Estimates

Estimates of individual terms in the ANOVA model are obtained by the method of least squares. If one makes the assumption that errors (ϵ_{ijkgr} in equation (14.1)) are normally distributed, then the least squares estimators are also maximum likelihood estimators. Least squares estimators may be sensitive to outliers in the data, and alternative methods of parameter estimation could prove to be more robust. In our experience, least squares estimators behave well, and we have not implemented robust algorithms in the current version of MAANOVA.

The usual method of fitting ANOVA models by least squares involves calculating the inverse of the *design matrix*. For microarray experiments, this matrix may have dimensions in the tens of thousands, and direct inversion is not practical. However, we note that the matrix has a regular structure that we can take advantage of to decompose the problem into many smaller calculations. The approach that we employ involves fitting the model in two stages. We first fit the *normalization* model

$$Y_{ijkgr} = \mu + A_i + D_j + AD_{ij} \tag{14.2}$$

to obtain residuals r_{ijkgr}. The rest of the model is fit iteratively on a per-gene basis. For each gene g (subscript suppressed), we fit the model

$$r_{ijkr} = G + VG_k + DG_j + AG_i + S_{r(i)} + \epsilon_{ijkr}. \tag{14.3}$$

In a typical microarray experiment, the same set of genes is assayed on all of the arrays. Thus, the factor G is said to be *balanced* with respect to factors A and D, and it is a consequence of this balance that the estimates obtained by fitting the model in two stages are identical to those that would be obtained by fitting the whole model (14.1) in a single step.

Numerically stable algorithms for least squares utilize the QR decomposition of the inner product of the design matrix (Seber, 1977). In the two-stage fitting algorithm, the same gene-specific model is fit many times. Furthermore, randomization tests may require that the whole process be repeated thousands of times. We achieve tremendous computational efficiency by precomputing the QR decomposition of the gene-specific design matrix once and storing it.

14.2.6   Statistical Inference

Parameters obtained by fitting the ANOVA model are estimates and as such are subject to error and uncertainty. In order to ensure that we are not misled into overinterpreting (or underinterpreting) the results, we appeal to methods of statistical inference.

Models

Statistical inference requires the specification of a model for the data. For hypothesis-testing inferences, it is necessary to specify two models, a null model and an alternative model. As noted above, each RNA sample in an experiment is associated with a label, the variety. Typically, RNA samples from a common source will share a common label. However, different labelings of the data can be applied to express different hypotheses about the data. The user of MAANOVA must input a vector of variety identifiers that serve as labels for the samples corresponding to each column of the intensity data. The varid is an integer vector, and its elements should be chosen from a set of consecutive integers (1, 2, 3, ...) with a unique integer for each distinct variety in the model. The structure of varid reflects the design of the experiment. For example a dye swap is specified as [1 2 2 1], a loop of five samples could be [1 2 2 3 3 4 4 5 5 1], and a reference design could be [1 6 2 6 3 6 4 6 5 6]. Under the null hypothesis of no differential expression, all of the samples would be considered to be one variety (i.e., there are no differences). For the dye swap and loop experiments, the varid is simply a vector of ones, [1 1 1 1] or [1 1 1 1 1 1 1 1 1 1]. For the reference design, the test samples are considered to be identical, but the reference sample is allowed to be distinct. The null model has two varieties, and the varid would be [1 2 1 2 1 2 1 2 1 2]. A more elaborate example is provided below.

Randomization Methods

Traditionally, statistical inference methods for ANOVA models have appealed either to normality of errors or to large sample theory to establish significance thresholds and/or p-values using tabulated distributions such as χ^2, t, or F. Microarray data often display dramatically nonnormal error distributions, and samples sizes (on a per-gene basis) are usually quite small. Thus, we have preferred to use randomization methods, permutation, and bootstrapping to establish distribution-free significance levels for statistical tests and confidence intervals. For hypothesis-testing applications in MAANOVA, we have implemented permutation methods that shuffle the residuals from fitting a null hypothesis ANOVA model to data. For computing confidence intervals, we employ bootstrapping methods that shuffle residuals obtained under an alternative model. Either procedure may be applied globally or restricted to shuffling within each gene. In bootstrapping applications, an inflation factor is applied to residuals to achieve the correct variance. The shuffling styles available in MAANOVA include restricted or unrestricted shuffling of the model residuals and sample shuffling. Restricted shuffling will shuffle the residuals within genes, and unrestricted shuffling will shuffle the residuals globally. Unrestricted shuffling assumes that error terms have a common variance. Restricted shuffling does not require this assumption, but it is only practical in large experiments where the number of residuals for each gene is sufficiently large. A third option, sample shuffling, which also requires a large experiment but makes few assumptions, is to shuffle whole arrays. Sample shuffling will freely exchange arrays that have the same (ordered pair of) variety identifiers. Like restricted shuffling, it is relatively assumption-free, but it is ineffective for small experiments. Enhancements to enumerate all possible permutation or bootstrap samples for moderately sized datasets are being considered, but the current implementation simply generates a random shuffle. Randomization can be time-consuming but, in light of the dramatic nonnormality of microarray data, we consider it to be worth waiting for. The standard tabulated p-values are also available in MAANOVA. The user can expect that the stringency of tests computed under various shuffling options will vary, and some judgment is required on a case-by-case basis to make an appropriate choice among these methods.

Hypothesis Tests

The MAANOVA package offers three test statistics (called F_1, F_2, and F_3) for hypothesis testing. We routinely compute all three types of tests because each one reveals different aspects of the data. All three test statistics are based on the gene-specific residual sums of squares, denoted by rss_g, and the residual degrees of freedom, denoted by df. Both quantities are model-dependent and are available in MAANOVA data objects after a model is fit to the data. Hypothesis testing involves the comparison of two models, and test statistics are computed on a gene-by-gene basis. Thus, we can suppress the subscript g and use the notation rss_0, df_0 for the null model and rss_1, df_1 for alternative model residual sums of squares and degrees of freedom, respectively.

The statistic F_1 is the usual F statistic that one would compute if data were available for only a single gene,

$$F_1 = \frac{(rss_0 - rss_1)/(df_0 - df_1)}{rss_1/df_1}. \tag{14.4}$$

It generalizes the t-test approach that is widely used in microarray analysis (Dudoit et al., 2002). Significance levels can be established by reference to the standard F distribution or by permutation analysis. This test does not require the assumption of common error variance. However, it has low power in typical microarray experiments because of small sample sizes, and it can be sensitive to variations in the estimates of residual variance, rss_1.

The test F_3 explicitly assumes a common error variance across all genes. The test statistic is

$$F_3 = \frac{(rss_0 - rss_1)/(df_0 - df_1)}{s^2_{pool}}, \tag{14.5}$$

where $s^2_{pool} = \frac{1}{N}\sum_{g=1}^{N} rss_{1g}/df_1$ is the estimated common variance. When testing the null hypothesis of no differential expression, the numerator in equation (14.4) (also in equations (14.5) and (14.6)) is equivalent to $\sum \widehat{VG}^2/df$. The F_3 statistic uses the same denominator for each gene. Thus, we are effectively testing based on the magnitude of relative expression values. In the case of two samples, this is equivalent to ranking genes by their log ratio. The F_3 test is powerful and can be applied to small experiments. However, it does assume a common variance, and we recommended checking this assumption (e.g., by inspecting residual plots). It may be necessary to apply a variance-stabilizing transformation such as `linlog`.

The test F_2 is a hybrid of the two other tests. The denominator of F_2 uses a gene-specific estimate of variance that is shrunken toward the global average variance,

$$F_2 = \frac{(rss_0 - rss_1)/(df_0 - df_1)}{(s_{pool}^2 + rss_1/df_1)/2}. \tag{14.6}$$

Although it is somewhat ad hoc in nature, we find that this test performs well in independently replicated experiments (much better than F_1 and slightly better than F_3). The motivating idea was to stabilize the gene specific variance estimates. The approach is similar to SAM t-tests (discussed in Storey and Tibshirani, Chapter 12, this volume) as well as Bayesian approaches employed by Baldi and Long (2001) and by Lönnstedt and Speed (2002). Properties of these "regularized" test statistics are an active area of investigation, and it is likely that we will expand the options available in MAANOVA.

Confidence Intervals

Formal hypothesis testing for differential expression produces p-values. A p-value for a test of differential expression summarizes the statistical significance of the test statistic, which is based on the variation in gene expression and the error variance. Oftentimes, p-values will be considered too concise in that they summarize the *statistical significance* of the data but do not give any information to evaluate the *biological significance*. Very small measured differences in gene expression may be statistically significant if the standard error is small but may be of no interest to a biologist. Confidence intervals allow one to gauge both the statistical significance and the potential biological significance of the result by providing the precision and magnitude of the changes in relative expression.

Confidence intervals are computed via bootstrapping to avoid normality assumptions. Currently, two kinds of confidence intervals can be computed. One kind is based on the assumption of a global error variance, analogous to the F-test F_3, and produces confidence intervals of uniform width for all genes. A second kind is based on the assumption of a gene-specific error variance, analogous to the F-test F_1. Confidence interval methods are another area of active investigation (see, for example, Kerr et al., 2002) where we anticipate advances that will be incorporated in MAANOVA.

Multiple Test Adjustment

When we compute test statistics or confidence intervals for differential expression we are simultaneously conducting thousands of inferences—one for each gene on our arrays. A well-recognized problem of this multiplicity is that the chances of obtaining a positive result become high even if all null hypotheses are true. The most common approach to the multiple-testing problem is to control the *familywise error rate*. The tests are done at a level of stringency so that the probability of making one or more type I errors is smaller than some nominal *alpha* level. Many scientists find this kind of control to be overly conservative for microarray studies. If the goal of an experiment is to generate a list of interesting genes, a certain number of false-positive results may be tolerable.

An alternative approach to multiple-test adjustment is with *false-discovery rates* (FDR) (Benjamini and Hochberg, 1995). The false-discovery rate is defined to be the expected proportion of false-positives among all rejected hypotheses. Tusher et al. (2001) incorporate a method for estimating the FDR in the SAM methodology for microarrays. However, FDR estimation is separable from the other aspects of the SAM methodology. The major requirement of the general methodology is that the null versions of the test statistics can be simulated. In the context of ANOVA methodology for microarrays, null test statistics can be simulated by permuting the data to preserve the experimental design except for the variety identifiers. Obviously, a certain amount of replication in the experimental design is required for this to be effective. For example, if the design is a loop with four slides, then there are only $4! = 24$ permutations of the array data, but many of these will be equivalent relative to the F statistic. There are really only $24/4 = 6$ permutations, hardly sufficient to simulate a distribution.

The current implementations of F-tests and confidence intervals in MAANOVA include the one-step adjustment method of Westfall and Young (1993) to control familywise error rates. The next release will include an implementation of the Westfall and Young step-down method for adjusted p-values and methods for estimating false-discovery rates.

14.2.7   Cluster Analysis

The estimated relative expression values $(\widehat{VG})$ obtained from fitting the ANOVA model capture the "profile" of expression across the samples in an experiment. We can use these quantities to cluster the genes or samples. By "clustering" we mean organizing the genes or samples into a hierarchical or grouped structure that represents the degree of similarity among profiles. The clustering structure is denoted by C. Prior to clustering, the user may want to filter out the insignificant genes using F-test results. The current implementation of MAANOVA includes a variety of options for hierarchical clustering as well as a K-means function. These functions are useful for organization and interpretation of long lists of significant genes.

Bootstrap Clustering

Our approach to clustering with expression data involves fitting the ANOVA model to obtain a derived dataset of relative expression values. A standard clustering algorithm is applied to them. Schematically, the process is $y \to VG \to C$. A cluster analysis will always produce a clustering C, but often there is no indication of how reliable it is.

The MAANOVA software includes features to assess the reliability of C by bootstrapping (Kerr and Churchill, 2001c). The steps are as follows:

1. Generate bootstrap dataset $y^\star$ by residual shuffling.
2. Fit the ANOVA model to $y^\star$ to obtain $VG^\star$.
3. Generate a clustering $C^\star$ from the $VG^\star$ matrix.
4. repeat steps 1–3 N times.

This process will yield N clusterings $C^\star$, and it will be necessary to obtain a summary. MAANOVA includes functions to generate and report the stable sets (i.e., sets of genes (or samples) that are grouped together in at least proportion p of all clusterings $C^\star$). For cluster analysis methods that partition a set of objects into a fixed number of groups, the stable sets are trivial to summarize. For hierarchical clustering, we use a consensus tree approach.

Consensus tree methods are widely used in phylogenetic research to summarize the common features among a set of trees (Felsenstein, 1985). Consensus trees are typically multifurcating, whereas the trees being summarized may all be bifurcating. Collapse of the bifurcating structure in the consensus trees indicates the absence of a consistent structure in the set of trees being summarized. The different types of consensus trees and algorithms for constructing them are described by Margush and McMorris (1981).

A clade is a grouping of objects defined by a tree. For each branch in a tree, there is a unique clade consisting of all tips below the branch. A majority rule consensus tree requires that a clade be included if and only if the same clade occurs in at least half of the trees. A generalization of the majority rule consensus tree requires that the clade occur in a proportion p of the trees, where $p > 1/2$ to ensure that self-contradictory clades do not appear in the consensus tree. The algorithm for consensus tree building used in our software is quick and straightforward. We store all the clades for all the trees in the bootstrap set, count the total occurrence of each, and then construct a list of clades in the consensus tree.

14.3   Software

14.3.1   Availability

MAANOVA functions have been developed and tested in Matlab Release 12 for Windows and for Linux RedHat 7.0. Some functions are written in C for speed. Executable software and source code can be downloaded from http://www.jax.org/research/churchill/software/ANOVA/. The

MAANOVA functions can be called from within the Matlab environment as part of an interactive data-analysis session. Alternatively, they may be run as scripts or incorporated in other user-defined Matlab functions. The user has access to standard Matlab functions and complete freedom to manipulate data objects in the analysis environment. We find this kind of freedom and flexibility to be appealing and preferable to compiled applications that limit the analyst's control over data objects. However, with freedom comes a certain degree of responsibility for knowing what you are doing. Some example datasets and scripted analyses are provided at the Web site listed above.

This section outlines the primary functions of MAANOVA in groups according to their standard usage. MAANOVA functions are designed to operate on special data structures (objects), and we provide their descriptions here. For detailed information on function syntax, the user can type "help [functionname]" in the Matlab environment. The fields contained in any data object can be listed simply by typing the object name. A user's manual is in preparation and will be regularly updated to reflect changes in the code. Lastly, the source code of each function is available and can be consulted or modified as needed by the user.

In parallel with our development of the Matlab version of MAANOVA, we have created an implementation of MAANOVA in the R programming environment. There are minor differences in syntax (e.g., underscore is a reserved character in R) but the functions and data structures are essentially identical. The R environment is freely available, and a number of other microarray analysis packages are being developed in R (see Parmigiani et al., Chapter 1, this volume). We will maintain the R version, and we plan to provide functions for creating and converting data objects to facilitate interoperability with SMA and other R-based packages.

14.3.2 Functionality

The MAANOVA package continues to grow. Rather than list all of the available functions here, we will highlight some of the most useful or important functions. A great strength of MAANOVA, in both the Matlab and R versions, is the capability for users to define their own functions and carry out analyses that could not be anticipated by the developers. Details regarding the syntax and the input and output data structures for MAANOVA commands are available through help functions in both Matlab and R environments.

Importing Data

Data can be read into the Matlab environment using one of several built-in functions such as load, tblread, dlmread, or textread. The raw intensity data should be formatted as a numerical data matrix with alternating columns representing the untransformed R and G values. Each array is represented by an adjacent pair of columns. Rows of the data matrix correspond to spots on the array, and if there are duplicated spots, they should be arranged in adjacent rows and should be identical in number for each spot. It is not possible to work with unequal numbers of duplicates in the MAANOVA model, and in some cases data may have to be discarded or excess replicates treated as separate genes. The arrangement of data is best done outside of the Matlab environment using a spreadsheet application. In addition to the raw intensity data, the user may wish to input gene identifiers and information about the array layout (e.g., meta-row, meta-column, row, and column positions for each spot).

Creating the Data and Model Objects

After reading in the raw data, the user can create several important objects. The function `createData` is used to create the data objects, and the function `makeModel` is used to create model objects. The data and model objects contain all of the information of the experiment design and are required inputs for most MAANOVA functions. The data object and the model object are created and maintained separately to allow the user to apply more than one model to a given dataset.

The function `createData` takes raw intensity data and the number of replicates as command-line inputs and creates a data object with these fields:

narrays: total number of arrays in the experiment;

ngenes: total number of genes in the experiment;

nspots: number of spots for each gene in total across all arrays;

nreps: the number of replicates of each gene on one array;

data: the raw intensity data;

adjdata: transformed data;

colmeans: column means for the transformed data;

offset: offset values from `shift`, `lowess`, or `linlogShift` function;

method: a string to indicate the data transformation method used.

By default, the `adjdata` field contains log-transformed data. This can be changed by calling one of the data transformation functions. The type of transformation applied and auxiliary information are stored in the fields `method` and `offset`.

The `makeModel` function takes a microarray data object, a `varid` vector, and a model indicator as input. The model indicator is an integer array with three elements to indicate whether to fit VG, DG, and AG effects or not. For example, if a user wants to fit only VG and AG effects and leave out the DG effect, this variable should be [1 0 1]. The model object has the following fields:

[fitVG fitDG fitAG]: flags to include terms in model;

varid: vector of variety identifiers;

nvars: total number of distinct varieties in the model;

varcount: counts of the occurrence of each variety in varid;

even_flag: indicates that the design is even;

latsq_flag: indicates that the design is a Latin square;

VDCon_flag: indicates that terms V and D are completely confounded;

X: the design matrix for a single gene in the experiment;

A: contrasts for zero-sum constraints on the design matrix;

Q, R: QR decomposition of the design matrix.

The model object stores several derived quantities that are used to speed other computations, and it is not advisable to modify this object once it is created.

Data Transformation and Visualization Functions

The first steps of an analysis often involve getting the data onto the right scale. An example of visualization tools used in conjuction with a data transformation is shown in Figure 14.4 (see color insert). MAANOVA provides a variety of functions for data transformations, including shift, lowess, and linlog. The lowess function includes options for spatial smoothing and for linear or quadratic fits. Results of applying a transformation function are stored in the adjdata field of the data object. The method of transformation is recorded along with auxiliary information such as offsets for shift-log transform. A data object can only store one type of transformed data. If more than one transformation is required, the user should create multiple data objects.

The function riplot takes a data object as input and generates a set of scatterplots of log ratio by log total intensity. The user can highlight selected subsets of genes that can be useful for diagnosis of problems with data or simply for viewing interesting subsets of data. The function arrayview can be used to display any function of the data as a color scale on the grid coordinates of an array. Used in conjunction with the function make_Ratio, it can be used to assess spatial heterogeneity in log ratios.

Matlab and R environments provide powerful and flexible graphical functions that can be used to view data. Some examples are provided in our online scripts. The user is encouraged to explore the possibilities and be creative.

Model Fitting and Diagnostics

The least squares fit of an ANOVA model to a dataset is achieved by a call to the function fitmaANOVA, which takes a data object and a model object as arguments. There is an optional flag to suppress computation of the sums of squares for the ANOVA table. This can be useful when things get slow. The output of fitmaANOVA is an ANOVA object with fields:

yhat: matrix of fitted value;

rss: residual sum of squares for each gene;

G: estimated gene effects;

VG: estimated relative expression values;

DG: estimated gene-specific dye effects;

AG: estimated spot effects;

model: a vector of flags [fitVG, fitDG, fitAG] that indicate which model terms were fit to the data;

table: cell array containing the ANOVA table.

Fitting of the ANOVA model is the core function of our software package. It is written in C and may be called repeatedly by other analysis functions.

After fitting the ANOVA model, it is recommended to check results by generating graphical diagnostics. The function resiplot will generate a standard residual plot. A variety of different scatterplots can be useful, and examples are provided in our online scripts. It is most effective to use built-in graphics functions for constructing these plots. The arrayview function can be used to visualize the spatial patterns of VG, AG, DG, and residuals.

Detecting Differentially Expressed Genes

The function make_Ftest will compute test statistics and carry out a permutation analysis of their distribution. It takes a data object and two model objects (a null model and an alternative model) as arguments. In addition, the user can set the number of permutations, the significance level(s), the type of shuffling, and the type(s) of F-test method to use. The output of make_Ftest is an object with the following fields:

`SigLevel`: user-specified significance levels;

`nIteration`: number of permutations;

`shuffle_flag`: indicates the shuffling method used;

`method`: type of F-test computed, may be any combination of [123];

`NullModel`: ANOVA terms fit under the null model;

`AltModel`: ANOVA terms fit under the alternative model;

`NullVarid`: null model variety identifiers;

`AltVarid`: alternative model variety identifiers;

`F1`, `F2`, `F3`: structure arrays for F-test result. Each contains the following fields:

> `Fobs`: observed F values;
>
> `Fcritpg`: calculated F critical value per gene;
>
> `Pvalpg`: calculated permutation P value per gene;
>
> `Fcritmax`: multiple-test adjusted F critical value;
>
> `Pvalmax`: multiple-test adjusted P value;
>
> `Ptab`: tabulated p value from F distribution (F_1 only).

A "volcano" plot provides a graphical summary of the simultaneous results from all three F-tests. The function `volcano` takes an ANOVA object, an F-test object, and significance thresholds for each test as input arguments. The y-axis of the volcano plot is the $-\log_{10}$ (tabulated p-value) for the F_1 test. If the experiment has only two samples, the x-axis is shown as the difference in estimated relative expression values, $\log_2(\widehat{VG}_{1g} - \widehat{VG}_{2g})$. The resulting figure has the appearance of an erupting volcano. When there are more than two samples, the x-axis is the root mean square of the relative expression values, $\sqrt{(\sum_{i=1}^{k}(\widehat{VG}_{kg})^2)}$. Although the volcano-like appearance is lost in the modified version, we kept the catchy name. A horizontal line on the plot represents the significance threshold of the F_1 test. Vertical lines represent thresholds for the F_3 test, and red color is used to indicate genes selected by the F_2 test. Genes in the upper-right (and upper-left) corner(s) of the figure are significant by all three criteria, and their indices are returned by the `volcano` function.

Confidence Intervals

The function `make_CI` is used to construct confidence intervals for user-defined contrasts of the relative expression values. Input arguments are a data object, a model object, a set of contrasts, and parameters to specify significance levels and the mode of shuffling. The default set of contrasts is all pairwise differences (i.e., all possible log ratios). Bootstrapping can be carried out at the individual gene level or with a multiple-testing adjustment. The output of this function is an object with the following fields:

`shuffle_flag`: shuffle method;

`output_selection`: indicates type of multiple-test adjustment;

`SigLevel`: significance level;

`Contrast`: contrast set;

`nIterations`: number of bootstrapping iterations;

`CI`: confidence interval for each gene;

`CImaxLo, CImaxHi`: limits for adjusted confidence intervals.

Cluster Analysis

Cluster analysis can be carried out using built-in functions in Matlab and R. In Matlab, we call built-in functions `pdist` and `cluster` from the Statistics Toolbox. In R, we call finctions `dist`, `hclust`, and `kmeans` from the `mva` package. The MAANOVA function `boothc` will run bootstrap analysis on a hierarchical clustering. It takes a data object, a model object, the index of the genes to be used in clustering, a flag for clustering genes or samples, and the number of bootstrap iterations as input arguments. It returns an object, `nodeobj`, that lists all of the clades in the bootstrap set and their frequencies. The function `consensus` operates on the node object to create a consensus tree. It calls the `drawconsensus` function to draw the tree and outputs an object to represent the consensus tree. The function `writephylip` can output the consensus tree to a text file in PHYLIP format (`http://evolution.genetics.washington.edu/phylip.html`) for use with other tree-drawing software.

The MAANOVA package includes a function `kmeans` to compute K-means cluster analysis of (selected) genes from their relative expression values. The function `fom` is useful for determining the number of groups for K-means analysis, and the function `bootkmean` will run a bootstrapping analysis. The results of bootstrapping a K-means cluster analysis can be visualized with the `VGprofile` function.

14.4   Data Analysis with MAANOVA

In this section, we will demonstrate data analyses using MAANOVA. The illustration is brief and is intended to highlight unique or commonly used

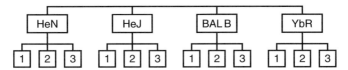

FIGURE 14.5. Experimental design for tumor survey data. Three independent tumor samples were obtained from each of four strains of mice. There are 12 independent RNA samples in total. The arrangement is a standard one-way layout with three replicates per group. The tumor samples were assayed with 24 microarrays using the dye-swap reference design as shown in Figure 14.2b.

features of MAANOVA. A complete script and data for this analysis are available on our Web site.

The design of the experiment at the level of the biological samples is shown in Figure 14.5. Three independent mammary tumor samples were obtained from each of four different strains of mice. These samples were compared using 24 microarrays. Each test sample was compared to a reference RNA using a dye-swap arrangement as in Figure 14.2b. Each microarray contained approximately 30,000 spots representing 15,000 genes printed in duplicate.

The raw intensities were read into the Matlab environment and named pmt, and there are two duplicated spots per gene. The row and column locations of each spot are stored in vectors grow and gcol. Function calls

```
> data = createData(pmt,2);
```

```
> data = malowess(data, grow, gcol, 'lowess2');
```

were used to create a data object and to apply the spatial lowess transformation.

We consider three models for the data. Model 0 assumes that all samples are the same (i.e., no differential expression of genes). Model 1 assumes that samples are the same within a strain but may vary between strains. Model 2 allows each sample to be unique. We first specify the variety identifiers.

```
> varid0 = [2 1 1 2 1 2 2 1 2 1 1 2 1 2 2 1 2 1 1 2 2 1 1 2 ...

2 1 1 2 1 2 2 1 2 1 1 2 2 1 1 2 2 1 1 2 2 1 1 2];

> varid1 = [5 1 1 5 1 5 5 1 5 1 1 5 2 5 5 2 5 2 2 5 5 2 2 5 ...

5 3 3 5 3 5 5 3 5 3 3 5 5 4 4 5 5 4 4 5 5 4 4 5];

> varid2 = [13 1 1 13 2 13 13 2 13 3 3 13 4 13 13 4 13 5 5 ...

13 13 6 6 13 13 7 7 13 8 13 13 8 13 9 9 13 13 10 10 13 13 11

11 13 13 12 12 13];
```

Then, we fit the ANOVA models.

```
> model0 = makeModel(data,varid0,[1 1 1]);
```

```
> ANOVA0 = fitmaANOVA(data,model0);
```

The argument "[1 1 1]" in makeModel indicates that all of the terms VG, DG, and AG are included in the model. Similar commands with varid1 and varid2 in place of varid0 are issued to fit models 1 and 2.

We computed F-tests to compare the various models. To compare models 0 and 1, issue the command

```
> Ftest0 = make_Ftest(data, model0,model1,0.95,500);
```

The last two arguments specify the confidence level and the number of permutations, respectively. The results of this F-test analysis are summarized as a volcano plot in Figure 14.6 (see color insert). Twenty genes were identified as being significant. The function call to generate the volcano plot is

```
> idx0= volcano(ANOVA1,Ftest0, 0.001, 0.05, 0.05);
```

where the last three arguments are significance levels for tests F_1, F_2, and F_3, respectively.

The 20 significant genes identified by Ftest0 were clustered using the K-means algorithm. The function call to generate the K-means analysis is

```
> [class, grp0] = bootkmean(data, model2, idx0,
    (1:12), 10, 500, 0.8, ...
```

```
'gene', 'VG');
```

Selected genes are indexed by idx0, selected samples are 1 through 12, the number of groups to fit is 10, the number of shuffles is 500, and results will be accepted at 0.8 confidence. The argument 'gene' specifies that we are clustering genes, not samples, and 'VG' specifies the sample-shuffling option. The function call

```
> VGprofile(VGdiff,grp0)
```

will generate the clustered relative expression profiles shown in Figure 14.7. Note that the testing and selection of genes were based on model 1 but the cluster analysis uses the unrestricted estimates of relative expression in model 2.

An objective of this study was to classify the samples according to their observed pattern of expression. For this purpose, we identified genes that were expressed differently without regard to strain of the sample by conducting a test of model 2 compared to model 0. Indices of significant genes

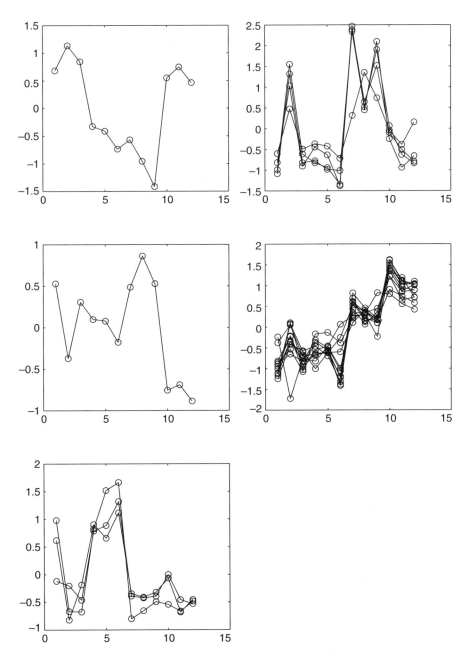

FIGURE 14.7. A K-means analysis of expression profiles from the tumor survey. Profiles are shown for the 20 significant genes selected by F-tests using the tumor survey data (Figure 14.6). The profiles cluster into five groups, and all of the assignments are supported at the 80% confidence level by bootstrap analysis. In each panel, the x-axis is the sample index and the y-axis is the relative expression value.

according to this test were stored in `idx2`, and a hierarchical cluster analysis of the samples was done. The function call

```
> nodeobj = boothc(data, model2, idx2, (1:12),500,'seuclid', ...
```

```
'sample', 'VG');
```

will generate 500 bootstrap trees using the standardized Euclidean metric to compute the distance. The last two arguments indicate that we are clustering the samples and using the sample-shuffling method.

The function call

```
> ctobj = consensus(nodeobj);
```

will generate the cluster diagram shown in Figure 14.8. The hierarchical cluster analysis was assessed using a sample bootstrap, and support values are shown on the tree. It is interesting to note that the clusters are largely concordant with strain but that there is an outlier that is strongly supported by the bootstrap analysis.

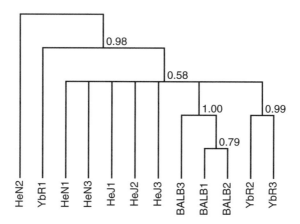

FIGURE 14.8. A majority rule consensus tree of tumor samples. A consensus was constructed using 500 bootstrap samples (residual method) from the tumor survey data. Strain and sample number of the tumors are indicated on the tips of the tree. Numbers on the branches indicate the proportion of bootstrap samples that support the clade, the grouping of samples below each branch. Branches that are supported at less than 50% are collapsed to a multifurcation in the consensus tree. The pattern of clustering is largely consistent with the strain origins of the tumors. A notable exception is HeN2, which appears to be a novel variant.

14.5   Discussion

There are several directions in which we plan to further the development of MAANOVA. Of these, the most substantial is the extension to mixed-effects ANOVA. As a first step, we plan to mimic the approach of Wolfinger et al. (2001). We are investigating alternative decompositions of the sums of squares in order to expand the scope of options for randomization tests in MAANOVA, and we are developing a more general approach to model specification in order to address hypothesis testing in a mixed-model context. In a complex experiment where the RNA samples have a nontrivial experiment design structure, it may be desirable to include both fixed and random terms in the VG component of the model. Mixed models can admit more general covariance structure and will provide shrinkage of estimated effects that can reduce bias.

Inevitably, there are missing data in microarray experiments. Missing data occur for any number of reasons but are often due to a technical failure of some spots on the array due to dust, scratches, or printing error. The algorithm employed to fit ANOVA models in two stages requires that the same set of genes be present on all arrays. Thus, if a gene is missing on just one array (or a few arrays) in a large experiment, it must be removed from the whole. In some cases, when a modest number of genes are missing at random from a large number of arrays, a substantial portion of the whole experiment may have to be discarded. This is clearly undesirable. The development of missing-data techniques that would allow us to retain useful information on genes with partial data would be desirable.

It seems likely that additional dyes will be developed for use with microarrays. Already, there are scanners on the market that can accommodate as many as five wavelengths. The basic structure of the ANOVA model is not affected by the use of multiple dye labels on the same array. The use of multiple dyes promises to substantially improve the accuracy of microarray experiments by effectively increasing the *block size* from two to three or more.

MAANOVA was originally developed for in-house use as a development and testing platform for microarray analysis. For these purposes, the interactive environments provided by Matlab or R are ideal. We use the same environment to carry out data analysis tasks. However, not everyone is comfortable with command-driven environments, and a different interface may be useful. Toward this end, we are actively discussing the form and function of a graphical user interface for MAANOVA.

References

Baldi P, Long AD (2001). A Bayesian framework for the analysis of microarray expression data: Regularized *t*-test and statistical inferences of gene changes. *Bioinformatics*, 17:509.

Benjamini Y, Hochberg Y (1995). Controlling the false discovery rate: A practical and powerful approach to multiple-testing. *Journal of the Royal Statistical Society, Series B*, 57:289.

Callow MJ, Dudoit S, Gong EL, Speed TP, Rubin E (2000). Microarray expression profiling identifies genes with altered expression in HDL deficient mice. *Genome Research*, 10:2022.

Chu S, DeRisi J, Eisen M, Mullholland J, Botstein D, Brown PO (1998). *Science*, 282:699.

Cui XQ, Kerr MK, Churchill GA (submitted for publication). Data transformations for normalization of cDNA microarray data.

Dudoit S, Yang YH, Speed TP, Callow MJ (2002). Statistical methods for identifying differentially expressed genes in replicated cDNA microarray experiments. *Statistica Sinica*, 12:111–139.

Felsenstein J (1985). Confidence limits on phylogenies—An approach using the bootstrap. *Evolution* 39:783.

Jin W, Riley RM, Wolfinger RD, White KP, Passador-Gurgel G, Gibson G (2001). The contribution of sex, genotype and age to transcriptional variance in *Drosophila melanogaster*. *Nature Genetics*, 29:389.

Kerr MK, Churchill GA (2001a). Experimental design for gene expression microarrays. *Biostatistics*, 2:183.

Kerr MK, Churchill GA (2001b). Statistical design and the analysis of gene expression microarray data. *Genetical Research*, 77:123.

Kerr MK, Churchill GA (2001c). Bootstrapping cluster analysis: Assessing the reliability of conclusion from microarray experiments. *Proceedings of the National Academy of Sciences USA*, 98:8961.

Kerr MK, Martin, M., Churchill GA (2000). Analysis of variance for gene expression microarray data. *Journal of Computational Biology*, 7:819.

Kerr MK, Leiter EH, Picard L, Churchill GA (2002). Sources of variation in microarray experiments. In: *Computational and Statistical Approaches to Genomics*, Zhang I, Shmulevich I (Eds), p 41. Kluwer Academic Publishers: Amsterdam.

Lönnstedt I, Speed TP (2002). Replicated microarray data. *Statistica Sinica*, 12:31.

Margush T, McMorris FR (1981). Consensus n-trees. *Bulletin of Mathematical Biology*, 43:239.

Pritchard CC, Hsu L, Delrow J, Nelson PS (2001). Project normal: Defining normal variation in mouse gene expression *Proceedings of the National Academy of Sciences USA*, 98:13266.

Rocke DM, Durbin B (2001). A model for measurement error for gene expression arrays. *Journal of Computational Biology*, 8:557.

Schena M (Ed) (2000). *DNA Microarrays: A Practical Approach*. Practical Approach Series 205. Oxford University Press: Oxford.

Seber GAF (1977). *Linear Regression Analysis*. Wiley: New York.

Tanner JM (1949). Fallacy of per-weight and per-surface area standards, and their relation to spurious correlations. *Journal of Applied Physiology*, 2:1.

Tusher VG, Tibshirani R, Chu G (2001). Significance analysis of microarrays applied to the ionizing radiation response. *Proceedings of the National Academy of Sciences USA*, 98:5116.

Westfall PH, Young SS (1993). Resampling-based multiple testing: Examples and methods for p-value adjustment. Wiley Series in Probability and Mathematical Statistics, Wiley: New York.

Wolfinger RD, Gibson G, Wolfinger ED, Bennett L, Hamadeh H, Bushel P, Ashfari C, Paules RS (2001). Assessing gene significance from cDNA microarray expression data via mixed-models. *Journal of Computational Biology*, 8:625.

Yang YH, Buckley MJ, Dudoit S, Speed TP (2002). Comparison of methods for image analysis on cDNA microarray data. *Journal of Computational Graph Statistics*, 11:108.

Yang YH, Dudoit S, Luu P, Lin DM, Peng V, Ngai J, Speed TP. (2001). Normalization for cDNA microarray data: A robust composite method addressing single and multiple slide systematic variation. *Nucleic Acids Research*, 30:e15.

15

GeneClust

KIM-ANH DO
BRADLEY BROOM
SIJIN WEN

Abstract

Two-way clustering techniques—such as hierarchical clustering, K-means clustering, tree-structured vector quantization, self-organizing maps, and principal components analysis—have been used to organize genes into groups or "clusters" with similar behavior across relevant tissue samples or cell lines. However, these procedures seek a single global reordering of the samples or cell lines for all genes, and although they are effective in uncovering gross global structure, they are much less effective when applied to more complex clustering patterns (for example, where there are overlapping gene clusters). This chapter describes *gene shaving* (Hastie et al., 2000), a simple but effective method for identifying subsets of genes with coherent expression patterns and large variations across samples or conditions. After summarizing the gene-shaving methodology, we describe two software packages implementing the method: a small package written in S (usable in either S-Plus or R) and a considerably faster, mixed-language implementation with a graphical user interface intended for more applied use. The package can perform unsupervised, fully supervised, or partially supervised gene shaving, and the user is able to specify various parameters pertinent to the algorithm. The package outputs graphical representations of the extracted clusters (as colored heat maps) and diagnostic statistics. We then demonstrate how the latter tool can be used to analyze two published datasets (the Alon colon data and the NCI60 data).

15.1 Introduction

Two-way clustering techniques have been explored by many researchers in the field of bioinformatics to organize genes into groups or "clusters" with similar behavior across relevant tissue samples or cell lines. Such methods include hierarchical clustering, K-means clustering, tree-structured vector quantization, self-organizing maps, and principal components analysis, to name a few. These procedures seek global organization of genes and samples and are effective in uncovering gross global structure by seeking a single reordering of the samples or cell lines for all genes. However, the power of these methods is challenged by more complex clustering patterns. For example, distinct gene groups may cluster the samples in different ways, or there may exist overlapping gene clusters where the genes in cluster C_1 may suggest discrimination between cancer groups G_1 and G_2, say, while the genes in cluster C_2, possibly including some genes in C_1, may suggest a different way of distinguishing between cancer groups G_3 and G_4.

Gene shaving (Hastie et al., 2000) is a simple but ingenious method that was proposed with the aim of resolving these issues. The development of the gene-shaving methodology was motivated by the research goal of identifying distinct sets of genes whose variation in expression could be related to a biological property of the tissue samples.

15.2   Methods

Let $X = x_{ji}$ be a row-centered $J \times I$ matrix of real-valued measurements representing the gene expression matrix, assuming no missing values. The rows are genes, the columns are tissue samples or cell lines, and x_{ji} is the measured (log) expression relative to a baseline.

15.2.1   Algorithm

Gene shaving is an iterative algorithm based on the principal components or the singular value decomposition (SVD) of the data matrix. It starts with the entire microarray gene expression matrix X and seeks a function F of the genes in the direction of maximal variation across the tissue samples. The general algorithm for gene shaving may be described in the following steps:

- *Step 1:* Calculate the simplest form of the function F as a normalized linear combination of the genes weighted by its largest principal component loadings, referred to as the *super gene*. The genes may be sorted according to the principal component weights.
- *Step 2:* A fraction α of the genes having lowest correlation (essentially the absolute inner product) with the *super gene* are then *shaved* off (discarded) from the original data matrix.
- *Step 3:* The process of calculating the leading principal component and shaving off some genes is iterated on the reduced data matrix until only two genes remain. This iterative top-down process produces a sequence of nested gene blocks of sizes ranging from the full set of J genes down to the final block consisting of just two genes.

- *Step 4:* Select the optimal cluster size based on a quality measure. In particular, to select the optimal number of genes in the cluster, we use a *Gap* function, which is based on the variances between and within the gene blocks computed from the raw data matrix and its permutation. The number of permutations should be specified a priori.
- *Step 5:* Remove the effect of genes in the optimal cluster, C_1 say, from the original matrix X. By computing the average gene or the vector of column averages for C_1, denoted by $\bar{C}_1$, we can remove the component that is correlated with this average. This is equivalent to regressing each row of X on the average gene row $\bar{C}_1$ and replacing the former with the regression residuals. Such a process is referred to as *orthogonalization* (Hastie et al., 2000), from which a modified data matrix X_{ortho} is produced.
- *Step 6:* With X_{ortho}, the whole process of calculating the leading principal component, producing another nested sequence of shaved gene blocks, applying the Gap statistic to obtain the next optimal cluster C_2, and orthogonalizing the current data matrix is repeated. This sequence of operations is iterated until M gene clusters $C_1, \ldots, C_M$ are found. The number of clusters to find, M, should be specified a priori.
- *Step 7:* For visual inspection, display graphically the derived M gene-shaving clusters and corresponding variance and Gap plots.

To allow negatively correlated genes to be included in a cluster, the average gene is actually a *signed mean gene*; that is, if a gene row has a negative principal component weight, then the signs of the expression values are flipped before the average is calculated. The shaving process requires repeated computation of the largest principal component of a particular data matrix X or its subset (after at least one step of shaving). This process is easily implemented using the singular value decomposition $U\Lambda V'$ of X, where U is a $J \times I$ matrix with orthonormal columns, $\Lambda = \text{diag}(\sqrt{\lambda_1}, \ldots, \sqrt{\lambda_I})$, where $\lambda_1 \geq \lambda_2 \geq \ldots \geq \lambda_r > \lambda_{r+1} = \ldots = \lambda_I = 0$, and V is an $I \times I$ orthogonal matrix. The columns of U are eigenvectors of XX', and the columns of V are the eigenvectors of $X'X$. Since the eigenvector corresponding to the largest eigenvalue of XX' is the first column of U, its elements form the loadings corresponding to the first principal component of XX'. At each step of the shaving process, it is not required to compute the complete SVD of the data matrix. Computational efficiency can be enhanced by deriving only the first column of U.

15.2.2   Choice of Cluster Size via the Gap Statistic

One important goal for any clustering technique is the ability to assess whether the extracted cluster is real; that is, we should be able to distinguish real patterns from random small clusters. Tibshirani and colleagues investigated a somewhat similar problem: estimation of the false-discovery rate in the context of detecting differential gene expression in DNA microarrays (Tusher et al., 2001); see also Storey and Tibshirani, Chapter 12 in this volume. In the context of gene shaving, the *Gap statistic* was devised for selecting a reasonable cluster size from the sequence of nested gene blocks (Hastie et al., 2000). Further theoretical and simulation results (Tibshirani et al., 2001) showed that the Gap statistic usually outperforms other methods proposed in the literature in its ability to estimate the actual number of clusters or groups in a set of data. The Gap test is an adaptation of the usual permutation test based on randomization and an appropriate definition of a quality measure, or test statistic, of each cluster. In other words, under the null hypothesis that the rows and columns of the data matrix are independent, the permuted version of X, obtained by permuting the elements within each row of X, should "look" the same as the original X. Thus, if we obtain many permutations X^* of X and compute any test statistic $T(X^*)$ repeatedly, we can obtain the *null distribution* of T. If the independence (or no structure) hypothesis were true, we would expect T to behave as we see in the repeated permutations. For our purposes here, the definition of T should reflect our ultimate goal of selecting clusters that simultaneously exhibit large variances between samples and high similarity between gene rows in the cluster. Established methods for a two-way analysis of variance can be adapted to our situation where the choice of a cluster size is governed by some function of the *between-to-within* variance ratio. After s shavings, let B_s denote the resulting gene block of k rows with elements x_{ji}. Define the *percent variance explained* as

$$R^2(B_s) = 100 \times \frac{V_B}{V_B + V_W} = 100 \times \frac{V_B/V_W}{1 + V_B/V_W}$$

where V_W, V_B and V_T denote the within, between, and total variances for the gene block B_s. The range of R^2 is over the interval $[0, 1]$, where values close to 0 imply no clustering evidence while values closer to 1 imply tight clusters of similarly expressing genes. R^2 for a gene block of size k_s can be explicitly computed by the following formulas:

$$V_W = \frac{1}{I} \sum_{i=1}^{I} \left[\frac{1}{k_s} \sum_{j \in B_s} (x_{ji} - \overline{x}_{.i})^2 \right], \qquad (15.1)$$

$$V_B = \frac{1}{I} \sum_{i=1}^{I} (\overline{x}_{.i} - \overline{x}_{..})^2, \qquad (15.2)$$

$$V_T = V_B + V_W = \frac{1}{k_s I} \sum_{j \in B_s} \sum_{i=1}^{I} (x_{ji} - \overline{x}_{..})^2. \qquad (15.3)$$

The *Gap function* for the gene block B_s of size k_s is defined as

$$Gap(k_s) = R^2_{k_s}(B_s) - \bar{R}^{2*}_{k_s}, \tag{15.4}$$

where $\bar{R}^{2*}_{k_s}$ denotes the averaged estimated value of the percent variance explained by blocks of size k_s computed from $NumPerm$ permutations.

The optimal cluster size is the value k_{opt} that maximizes the Gap statistic over all values of $k_s \in \{2, 3, \dots, N\}$. Implementation of the Gap statistic criterion is enhanced by plotting the percent variance curve $R^2_{k_s}(B_s)$ for the observed data matrix versus the corresponding averaged curve $\bar{R}^{2*}_{k_s}$ for the collection of permuted data matrices as a function of $k_s \in \{2, \dots, N\}$. Alternatively, one can also include a plot of the computed values of $Gap(k_s)$ against $k_s \in \{2, \dots, N\}$. Since the optimal cluster size usually assumes a small integer, these plots are more meaningful if depicted on the log scale.

The problem of estimating the number of clusters and finding the optimal cluster size in a dataset is a difficult one since there is no clear definition of a "cluster." Simulation studies have demonstrated that the Gap estimate is good for identifying well-separated clusters (Hastie et al., 2000). However, when data are not clearly separated into groups, different people might have different opinions about the number of distinct clusters. In practice, with DNA microarray data, the Gap curve plots often exhibit some flatness near the maximum or may have multiple peaks that are almost equivalent. A practicing biologist often wishes to explore further those genes next in line to the ones in the cluster corresponding to the estimated maximum Gap value. Therefore, graphing the Gap statistic with respect to increasing cluster size is important. Further to this aim of exploratory analysis, a more relaxed or flexible method for cluster size selection is to choose a cluster with a larger size than the optimal cluster but still maintaining a Gap statistic within a small percentage of the maximal Gap statistic, say 5%. In our implementations, we refer to this parameter as the *Gap tolerance*, where a Gap tolerance of 0 refers to cluster sizes corresponding to the maximum Gap estimate.

15.2.3 Supervised Gene Shaving for Class Discrimination

Either fully or partially supervised shaving with the ultimate aim of class discrimination can be carried out if the column (sample) grouping is known. Briefly, supervised shaving maximizes a weighted combination of column variance and an information measure that depends on the nature of the auxiliary information. Let B_k denote a generic cluster of k genes, where $B_k = x_{ji}(j = 1, \dots, k; i = 1, \dots I)$ with corresponding column average vector $\bar{x}_{B_k}$. Suppose that there are c classes or categories, where $c < I$. Let $Y = (y_1, \dots, y_I)$ denote a vector of integers indicating class membership for each sample. For example, if the data samples represent normals and tumors, then Y_j can take a value of 1 if the sample came from a normal tissue or a value of 2 if it was extracted from a tumor sample. Let class averages be denoted by $\overline{x}_{class} = (\overline{x}_{normals}, \overline{x}_{tumors})$, which represent the average expression of all genes in the gene block B_k summed over the class of normal or tumor tissues, respectively. Thus, a measure of class discrimination can be represented as $\mathrm{Var}(\overline{x}_{class})$. Under principal component shaving, it can be proven that

$$\mathrm{Var}(\overline{x}_{B_k}) = \mathbf{w}(X^\dagger X^{\dagger T})\mathbf{w},$$

where $\mathbf{w} = (w_1, \ldots, w_J)$ represent the principal component loadings associated with each gene in the original data matrix, and $X^\dagger$ is a $J \times c$ matrix whose rows are standardized versions of the class means for each gene. Define an information measure $\mathcal{J}_Y(\overline{x}_{B_k})$ to be an appropriate quadratic function of the column averages $\overline{x}_{B_k}$; then, the general supervised gene-shaving method is based on maximizing a weighted combination of the column means variance and the information measure

$$(1 - \omega)\mathrm{Var}(\overline{x}_{B_k}) + \omega\mathcal{J}_Y(\overline{x}_{B_k}) \qquad \omega \in [0, 1] \tag{15.5}$$

over all possible clusters of sizes ranging from 2 to J. Full supervision is equivalent to $\omega = 1$, while partial supervision is indicated by values of ω between 0 and 1. The equation (15.5) presents a compromise between supervised and unsupervised clustering. In the context of class discrimination, an indicator vector with the length of samples is needed to specify the external classification of the samples; the information measure is then taken as the sum of squared differences between the class averages. For a more detailed discussion of the methodology development and extension in the context of general supervision, see Hastie et al. (2000).

15.3   Software

15.3.1   The GeneShaving Package

We have implemented the gene-shaving method, for both unsupervised and supervised analyses, in an S language package we call GeneShaving. The source code is available from the StatLib S archive collection at http://lib.stat.cmu.edu/S/.

To illustrate the gene-shaving methodology, and in particular the use of the GeneShaving package, we present below a simple example involving simulated data. We generated a data matrix X of J genes and I samples. Each element x_{ji} consists of a simple signal s_{ji} that can take values $-H, 0, +H$ and noise $\epsilon_{ji} \sim N(0, 1)$. Specifically, for K an even integer less than J, let

$$x_{ji} = s_{ji} + \epsilon_{ji}; \tag{15.6}$$

$$\text{for } 1 \le j \le K/2 : s_{ji} = \begin{cases} -H & \text{if } i \le I/2 \\ +H & \text{if } i > I/2 \end{cases} \qquad (15.7)$$

$$\text{for } K/2 < j \le K : s_{ji} = \begin{cases} -H & \text{if } i \text{ odd} \\ +H & \text{if } i \text{ even} \end{cases} \qquad (15.8)$$

$$\text{for } j > K : s_{ji} = 0. \qquad (15.9)$$

For example, suppose that a data frame `demo.dat` is generated by taking $H = 2.5$, $J = 100$, $K = 40$, and $I = 40$.

```
> dim(demo.dat) [1] 100  40
# 100 genes (rows) and 40 samples (columns)
> demo.dat[(1:5), ]
# part of the data frame (first 5 genes)
```

```
          S1         S2         S3         S4         S5
G1 -2.207234 -1.480847 -2.177457 -4.260801 -3.407388
G2 -3.793671 -2.979609 -3.069576 -3.474391 -1.611161
G3 -2.108358 -3.166314 -3.811151 -4.053201 -3.869151
G4 -3.112067 -2.992520 -2.024178 -3.646037 -4.109120
G5 -4.371155 -5.390998 -1.957976 -3.270669 -2.245929
```

```
          S36      S37      S38      S39      S40
     5.052145 4.323419 2.703080 2.548090 4.489734
     1.713629 3.588597 2.443339 3.032010 2.852629
. . . 2.743546 3.445167 4.670663 2.045603 2.986980
     2.964064 1.041393 2.827109 3.781293 2.153697
     2.241520 3.447944 3.595235 4.946946 2.183231
```

Its heat map on the gray scale, Figure 15.1, was produced by the `image()` function in S-Plus.

The S-Plus function `shave()` implements *unsupervised gene shaving*. It has three parameters:

- **X**: the gene expression microarray data. It is a data frame with gene names (rows) and sample names (columns).
- **NumClusts**: the number of clusters to extract.
- **NumPerm**: the number of permutations used to select the optimal cluster size by the *Gap* function.

For instance, to extract two gene clusters using five permutations, the inputs of `shave()` are

1. the S-Plus data frame `demo.dat`,
2. the number of clusters (i.e., 2), and
3. the number of permutations (i.e., 5).

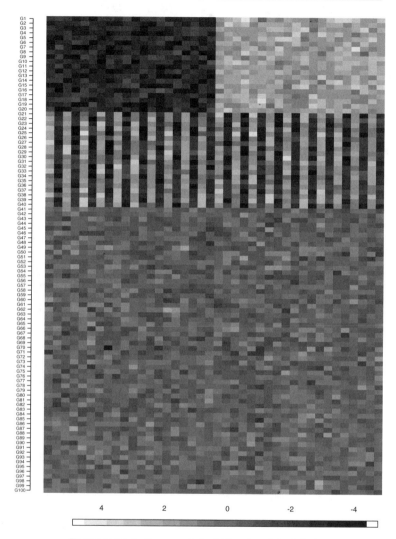

FIGURE 15.1. Image plot of the simulated data.

The shave() function returns the gene labels belonging to each extracted cluster.

```
> shave(demo.dat,2,5)
## this is the output of gene labels for cluster #1:
 [1] "G1"  "G2"  "G3"  "G4"  "G5"  "G6"  "G7"
 [8] "G8"  "G9"  "G10" "G11" "G12" "G13" "G14"
[15] "G15" "G16" "G17" "G18" "G19" "G20"
```

```
## this is the output of gene labels  for cluster #2:
 [1] "G21" "G22" "G23" "G24" "G25" "G26" "G27"
 [8] "G28" "G29" "G30" "G31" "G32" "G33" "G34"
[15] "G35" "G36" "G37" "G38" "G39" "G40"
```

The shave() function also outputs, for each extracted cluster, three plots:

* the percent variance explained by the actual and permuted data,
* a gap plot,
* a cluster plot (heat map) of the genes.

Figure 15.2 shows the plots output by shave() for the two clusters extracted from the simulated data.

Suppose that instead of unsupervised shaving we wish to perform a supervised shaving procedure. Perhaps the 40 samples (columns) in demo.dat are classified into two groups, with the first 20 samples in the first group and the last 20 samples in the second group. Thus, an indicator vector can be constructed and is coded 1 for the first 20 samples (columns) and coded 2 for the last 20 samples (columns):

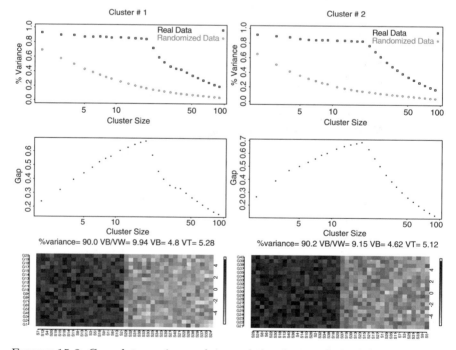

FIGURE 15.2. Gap plots, variance plots, and the clusters derived from applying unsupervised gene shaving to the simulated data.

```
>clf <- c(rep(1,20),rep(2,20)) #indicator vector
> clf
 [1] 1 1 1 1 1 1 1 1 1 1 1 1 1 1 1 1 1 1 1 1
[11] 2 2 2 2 2 2 2 2 2 2 2 2 2 2 2 2 2 2 2 2
```

Since there are only two classes of samples, we should only try to find a single cluster. In the general scenario of more $g > 2$ classes, the number of clusters to consider can be greater than g to cover both the main effects and the possible interactions between the different classes.

In the GeneShaving package, the S-Plus function super.part() implements supervised gene shaving. The outputs for super.part() are the same as those in shave(); however, the inputs are slightly different:

- **X**: gene expression microarray data. It is a data frame in S-Plus with gene names (rows) and sample names (columns).
- **clf**: an indicator vector to indicate the sample classification.
- **lam**: supervision parameter—fully supervised shaving is carried out if lam = 1; partially supervised shaving is carried out if lam takes a value between 0 and 1.
- **NumClusts**: the number of clusters to extract.
- **NumPerm**: the number of permutations for selecting the optimal cluster by the Gap function.

```
## fully supervised shaving with 5 permutations
## 2 clusters to extract

> super.part(demo.dat,clf,1,2,5)

## this is the output of gene labels for cluster #1:
 [1] "G1"  "G2"  "G3"  "G4"  "G5"  "G6"  "G7"  "G8"
 [9] "G9"  "G10"  "G11" "G12" "G13" "G14" "G15" "G16"
[17] "G17" "G18" "G19" "G20"
## this is the output of gene labels  for cluster #2:
 [1] "G21" "G23" "G30" "G40" "G69"
```

The outputs including Gap plots, variance plots, and heat maps of the derived clusters under full supervision are shown in Figure 15.3. Observe that the first cluster gives a perfect discrimination of the samples (samples S1–S20 versus samples S21–S40). The second cluster is orthogonal to the first derived cluster and groups the even-numbered samples together versus the odd-numbered samples. Since full supervision was employed, the whole process was dominated by the external classification pattern (captured by cluster 1), and only the strongest genes for the alternative clustering pattern emerged in cluster 2, suggesting that further exploration is required here, such as a partially supervised analysis.

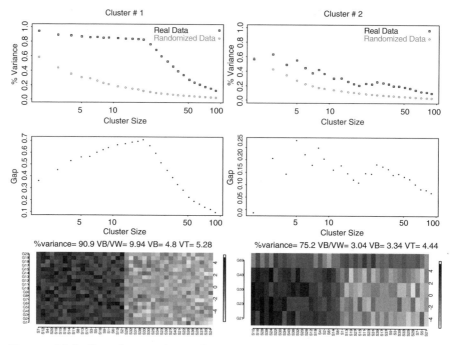

FIGURE 15.3. Gap plots, variance plots, and the clusters derived from applying fully supervised gene shaving to the simulated data.

15.3.2   GeneClust: A Faster Implementation of Gene Shaving

The GeneShaving package described above has two significant limitations that prevent its routine use in clinical data analyses:

- it is far too slow, and
- it is only usable by S programmers.

We have developed the GeneClust software package to address these issues.
GeneClust has a graphical user interface (GUI) written in Java. Figure 15.4 shows the initial GUI. The GUI allows users to

- perform a simple one-way hierarchical clustering by genes or by samples;
- perform unsupervised, fully supervised, or partially supervised gene shaving; if the latter is chosen, the user can also select the amount of partial supervision;
- specify the number of clusters to extract, the percent to shave for each iteration, the number of permutations used to calculate the Gap statistic, and the level of Gap tolerance.

GeneClust may be used either to analyze a real dataset by selecting the *Raw Data* mode or to investigate the performance of gene shaving for simulated

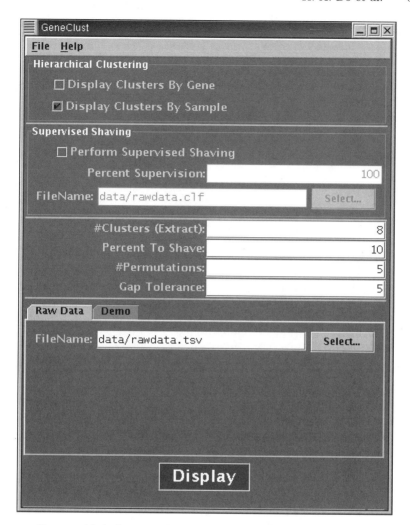

FIGURE 15.4. A screen dump of the GeneClust GUI template.

datasets that are variations of the model (15.6) by selecting the *Demo* mode.

When the user starts the gene-shaving procedure, the GUI invokes the back-end statistical analysis process. Because this is an S-Plus (or R) application, with which the GUI communicates using a pseudoterminal, either S-Plus or R must be installed (see Parmigiani et al., Chapter 1, this volume). The computationally intensive gene-shaving algorithm is implemented using C and is dynamically loaded into S-Plus (or R) to perform the analysis. After the clusters have been extracted, the S-Plus (or R) application presents the analysis results graphically.

The Geneclust software has been implemented for the Solaris and Linux operating systems and for the S-Plus and R statistical programming environments. For exact details and continuous updates, check the Web site

http://odin.mdacc.tmc.edu/~kim/geneclust.

15.4 Applications

We demonstrate how GeneClust can be employed in the analyses of two real published datasets: a colon dataset (Alon et al., 1999) and the human tumor NCI60 dataset (Ross et al., 2000; Scherf et al., 2000).

15.4.1 The Alon Colon Dataset

Alon and colleagues used Affymetrix oligonucleotide arrays to monitor absolute measurements on expressions of over 6500 human gene expressions in 40 tumor and 22 normal colon tissue samples (Alon et al., 1999). These samples were taken from 40 different patients so that 22 patients supplied both a tumor and normal tissue sample. They focused on the 2000 genes with the highest minimal intensities across the samples. There was a change in the protocol during the conduct of the microarray experiments (Getz et al., 2000). Tumor and normal tissue samples were taken from the first 11 patients using a poly detector, while the remaining tumor and normal tissue samples were taken from the remaining 29 patients using total extraction of RNA.

Before performing any clustering of this set, we processed the data by taking the (natural) logarithm of each expression level in X. Then, each column of this matrix was standardized to have mean zero and unit standard deviation. Finally, each row of the consequent matrix was standardized to have mean zero and unit standard deviation.

We then selected a 446-gene reduced Alon dataset, as described in McLachlan, Bean and Peel (2002), by selecting a subset of relevant genes based on the likelihood ratio statistic $-2 \log \lambda$ calculated from fitting a single t distribution versus a mixture of two t components. The relevance of each of the J genes was assessed by fitting one- and two-component t mixture models to the expression data over the I tissues for each gene considered individually. If $-2 \log \lambda$ is greater than a specified threshold b_1,

$$-2 \log \lambda > b_1, \tag{15.10}$$

then the gene is taken to be relevant provided that

$$s_{min} \geq b_2, \tag{15.11}$$

where s_{min} is the minimum size of the two clusters implied by the two-component t mixture model and b_2 is a specified threshold. Following McLachlan et al. (2002), we chose $b_1 = 8$ and $b_2 = 8$, thus retaining the reduced Alon microarray data matrix X that has $J = 446$ rows and $I = 62$ columns. We have rearranged the data consecutively so that the first 40 columns correspond to tumor samples followed by 22 columns of normal tissues. Further, the first 11 tumor columns are matched with the first 11 normal columns by patient. Figure 15.5 displays the observed gene expression matrix X of 446 genes.

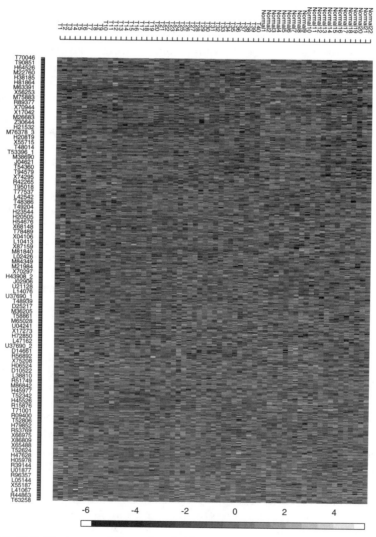

FIGURE 15.5. Heat map of the reduced Alon colon expression array of 446 genes.

We investigated the performance of unsupervised gene shaving when applied to the reduced Alon dataset of 446 genes. We applied GeneClust with a Gap tolerance of 5%, which allows us to pick larger cluster sizes than those dictated by the maximum Gap criterion. The first four orthogonal unsupervised clusters along with their corresponding variance and Gap plots are presented in Figures 15.6 (see color insert) and 15.7. The columns in the heat maps are sorted according to the column averages. Estimated eigenvalues, R^2, and between-to-within variance ratios are also printed beneath the heat map for each cluster. The first cluster corresponds to a relatively small between-to-within variance ratio of 1.72 and does not seem to reflect any pattern resembling the normal versus tumor or change in paradigm external classification. It can be seen, however, that cluster 2 (19 genes) consists of two negatively correlated subclusters that capture the normal versus tumor structure quite well, showing genes that are either underexpressed or overexpressed for most of the normal tissues (clustered to the right). In contrast, the change in paradigm structure is reflected in cluster 3 (seven genes), where most of the normal and tumor tissues corresponding to the first 11 patients are clustered to the right. A number of muscle-specific genes have been identified as being characteristic of normal colon samples (Ben-Dor et al., 2000). We note that two of these genes (J02854, T60155), along with two suspected smooth-muscle genes (M63391, X74295), are included in the second cluster. Clusters 2 and 3 also correspond to high R^2 estimates of around 70% or more and large values of V_B/V_W (2.3 and 5.6, respectively). Inspection of the variance and Gap curves can also be informative. The variance curves are well-separated for all four clusters, indicating that the clusters found are not entirely random. Except for cluster 3, with a clear maximum at a cluster size of 9, the Gap curves for the other clusters exhibit some flatness to the right of the maximum, suggesting that it is worth exploring additional genes in slightly larger cluster sizes than those presented here.

15.4.2 The NCI60 Dataset

The second dataset that we analyzed is the cDNA microarray gene expression data collected by the National Cancer Institute's Developmental Therapeutics Program (DTP) (Ross et al., 2000). The gene expression represents 60 cancer cell lines (the NCI60) and contains 9703 spotted cDNA sequences. The cell lines are derived from tumors with different sites of origin: 7 breast, 5 central nervous system (CNS), 7 colon, 6 leukemia, 8 melanoma, 9 non-small-cell lung carcinoma (NSCLC), 6 ovarian, 2 prostate, 9 renal, and 1 unknown. The fluorescent cDNA targets were prepared from an mRNA sample using red dye Cy5, while the reference sample used green dye Cy3. All hybridizations were prepared by pooling equal mixtures of mRNA from 12 of the cell lines. Independent microarray experiments using a leukemia cell line (K562) and a breast cancer cell line (MCF7) were each grown in

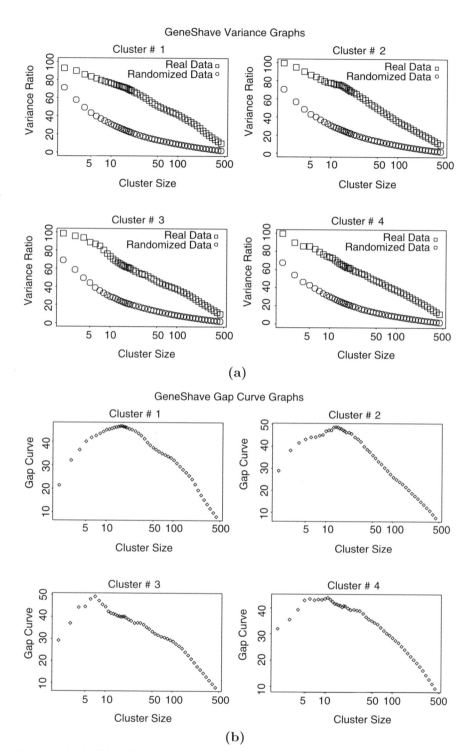

FIGURE 15.7. Alon data set. (a) Variance plots for the original and randomized data, with the percent variance explained by each cluster both for the original data and for an average over 20 randomized versions. (b) Gap estimates of cluster size. The Gap curve corresponds to the difference between the pair of variance curves.

independent cultures to investigate the reproducibility of the entire experiment. As in Dudoit et al. (2001), we performed some preprocessing steps including: (i) from the 60 tumor cell lines, we excluded the two prostate cancer cell-line observations and the unknown cell-line observation; the remaining are labeled BREAST-BREAST7, CNS-CNS5, COLON-COLON7, LEUK-LEUK6, MELA-MELA8, NSCLC-NSCLC9, OVAR-OVAR6, and RENAL-RENAL9; (ii) retained four of the independent cell-line experiments (labeled K562A, K562B), (iii) screened out genes with more than two missing data points; (iv) for genes with two or fewer missing data points, imputation was performed by replacing the missing entry with the average of the corresponding entries from five nearest-neighboring genes (in terms of correlation). Our final matrix X thus consists of 5244 genes (rows) and 61 samples (columns), where x_{ji} denotes the natural logarithm of the red/green fluorescence ratio for gene j in sample i. Figure 15.8 displays the observed gene expression matrix X.

A fully supervised gene-shaving analysis of the NCI dataset was performed. The first four supervised clusters obtained from running GeneClust with a 5% Gap tolerance based on five permutations are presented in Figure 15.9 (see color insert). Cluster 1 (86 genes) exhibits two subclusters that group most of the leukemia samples to the left while the majority of the renal tissue samples cluster at the rightmost end of the heat map. Cluster 2 (33 genes), orthogonal to cluster 1, found a different set of genes that specifically overexpress or underexpress for all melanoma tissues and several breast tissues. Five genes that specifically overexpress for colon and NSCLC cancers are displayed in cluster 3. The fourth cluster consists of only three genes that seem to express highly for the two breast cell lines and several breast tissues.

15.5 Discussion

We have presented a brief description of the methodological underpinnings and implementation details of our software GeneClust for the clustering of microarray gene expression data by the gene-shaving approach. We illustrated the simplicity of usage of our software via some simple tutorial exercises based on simulated data as well as its application to two real published datasets. In practice, scientific investigators often also wish to evaluate how much of the dimensionality of the gene expression is captured by the first few clusters derived from any specific method used. For gene shaving, the expression profile for each gene in the original complete dataset can be explained as a linear combination of the super genes from each gene-shaved cluster. Alternatively, one can also compute the relative sum of squares total for each cluster relative to that of the complete dataset as long as the number of overlapping genes between clusters is small. The percent variance explained by the first K clusters from gene-shaving can

FIGURE 15.8. The NCI60 human tumor expression array.

be compared with those obtained from other methods, such as a full principal component analysis, by appropriate plots (Hastie et al., 2000). In the group discrimination context, the error rates can also be compared between methods. The gene-shaving method finds clusters of genes in which the gene expression varies greatly over the tissue samples while maintaining a high level of coherence between the gene expression profiles. The extracted clusters are orthogonal to each other and are of varying sizes so that once a specific structure is captured in one cluster, the same structure will no longer be captured in subsequent clusters. However, overlapping genes between clusters are allowed if such genes induce different groupings of the columns (tissue samples). Gene shaving is also flexible in allowing the user to apply any amount of supervision required during the data analysis process. GeneClust is implemented with this flexibility in mind and allows interaction with the user to choose the Gap tolerance level as well as the amount of supervision. GeneClust is at least a hundredfold faster than the GeneShaving package implemented fully in S and hence is recommended for real-time interactive analyses of large microarray data.

Acknowledgments. We would like to thank the programmers, Paul Roebuck and Rumiana Nikolova, for programming assistance and the development and maintenance of the GeneClust Web site. Keith Baggerly provided many insightful discussions. This work was partially funded by the National Institute of Health via the University of Texas SPORE in Prostate Cancer (CA90270) and the Early Detection Research Network (Grant CA99007).

References

Alon U, Barkai N, Notterman DA, Gish K, Ybarra S, Mack D, Levine AJ (1999). Broad patterns of gene expression revealed by clustering analysis of tumor and normal colon tissues probed by oligonucleotide arrays. *Proceedings of the National Academy of Sciences USA*, 96:6745–6750.

Ben-Dor A, Bruhn L, Friedman N, Nachman I, Schummer M, Yakhini Z (2000). Tissue classification with gene expression profiles. *Journal of Computational Biology*, 7:559–584.

Dudoit S, Fridlyand J, Speed TP (2001). Comparison of discrimination methods for the classification of tumors using gene expression data. *Journal of the American Statistical Association*, 97(1):77–87.

Getz G, Levine E, Domany E (2000). Coupled two-way clustering analysis of gene microarray data. *Cell Biology*, 97:12079–12084.

Hastie T, Tibshirani R, Eisen MB, Alizadeh A, Levy R, Staudt L, Chan W, Botstein D, Brown P (2000). 'Gene shaving' as a method for identifying distinct sets of genes with similar expression patterns. *Genome Biology*, 1:research0003.1–0003.21.

McLachlan GJ, Bean RW, Peel D (2002). EMMIX-GENE: A mixture-model based approach to the clustering of microarray expression data. *Bioinformatics*, 18:413–422.

Ross DT, Scherf U, Eisen MB, Perou CM, Reese C, Spellman P, Iyer V, Jeffrey SS, de Rijn MV, Waltham M, Pergamenschikov A, Lee JCF, Lashkari D, Shalon D, Myers TG, Weinstein JN, Botstein D, Brown PO (2000). Systematic variation in gene expression patterns in human cancer cell lines. *Nature Genetics*, 24:227–235.

Scherf U, Ross DT, Waltham M, Smith LH, Lee JK, Tanabe L, Kohn KW, Reinhold WC, Myers TG, Andrews DT, Scudiero DA, Eisen MB, Sausville EA, Pommier Y, Botstein D, Brown PO, Weinstein JN (2000). A gene expression database for the molecular pharmacology of cancer. *Nature Genetics*, 24:236–244.

Tibshirani R, Walther G, Hastie T (2001). Estimating the number of clusters in a dataset via the gap statistic. *Journal of the Royal Statistical Society B*, 63:411–423.

Tusher VG, Tibshirani R, Chu G (2001). Significance analysis of microarrays applied to the ionizing radiation response. *Proceedings of the National Academy of Sciences USA*, 98(9):5116–5121.

16

POE: Statistical Methods for Qualitative Analysis of Gene Expression

ELIZABETH S. GARRETT
GIOVANNI PARMIGIANI

Abstract

In many gene expression studies, the goals include discovery of novel biological classes and identification of genes whose expression can reliably be associated with these classes. Here we present a statistical analysis approach to facilitate both of these goals. The key idea is to model gene expression using latent categories that can be interpreted as a gene being turned "on" or "off" compared to a baseline level of expression. This three-way categorization is used for defining a reference in the unsupervised setting, for removing noise prior to clustering, for defining molecular subclasses in a way that is portable across platforms, and for defining easily interpretable probability-based distance measures for visualization, mining, and clustering.

16.1 Introduction

Molecular class discovery can be defined as investigating a set of previously unclassified subjects from a population to find subgroups that share similar molecular profiles (Duggan et al., 1999). The most common application to date is to data consisting of sets of tumors that are not distinguishable morphologically (Golub et al., 1999; Alizadeh et al., 2000; Bittner et al., 2000; Bhattacharjee et al., 2001; Garber et al., 2001). Subclasses could be defined using genome-wide profiles, but it is often more practical to identify a set of genes that show differential expression across classes. This allows for biological characterization of the classes, for out-of-sample prediction across platforms, and for confirmatory work using alternative gene expression measurement techniques such as RT-PCR. In describing genomic variability at the population level, it can also be important to define subclasses in terms of differential expression of these genes compared to the typical level in the population.

This chapter describes a statistical approach and software tools to support this type of unsupervised molecular classification. The tools were developed with cancer-related applications in mind but are more generally applicable. Our starting point is a probabilistic definition of differential expression for the unsupervised setting. This is used to define molecular profiles and assess quantities of potential use in classification, such as the probability that an individual belongs to a given profile and the probability that a group of individuals has the same given profile. The detailed context, motivation, and application to the classification of ductal breast cancers using this approach are given by Parmigiani et al. (2002). Successful molecular classification so far has been characterized by a combination of visualization, formal quantitative analysis, and informal a priori information on gene function. Our analysis tools are not intended to replace, but to facilitate, the extensive manual work required to develop novel subclasses.

Exploration of molecular profiles using the tools described in this chapter proceeds along the following lines.

- Identification, for each gene, of a baseline or modal class, using statistical modeling.

- Estimation, for each gene-subject combination, of whether gene expression is at baseline, overexpressed, or underexpressed and derivation of probabilities of these differential expressions. These probabilities can offer an effective way to stabilize the measurements by eliminating a large portion of the noise and the hard-to-cluster variation associated with extreme gene expression values. At the same time, they provide an interpretable scale for classification of tumors to patterns.

- Mining for candidate genes to predict classes using the probability that they follow a pattern of specified characteristics. Mining produces potentially overlapping groups of genes with similar patterns.

- Visualization of probabilities that subjects belong to a specified profile, defined in terms of a small set of interpretable genes. Not all samples need to be assigned to a class.

Once a small set of s genes is identified, a molecular profile can be defined as an s-dimensional vector, with each coordinate taking one of three values (overexpressed, underexpressed, or expressed as baseline). This allows for a crisp and portable definition of a molecular profile. Also, in some datasets, the definition of a baseline or modal class allows us to model, and potentially remove, a noise component.

Model estimation is implemented using a Bayesian approach with Markov chain Monte Carlo (MCMC) evaluation of posterior quantities. We describe R software to carry out the estimation. The functions for performing the analysis require only a data matrix of expression values and a vector of gene identifiers. Functions are available for estimating the mixture model described below, producing graphical displays of the data and the model, and mining for candidate genes to predict classes.

16.2   Methodology

In this section, we present the gene expression mixture model and our Bayesian estimation approach. The basic assumption of the model is that for each gene that is related to tumor subtypes, gene expression will be different for tumors of different subtypes. For example, if there are k classes of tumors and gene j^* is related to tumor class, then the distribution of expression in gene j^* is different in at least one of the k classes as compared to the other classes. A general case that can be estimated for varying k assumes three qualitative categories of gene expression: a baseline level, underexpression, and overexpression. After assessing expression categories, genes and tumors are examined based on their patterns in terms of the qualitative categories. Visualization can be done in a variety of ways. We propose one method later in this chapter, but functions from other software packages can be used as well (Quackenbush, 2001). In this context, we assume that there is no phenotypic classification of samples. However, we are currently developing complementary software that fits models where either baseline data or phenotype data are available.

16.2.1   Mixture Model for Gene Expression

Our discussion begins with a mixture model for qualitative categories of gene expression. We describe this in the context of tumor classification (i.e., subjects are tumors) to make the notation and terminology consistent with our related work. Our gene expression data matrix, X, has J rows and I columns, corresponding to J genes and I tumors, $j = 1, \ldots, J; i = 1, \ldots, I$. Element x_{ji} represents the measured intensity or abundance of gene j in tumor i or perhaps a transformation of this quantity. For example, in two-channel cDNA arrays, logs of ratios of expression measure in tumors compared to a reference would be indicated. The matrix X is assumed to have been normalized and preprocessed so that experimental artifacts have been removed. Discussion of normalization are provided in Dudoit and Yang in Chapter 3 and Irizarry et al. in Chapter 4 of this volume.

We define three categories from which x_{ji} can arise and use e_{ji} to represent them:

$$e_{ji} = -1 \text{ gene } j \text{ has abnormally low expression in tumor } i;$$

$$e_{ji} = 0 \text{ gene } j \text{ has baseline expression in tumor } i;$$

$$e_{ji} = 1 \text{ gene } j \text{ has abnormally high expression in tumor } i.$$

Defining "baseline expression" in the unsupervised case is challenging because there may not be a biological reference for comparison. Our approach seeks to identify a class that includes a large number of cases and presents relatively low variability.

For each gene j, the distribution of x_{ji} given e_{ji} follows probability distribution $f_{e,j}$; that is,

$$x_{ji}|(e_{ji} = e) \sim f_{e,j}(\cdot), \qquad e \in \{-1, 0, 1\}.$$

We define the population proportion of overexpressed tumors in gene j to be π_j^+ and that of underexpressed tumors to be π_j^-. The proportion of all differentially expressed tumors is $\pi_j = \pi_j^+ + \pi_j^-$. The model also assumes that conditional on the π_j's and f's, the e_{ji}'s are independent.

A critical aspect of our approach is the transformation of the observed expressions to probabilities of latent categories. Using Bayes's rule, we can estimate for each data point x_{ji} the probability that it is overexpressed (p_{ji}^+) and the probability that it is underexpressed (p_{ji}^-) by

$$p_{ji}^+ = P(e_{ji} = 1|x_{ji}, \omega)$$

$$= \frac{\pi_j^+ f_{1,j}(x_{ji})}{\pi_j^+ f_{1,j}(x_{ji}) + \pi_j^- f_{-1,j}(x_{ji}) + (1 - \pi_j^+ - \pi_j^-)f_{0,j}(x_{ji})}$$

$$p_{ji}^- = P(e_{ji} = 1|x_{ji}, \omega)$$

$$= \frac{\pi_j^- f_{-1,j}(x_{ji})}{\pi_j^+ f_{1,j}(x_{ji}) + \pi_j^- f_{-1,j}(x_{ji}) + (1 - \pi_j^+ - \pi_j^-)f_{0,j}(x_{ji})}.$$

We use ω as shorthand for the full set of unknown parameters.

Our current software implementation uses uniform distributions ($\mathcal{U}$) for $f_{1,j}$ and $f_{-1,j}$ and a normal distribution ($\mathcal{N}$) for $f_{0,j}$ and has been successful in modeling gene expression in both simulated and real data examples of different cancer tumors. More specifically, we use

$$f_{-1,j}(\cdot) = \mathcal{U}(-\kappa_j^+ + \alpha_i + \mu_j, \alpha_i + \mu_j)$$

$$f_{0,j}(\cdot) = \mathcal{N}(\alpha_i + \mu_j, \sigma_j)$$

$$f_{1,j}(\cdot) = \mathcal{U}(\alpha_i + \mu_j, \alpha_i + \mu_j + \kappa_j^-).$$

There are, however, many alternative choices for the distributions f that may achieve the same goals.

In the equations above, $\alpha_j + \mu_j$ is the center of the distribution of baseline expression levels for gene j in tumor i, with μ_j measuring the gene effect in gene j for normal expression and α_i measuring the sample effect for normal expression levels in sample j. The choice of the normal distribution for the center distribution has worked well in practice and is consistent with the idea that, for each gene, there is at least a portion of tumors with very similar expression values. In these similar tumors, the differences in observed expression are primarily the result of noise introduced at the experimental and preprocessing stages.

In cancer applications, the uniform distribution may reflect cases where abnormally altered expression of a gene results from the failure of a biological mechanism that controls the expression level. In the absence of a control mechanism, high or low expressions may take a broad range of values. Statistically, the uniform distribution is appealing because it provides stable estimates of mixture model parameters, and no values are assigned very low density in the differentially expressed cases. The upper and lower limits are notated by κ_j^+ and κ_j^-. We add the constraint that $\kappa_j^+ > r\sigma_j$ and $\kappa_j^- > r\sigma_j$ to ensure that the uniform distributions capture differential expression. When this constraint is relaxed, it is possible for the uniform component to be assigned very low values of κ_j^+ and κ_j^- to fit broad shoulders of the baseline distribution. In our applications, we have chosen $r \geq 5$. The supports of $f_{-1,j}$ and $f_{1,j}$ are nonoverlapping, which simplifies the estimates of p_{jt}^+ and p_{jt}^- given above.

Examples of normal/uniform mixtures for finding outliers and sparse clusters are discussed by Fraley and Raftery (1998). For other examples of mixture modeling applied to microarray data see Lee et al. (2000); McLachlan et al. (2002); Yeung et al. (2001).

16.2.2   Useful Representations of the Results

For each observed expression level x_{ji}, we have two parameters to describe its expression according to our model: p_{ji}^+ and p_{ji}^-. We can define a new measure, $p_{ji}^* = p_{ji}^+ - p_{ji}^-$, which ranges from -1 to 1, with negative numbers indicating underexpression and positive numbers indicating overexpression. Using the estimated mixture model parameters, we can also use an approximate Bayes estimator to remove the estimated experimental noise in the data (i.e., to "shrink" the data). In this way, an observation x_{ji} that shows strong evidence of being from the normal component of the mixture (i.e., has small p_{ji}) is estimated to be close to $\mu_j + \alpha_i$, while an observation that has a large value of p_{ji} is estimated to be close to the observed value. To develop these estimates, we reexpress the model by introducing unknown quantitative expression values η_{ji} and defining $x_{ji} \sim N(\eta_{ji}, \sigma_j)$. The normal class is then defined by $\eta_{ji} = \mu_j + \alpha_i$ and σ_j unknown, the overexpressed class is $\eta_{ji} - \mu_j - \alpha_i \sim \mathcal{U}(0, \kappa_j^+)$, $\sigma_j = 0$, and the underexpressed class is $\eta_{ji} - \mu_j - \alpha_i \sim \mathcal{U}(\kappa_j^-, 0)$, $\sigma_j = 0$. The posterior means z_{ji} of the η_{ji} can be used as estimates of expression values:

$$E(\eta_{ji}|x_{ji}, \omega) = z_{ji} \approx \mu_j + \alpha_i + (x_{ji} - \mu_j - \alpha_i)p_{ji}$$

$$= p_{ji}x_{ji} + (1 - p_{ji})(\mu_j + \alpha_i). \qquad (16.1)$$

These are similar to multiple shrinkage estimates (George, 1986) and can be used as a "denoised" version of the original gene expression data. See Clyde et al. (1998) for an application to wavelets.

16.2.3 Bayesian Hierarchical Model Formulation

There are several ways in which the model proposed above can be estimated. We have chosen a Bayesian hierarchical model where we use a Metropolis–Hastings MCMC approach to obtain posterior distributions of the parameters. We assume that sample-to-sample variation of gene expression is described by the model in the previous section, and that the variation of gene-specific parameters can follow additional probability distributions. The benefit of this approach is that it borrows strength across genes using the the entire genomic distribution instead of fitting a separate, independent model for each gene. In addition, it is sensible to assume that the technological aspects of acquiring gene expression data affect many genes in similar ways.

Hierarchical Bayesian models have been shown to have appealing properties in estimation of large vectors of related quantities. An introduction and review are provided by Berger (1985). In Chapter 11 of this volume, Newton and Kendziorski take a similar approach, where they estimate probabilities of differential expression using parametric and semiparametric hierarchical approaches.

In our estimation approach, we use the following hierarchical distributions to describe the variation of parameters across genes:

$$\mu_j | \theta_\mu, \tau_\mu \sim \mathcal{N}(\theta_\mu, \tau_\mu),$$
$$\sigma_j^{-2} | \gamma, \lambda \sim \mathcal{G}(\gamma, \lambda),$$
$$\kappa_j^+ | \theta_\kappa^+ \sim \mathcal{E}(\theta_\kappa^+),$$
$$\kappa_j^- | \theta_\kappa^- \sim \mathcal{E}(\theta_\kappa^-),$$
$$\text{logit}(\pi_j^+) | \theta_\pi^+ \sim \mathcal{N}(\theta_\pi^+, \tau_\pi^+),$$
$$\text{logit}(\pi_j^-) | \theta_\pi^- \sim \mathcal{N}(\theta_\pi^-, \tau_\pi^-),$$

where $\mathcal{G}$ is the Gamma distribution and $\mathcal{E}$ is the exponential distribution. We assume that gene-specific parameters are independent conditional on the hyperparameters on the right-hand side of the distributions above. Hyperparameters can be assigned dispersed, noninformative priors because the large number of genes allows for data-driven estimation. An advantage of the hierarchical model is that for genes that show little or no evidence of differential expression (i.e., $\pi_j \approx 0$), there is essentially no information in the data with which to estimate the parameters associated with the over-expression and underexpression distributions. Improper priors would likely lead to identifiability issues, and in the maximum likelihood setting, additional constraints would need to be implemented to achieve identifiability of all parameters. We avoid these problems by borrowing strength from the genomic distribution.

In the current implementation, the sample effects, α_i's, are assumed to be independent, identically distributed according to a normal, with user-specified parameters. In practice, a hierarchical model can be applied to the sample effects as well. In our experience, the two approaches yield similar results when the number of genes is large.

In the MCMC, the data were augmented with a trichotomous indicator, e_{ji} for each data point, similarly to Diebolt and Robert (1994), and West and Turner (1994). To facilitate sampling of the κ's, we used the sequence $[\kappa|\omega^*]$ $[e|\kappa, \omega^*]$ $[\omega^*|\kappa, e]$, where symbols refer to parameter vectors or matrices, brackets refer to posterior distributions, and ω^* is ω with κ removed. Given the class indicators (e_{ji}'s), the full-conditional distribution of the π_j's is a Dirichlet distribution, and the full-conditional distribution of the parameters of the normal component is conjugate, with the additional constraint that $\sigma > r\max(\kappa_j^+, \kappa_j^-)$.

16.2.4   Restrictions to Remove Ambiguity in the Case of Only Two Components

Our model assumes that there are three gene expression categories and that the inner component represents the "modal" class. However, there will certainly be genes that appear to be suitably described by a two component model. Consider a gene in which 51% of the samples appear to be underexpressed, 49% appear to be normally expressed, and no samples appear to be overexpressed. This appears to be a two-component mixture model, yet we have overparameterized it with three components and so the labels we give to the classes are relatively arbitrary. The same data could very likely lead to the conclusion that 51% of samples are normally expressed and 49% of the samples are overexpressed. When the data naturally fall into only two categories of expression, this ambiguity arises. To alleviate the arbitrariness of the labels, we enforce two restrictions: $\pi^+ < 0.50$ and $\pi^- < 0.50$. Additionally, when one of the components of the mixture is very small (i.e., $\pi^+ < 1/N$ or $\pi^- < 1/N$ where N is the number of samples) we collapse our three component mixture to two components. In the relatively rare case when the size of these two components are within 5% of each other (i.e., both are close to 50%), we label the two components so they represent normal expression and overexpression.

16.2.5   Mining for Subsets of Genes

A useful step in molecular classification is identifying a subset of candidate genes for subsequent validation. The results of the mixture analysis can contribute to this process by providing the basis for visualization and mining tools that are intuitive and robust to noise. Some of these visualizations will involve the whole profile. A useful alternative to that is the visualization of a selected subset of candidate genes in concert with other information such as functional annotations (see also Bouton et al., Chapter 8 in this volume). Unlike clustering, the latter can be an iterative process in which several subsets are successively identified and examined based on various mining criteria.

Our approach is to identify a subset of k (where k is usually between 5 and 20) distinct genes that show strong evidence of differential expression across samples and to find genes with patterns similar to these k genes. Similar genes are pooled into gene groups, so that we have a resulting set of k groups, each formed by iteratively choosing a seed gene, then identifying a set of similar genes. The goals of displaying groups are to reduce redundancy in the set of seed genes, to provide additional context to facilitate interpretations of subtypes, and to make it less likely to select isolated artifacts. It is not critical that the gene groups contain all of the genes that are actually involved in a coregulated pattern. Although such a result would be of great biological interest, it is often too difficult a task for the types of datasets at hand. Likewise, it is not critical that the gene sets be disjoint. We describe our approach in steps as follows.

1. Choose the differential expression pattern of interest. The idea is to state a priori how many samples within a gene are expected to be underexpressed and how many overexpressed. For example, the pattern $\{0.05, 0.20\}$ indicates that 5% of genes should be underexpressed and 20% of genes should be overexpressed. The remaining 75% would then be in the "normal" component of the mixture.

2. Sort genes according to consistency with patterns defined in step 1. For each gene, using the estimates of p_{ji}^{+} and p_{ji}^{-}, we can calculate the probability that the samples have a pattern of the type specified. We sort genes by this probability.

3. Calculate a $J \times J$ matrix of gene agreement, where r_{jm} represents agreement between genes j and m:

$$r_{gm} = \sum_{i=1}^{I} (p_{gi}^{+} p_{mi}^{+} + p_{gi}^{-} p_{mi}^{-} + (1 - p_{gi})(1 - p_{mi})). \qquad (16.2)$$

4. Define gene "coherence" as the diagonal of the agreement matrix, that is, coherence for gene j is r_{jj}. This is the probability that the gene would have the same true profile as another gene with identical observed expression and measures the reliability with which samples can be assigned to classes for that gene. Genes with low coherence are highly noisy or poorly fit by the model. Identify genes as potential "seed" genes if their coherence is above a prespecified cutoff.

5. Choose the gene with the largest score from step 2 and that is sufficiently coherent as the seed gene.

6. Choose genes that show substantial agreement with the seed gene, either as a fixed agreement cutoff or as a proportion of coherence of the seed gene. Add these genes to the gene "group" that is seeded by the gene chosen in step 5.

7. Remove the genes in the group from consideration as seeds. Repeat steps 5 and 6 for the remaining groups to be identified.

The gene groups obtained above represent distinct sets of genes, with each group showing different expression patterns across genes. The creation of these groups achieves our goal of gene selection. These sets are, however, conditional on the pattern chosen in step 1 of the algorithm. It is important to consider several types of patterns and varying proportions for overexpressed and underexpressed genes.

16.2.6 Creating Molecular Profiles

We envision this approach to be carried out interactively with the help of expert input. If successful, it will identify a relatively small number s with which we can create molecular profiles. For example, for $s = 4$, we get 81 possible marginal profiles:

	Gene A	Gene B	Gene C	Gene D
Profile 1	−1	−1	−1	−1
Profile 2	−1	−1	−1	0
...				
Profile 81	1	1	1	1

Marginal profiles are coarse classifications that are susceptible to further refinement when additional evidence becomes available or a different perspective is taken. Of course, other genes may also display differential expression. This approach to defining profiles has biological interpretability, is potentially independent of the platform used, and is independent of the classification algorithm used to assign tumors to profiles.

For each sample i, based on the estimates of p_{ji}^{+} and p_{ji}^{-}, we can estimate the probability that it belongs to each profile, generating a vector of length 3^s. We can then visualize this using an image so that the most likely patterns can be differentiated from the less likely patterns. This is useful in two ways. First, we can identify the molecular profile for each sample. In other words, we can describe the expression pattern for each sample. Second, we can identify subgroups of samples that are likely to belong to the same profile.

16.3   R Software Extension: POE

16.3.1   An Example of Using POE on Simulated Data

The name of the R extension that we developed for the mixture model described above is POE, for (for Probability Of Expression). The current code consists entirely of R scripts, so no compilation is necessary to install it. The full library, for use on all platforms on which R is available, can be accessed at http://astor.som.jhmi.edu/poe. In this section, we describe the suite of functions for version 0.1. This includes one main function for estimating the model and several tools for data exploration, model visualization, and gene mining. Table 16.1 lists the functions and their uses.

We begin our illustration using a simulated example, with 1000 genes ($j = 1, \ldots, 1000$) from 72 samples ($i = 1, \ldots, 80$). Genes 1 through 50 are overexpressed in the first one-third of samples ($i = 1, 2, \ldots, 24$), underexpressed in the following one-eighth of samples ($i = 25, 26, \ldots, 33$), and normally expressed in the remaining samples ($i = 33, 34, \ldots, 72$). Genes 51 through 100 are underexpressed in every third sample ($i = 1, 4, 7, \ldots, 70$), overexpressed in one-eighth of the samples ($i = 2, 5, 8, 11, 14, 17, 20, 23, 26$), and normally expressed in the remaining samples. Genes 101 through 150 are overexpressed in the first quartile of samples ($i = 1, \ldots, 18$), underexpressed in the third quartile of samples ($i = 37, \ldots, 54$), and normally expressed in the second and fourth quartiles of samples. Genes 151 through 1000 are normally expressed in all samples. Underexpressed, normally expressed, and overexpressed data were generated as follows:

$$x_{ji}|e_{ji} = -1 \sim \mathcal{N}(-4, 2^2),$$
$$x_{ji}|e_{ji} = 0 \sim \mathcal{N}(1, 1),$$
$$x_{ji}|e_{ji} = 1 \sim \mathcal{N}(4, 2^2).$$

TABLE 16.1. POE functions.

poe.fit	fits the gene expression mixture model
poe.bigpicture	graphically displays the raw data or the transformed data (p_{ji}^*) in an image plot
poe.onegene	makes a diagnostic plot for model fit for a selected gene
poe.mine	selects gene groups based on a target expression profile
poe.pattern	graphically displays the results from poe.mine
poe.profile	takes a small set of genes and calculates posterior probabilities of gene profiles for each sample

A graphical depiction of the simulated data is shown in the top panel of Figure 16.1 (see color insert). We store our data matrix with 1000 rows and 72 columns as the R matrix object Y.

16.3.2 Estimating Posterior Probability of Expression Using poe.fit

The poe.fit Function

The model described above can be applied to a dataset using the command poe.fit, which requires just one argument: a matrix with rows corresponding to genes and columns to samples. We assume that the data have been previously normalized. There are optional arguments that can be included, which are described below. To fit the model, we call the poe.fit function at the R prompt:

```
> my.output <- poe.fit(Y)
```

The function poe.fit will fit the mixture model described in the previous section using default values of the prior hyperparameters. Genes will be labeled with the numbers $1, \ldots, J$. Using defaults, a burn-in period of 100 iterations will be performed, after which every 5th iteration will be saved, and a total of 2000 iterations will be saved. The chain results will be saved to a file called poe.out.dput in the default directory. Using these defaults, a total of $100 + 5 \times 2000 = 10100$ iterations will be performed. Note that these are the default settings and there is no guarantee that the number of iterations is sufficient for burn-in or for convergence for any particular analysis. In this simulated data example, 10100 iterations took approximately 14 hours to run on an 866 Mhz processor. However, fewer iterations would probably have been sufficient, given the stable behavior of the chain. It is imperative to check the results of the chain to ensure that extreme values within genes do not fall outside limits of the uniform distributions. This can be done using the poe.ongene command defined in section 16.3.3. Running longer chains will correct this problem. While this has been rare in practice, it can affect inferences if left unnoticed.

Convergence diagnostics for MCMC are discussed by Cowles and Carlin (1996). We also refer the reader to the R version of the CODA library (output analysis and diagnostics for Markov chain Monte Carlo simulations)(http://www-fis.iarc.fr/coda/) for code for checking MCMC convergence. In our example, the posterior point estimates from the chain will be saved to the R object called my.output.

Optional Arguments to poe.fit

The full command line for the default poe.fit function is

```
poe.fit(Y, id=NULL, burnin=100, M=2000, skip=4, logdata=F,
    stepsize=0.5, outfile="poe.out.dput", domedians=T,
    savechain=T, centersample=F, centergene=F,
    generatestarts=T, startobject=NULL,kap.min=7,
    PR=list(alpha.mm=0, alpha.sd=100, mu.mm=0, mu.sd=100,
        pipos.mm=0, pipos.sd=0, pineg.mm=0, pineg.sd=0,
        kap.pri.rate=1, tausqinv.aa=1, tausqinv.bb=0.1))
```

The definitions of these options are as follows:

id: A numeric or character vector of length J giving gene identifiers. Preferably, a vector of accession numbers.

burnin: The number of MCMC iterations to be performed before starting to save results.

M: The number of saved iterations. $M \times$ skip+1 is the total number of iterations after burn-in.

skip: Every skip+1 iteration will be saved to the output file.

logdata: If T, then data are (natural) log-transformed before fitting the model.

stepsize: Multiplier applied to the standard deviations of the proposal distributions; acts on all the Metropolis steps in the chain.

outfile: Name of output file into which the draws from all saved iterations are written. Format is an R dput file.

domedians: Indicates whether or not to summarize point estimates from posterior distribution by the median. If domedians=F, then means are taken instead.

savechain: If savechain=F, then no dput file is written to outfile. If savechain=T, then M chain values are saved to outfile.

centersample, centergene: If centersample=T, then the column median will be subtracted from Y before fitting the model. If centergene =T, then the row median will be subtracted from Y before fitting the model.

generatestarts: If generatestarts=T, then initial values for the chain will be determined based on initial estimates. If generatestarts=F, then starting values for all parameters must be defined in startobject.

startobject: If generatestarts=F, then initial values need to be specified in startobject, a list object. startobject can be a list generated by the user or the last iteration of a previously run chain (created from the output file of a previously run chain).

TABLE 16.2. Values for a `poe.fit` object.

`phat.pos`	p_{ji}^+, $J \times I$ matrix
`phat.neg`	p_{ji}^-, $J \times I$ matrix
`alpha`	vector of α_i , $i = 1, \ldots, I$
`mug`	vector of μ_j, $j = 1, \ldots, J$
`sigmag`	vector of σ_j, $j = 1, \ldots, J$
`kappaposg`	vector of κ_j^+, $j = 1, \ldots, J$
`kappanegg`	vector of κ_j^-, $j = 1, \ldots, J$
`piposg`	vector of π_j^+, $j = 1, \ldots, J$
`pinegg`	vector of π_j^-, $g = 1, \ldots, J$
`AA`	(centered) input matrix, $J \times I$ matrix
`id`	accession or numeric values, length J

`kap.min:` Defines r, the minimum value for κ_j^+ and κ_j^-, in units of the standard deviation σ_j of the center (i.e., normal) component of the mixture.

`PR:` List of prior hyperparameters.

The most commonly used values of the object created by `poe.fit` can be found in Table 16.2. Some additional parameters (e.g., values of the hyperparameters for bookkeeping convenience) are also included in the object but not included here for brevity.

16.3.3 Visualization Tools

The results of model fitting make available (1) the raw data x_{ji}, the inputted data to `poe.fit`, after being centered if specified in `poe.fit`, (2) the POE scale data p_{ji}^*, which ranges from -1 to 1, and (3) the shrinkage estimates z_{ji}, denoised according to the mixture model results. Any one of these can be used as the input to a broad array of visualization and class discovery techniques. In addition, there are two visualization tools for the result of `poe.fit` in the POE library. The first, `poe.bigpicture`, produces a red and green image of the gene expression data from all genes and samples in the data matrix (Eisen et al., 1998). The gene expression data displayed can be of one of three forms. The `poe.bigpicture` will plot one of these matrices and, if specified, order genes and samples using divisive or agglomerative hierarchical clustering via the R functions `diana` or `agnes`, respectively, from the R library `cluster` (Rousseeuw et al., 1996). The syntax is as follows:

```
> poe.bigpicture(my.output)
```

With no additional arguments, an image of the x_{ji} matrix will be plotted with the original gene and sample order preserved. Options can be adjusted to see different representations and to choose how rows and columns of the data matrix are ordered. The full command is

```
poe.bigpicture(my.output,psout=F,jpout=F,outfile="bigpic",hz=T,
    plotphat=F,plotshrink=F,order.genes=F,order.samples=F,
    bw=F,filter=T,filterp=0.025,order.by="raw",
    cltype="diana",center.scale=F,genelabels=F,
    geneids=NULL,samplelabels=T,sampleids=NULL,divide.sd=T)
```

where we have arguments with the following usage:

psout, jpout: If psout=T or jpout=T, a PostScript or JPEG file with name outfile and appropriate extension will be created. Set no more than one of psout and jpout to T or errors will occur.

outfile: Defines the name of the PostScript or JPEG file that will be created if psout=T or jpout=T. The appropriate extension will be added automatically.

hz: If T and psout=T, then landscape view will be produced. If F, then portrait view will be produced.

cl.type: Indicates which hierarchical clustering method to use. Takes arguments diana for divisive and agnes for agglomerative.

plotphat, plotshrink: If both are set to F, then the raw data (i.e., x_{ji}) are plotted. Otherwise, p^*_{ji} are plotted if plotphat=T, or z_{ji} are plotted if plotshrink=T. Set no more than one of these to T.

order.genes, order.samples: These indicate whether or not hierarchical clustering should be performed for the samples and the genes. If so, genes and/or samples will be arranged accordingly.

bw: Indicates black and white versus color picture (bw=T makes grayscale).

filter, filterp: If filter=T, genes with low evidence of differential expression are removed before creating the image. filterp, which ranges from 0 to 1 on the scale of $\sum_i p_{ji}$, indicates the level of filtering, where 0 means no filtering and 1 removes all genes.

cluster.by: Takes one of the arguments phat, shrink, and raw. This indicates which measure of gene expression should be used for clustering genes and samples.

cltype: Indicates what method to use for ordering the data. diana performs divisive hierarchical clustering. agnes performs agglomerative hierarchical clustering.

center.scale: If T, forces the center of the color (or gray) scale to be zero and sets the extremes to the same absolute value.

genelabels, geneids: If genelabels=T and geneids=F, the gene numbers 1 through G will be included on the plot. If genelabels=T and
geneids is set to a vector of strings or numeric values (which are
contained in the geneids given in the poe.fit call), then only the
geneids in the vector will be plotted.

samplelabels, sampleids: See genelabels, geneids above.

divide.sd: If T, then when plotting the raw data, each row of data is
divided by the row standard deviation so that the overall scale per
gene is consistent.

We produced the two panels of Figures 16.1 using the following commands:

```
> poe.bigpicture(my.output,psout=T,outfile="figure1a",
        genelabels=T,geneids=c(15,84,101))
> poe.bigpicture(my.output,psout=T,outfile="figure1b",
        genelabels=T,geneids=c(15,84,101),
        plotphat=T,order.by="phat",
        order.samples=T,order.genes=T)
```

The other graphical tool for the output from poe.fit is poe.onegene,
which displays for a specified gene both (1) a picture of the fitted mixture
model with the empirical density and the expression data, and (2) a q-q
plot. These can be used to evaluate model fit. The command requires the
output from the poe.fit function and the gene number (i.e., $1, \ldots, J$) or
the accession number. To plot the 1st and the 53rd genes, we would type

```
> poe.onegene(my.output, gg=1, outfile="figure2a",psout=T)
> poe.onegene(my.output, gg=53, outfile="figure2b",psout=T)
```

The results from these genes are shown in Figure 16.2 (see color insert) at the top, while the results for genes 115 and 199 are shown in
Figure 16.2 at the bottom. There are few optional arguments to this command, as can be seen below, and most are familiar from the description of
the poe.bigpicture command.

```
poe.onegene(my.output, gg=1, qq=T, dens=T, xbounds=NULL,
        psout=F, jpout=F, outfile="onegene", bw=F, hz=T)
```

qq: Indicates whether or not to include the q-q plot in the figure.

dens: Indicates whether or not to include the mixture model and empirical
density plot.

xbounds: By default, the x-limits on the plot are chosen based on the data,
but others can be input here in the form of a vector with two entries
(the lower and upper limits of the x scale).

16.3.4   Gene-Mining Functions

To mine for interesting subsets of genes that show differential expression, we use the approach described in subsection 16.2.4. The command that implements the formation of gene groups is `poe.mine`, and it takes an object resulting from the `poe.fit` command.

```
> my.groups <- poe.mine(my.output)
```

The full command line and additional arguments are as follows.

```
poe.mine(my.output,k=10,cohfilter=.8,cutp=0.90,cutperc=T,
         perc=0.90,targetp=c(0.2,0.2),filter=F,
         filterp=0.025)
```

k: Number of groups.

cohfilter: Minimum value of coherence necessary for a gene to be a seed.

cutp: Cutoff for agreement between seed gene and genes admitted to gene group.

cutperc: If F, then agreement is determined by cutp. If T, then the agreement cutoff for the inclusion of genes to the group is the coherence of the seed gene times perc.

perc: See cutperc above.

targetp: Vector of length 2, giving target proportion of samples that are underexpressed and overexpressed, respectively.

filter, filterp: If filter=T, genes with low evidence of differential expression are removed before choosing genes. filterp, which ranges from 0 to 1, is the filter cutoff. 0 means no filtering and 1 removes all genes.

This function is designed to be used interactively, changing the options to explore the data. Specifically, we recommend considering different values for the agreement threshold (via cutp and perc), which will change the size of the groups. Also, we recommend several different vectors for targetp, which will change the seed genes and, as a result, the distribution of the samples within groups. The values in a `poe.mine` object are listed in Table 16.3.

To mine for genes with several different patterns and to display them, we use the `poe.mine` and `poe.pattern` functions in sequence:

```
> my.groups <- poe.mine(my.output,k=6,targetp=c(0.2,0.2),
                cutperc=T,cutp=0.80)
> poe.pattern(my.groups,my.output,psout=T,outfile="figure3",
                order.samples=F,bw=T)
```

TABLE 16.3. Values of a `poe.mine` object.

`C.id`	Id numbers (numeric), list object.
`C.fullid`	Id numbers, as entered into the original call to `poe.fit`. (could be either numeric or string), list object.
`C.ppm`	p_{ji} for genes in clusters, list object.
`seedgenes`	vector giving the numeric Id value for seedgenes.
`k`	number of gene groups.
`coherence`	vector of length J, giving coherence value for each gene.
`score`	vector of length J, giving score (as defined above) for each gene.

Results are illustrated in Figure 16.3. There are only a few optional arguments to `poe.pattern` that have not been described in the previous functions.

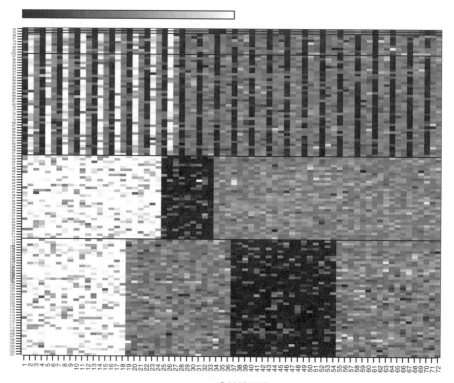

SAMPLES

FIGURE 16.3. Gene groups based on mining for profiles close to 20% overexpressed and 20% underexpressed.

```
poe.pattern(my.groups,my.output,fullid=T,cexx=0.5,cexy=0.5,
           order.samples=T,psout=F,jpout=F,bw=F,hz=T,
           outfile="grouppic",samplelabels=T,sampleids=NULL)
```

fullid: if T, the geneid from the input to poe.fit will be used. If F, numeric values will be used instead.

cexx,cexy: give cex parameter for x and y axes for plotting labels. Depending on the number of genes and samples, these can be adjusted to make legible labels.

16.3.5 Molecular Profiling Tool

By examining the gene groups identified in the previous section, we can identify a small set of genes (ideally, each from a different group) with which to derive molecular profiles and group samples. We implement this using the poe.profile command, which both calculates the molecular profiles based on a list of genes and produces an image plot of the profiles by samples. If we are interested in creating molecular profiles based on genes 13 and 51, then we would input the following:

```
> my.profile <- poe.profile(my.output,c(13,51),psout=T,
                 outfile="figure4")
```

Note that any number of genes can be used to create the profile. However, g genes lead to 3^g possible patterns so that profiles based on four or more genes are seldom required with the sample sizes and signal-to-noise ratios currently achievable. The only options refer to the plot that is produced. Most of them have been described in previous functions.

```
poe.profile(CS,gg,plot.it=T,psout=F,jpout=F,order.samples=T,
            order.pats=F,outfile="profilepic",bw=F,hz=T,
            samplelabels=T,sampleids=NULL)
```

order.pats: If T, patterns are arranged in an order determined by a divisive clustering algorithm.

The estimated molecular profiles based on the fit of POE can be visualized using the poe.profile call above. Results are shown in Figure 16.4 at the top. The true molecular profile of each sample is shown at the bottom. Comparing top and bottom, we see quite good agreement: the estimated molecular profiles are consistent with the actual profiles.

The values produced by the poe.profile function are listed in Table 16.4.

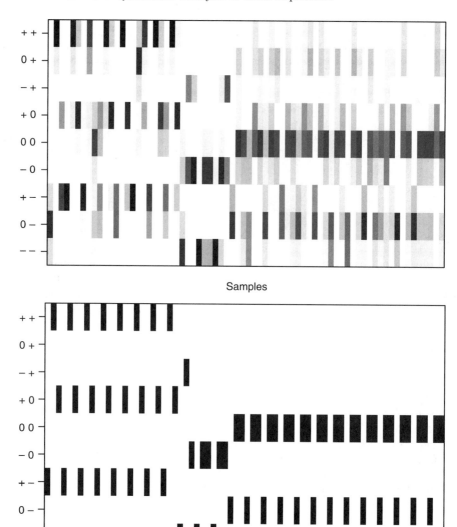

FIGURE 16.4. Molecular profiles based on results of POE (top) and true molecular profiles (bottom).

TABLE 16.4. Values for a `poe.profile` object.

classpro	$I \times 3^s$ matrix, probability of each molecular profile by sample
class	$3^s \times s$ matrix of profiles

16.4   Results of POE Applied to Lung Cancer Data

We also applied POE to gene expression array data from lung cancer samples (Bhattacharjee et al., 2001). The goal is to identify gene profiles that can differentiate between the lung cancer types. Another goal, which will not be addressed in this application, is class discovery within the lung adenocarcinoma samples. Some tumor samples are replicates from the same individual. For this descriptive example, we will ignore the likely within-individual correlation.

Affymetrix arrays were used to obtain gene expression data on 203 samples for 460 genes. This set of 460 genes is a subset of the original dataset of over 3000 genes. We selected the 460 genes in our analysis based on evidence of reproducibility between these genes and those in a similar cDNA microarray experiment of lung tissues. A POE analysis could have been carried out on the whole set. The specimens in this dataset are 139 lung adenocarcinoma ("adeno"), 20 pulmonary carcinoid (PC), 6 small-cell lung carcinoma (SCLC), 21 squamous-cell lung carcinoma (SQ) cases, and 17 normal lung (NL) specimens. In the graphical displays, the samples are differentiated by symbols: X = NL, + = PC, 0 = SCLC, and ∧ = SQ. The adenos labels are left blank due to the impossibility of seeing all of the labels simultaneously. Samples without labels are adeno samples.

More detailed information about the experimental processes can be found in Bhattacharjee et al. (2001). The data have been preprocessed to remove experimental artifacts, and a cube root transformation was performed to normalize expression levels within genes across samples. The R object `lungdata` is the 460×203 matrix representing gene expression.

An image of the raw data, ordered by genes and by samples using a divisive clustering method, is shown in Figure 16.5 at the top (see color insert). First, we estimate the Bayesian hierarchical model using the `poe.fit` command and view the estimated 460×203 p matrix using the `poe.bigpicture` command.

```
> lung.out <- poe.fit(lungdata, id=geneid, M=1000, skip=9,
                burnin=100)
> poe.bigpicture(lung.out,psout=T,outfile=''figure5b'',
                order.by=''phat'',plotphat=T,sampleids
                =lung.symbols,
                order.samples=T,order.genes=T)
```

The results are shown in Figure 16.5 at the bottom (see color insert). In comparing the two, we see that the image in the POE scale shows much stronger signals (i.e., brighter reds and greens) and much more clearly defined groups of similar genes and samples. Occasionally, investigators apply nonlinear transformations to the color scales of images to dampen the effect of outliers. POE provides a systematic and interpretable alternative to these manual changes. See also Vilo et al. (Chapter 6 of this volume) for additional discussion on manipulation of color scales.

We take a closer look at the fitted model by looking at the fit for several genes. Specifically, we examine the fit for genes AA894557 (CKB: creatinine kinase, brain) and AA487560 (CAV1: caveolin1, caveolae protein 22kD) by implementing the `poe.onegene` command. Expression levels are bimodal and the baseline level is closely approximated by a normal. The overexpressed levels are not uniformly distributed but the uniform distribution is adequate in capturing the fact that there is a departure from the baseline. Although it is common for the inner distribution to fit well, it is less common that the differentially expressed values will be uniformly distributed. This lack of fit of the uniform distribution is not necessarily problematic, as it often coexists with good estimates of the probabilities of differential expression.

```
> poe.onegene(lung.out,"AA894557", psout=T,outfile="figure6a")
> poe.onegene(lung.out,"AA487560", psout=T,outfile="figure6b")
```

Both genes show evidence of overexpression of a large portion of the samples. There is one notably underexpressed sample in gene CAV1. Our next step is to look for groups of genes that show different patterns of expression across samples. The `poe.mine` function is applied to the fitted model where we choose genes with a target expression pattern of 5% underexpressed and 15% overexpressed. The `targetp` and `cutperc` should be fine-tuned depending on the dataset and the profile of interest.

```
> lung.mine <- poe.mine(lung.out,k=4,targetp=c(0.05,0.15),
                  cutperc=T,perc=0.9)
> poe.pattern(lung.mine,lung.out,psout=T,outfile="figure7",
                  bw=T,sampleids=snames)
```

The results in Figure 16.7 show the four groups identified by `poe.mine`. The top group is a singleton, the second is somewhat heterogeneous and most genes show little classification ability. The bottom two groups are more homogeneous and show potential for further investigation. We choose two genes from each of the bottom two groups and study the resulting molecular profiles using the `poe.profile` command. In this example, we use the same two genes that were shown in Figure 16.6 (see color insert), CKB and CAV1. If the main goal is to identify novel subtypes, a small subset of genes may be sufficient for assigning samples to classes. But displaying groups of genes is useful in providing a framework for the interpretation of the genes selected for classification and for supporting the belief that a profile may reflect a biological mechanism.

```
> lung.profiles <- poe.profile(lung.out,
                  gg=c("AA894557","AA487560"),
```

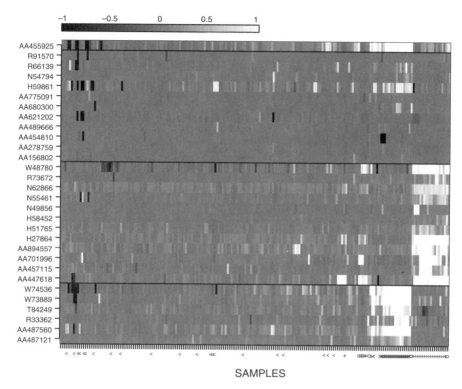

SAMPLES

FIGURE 16.7. Gene groups from lung cancer mined for 5% and 15%.

```
plotit=T,psout=T,outfile="figure8",
sample.labels=snames)
```

Figure 16.8 shows the resulting patterns. It appears that the adenos comprise most of the genes showing normal expression in both genes, but many of the 21 squamous tumors are consistent with this pattern also. The normal tissue samples tend to show overexpression in CKB and normal expression in CAV1. Almost all the pulmonary carcinoids and small cells together show overexpression in CAV1 and normal expression in CKB. Lastly, there is a subset of the squamous samples that tends to show underexpression on CKB and normal expression on CAV1. Based on this unsupervised analysis, we were able to define subsets of genes that are consistent with the known subtypes of lung cancer.

Both these genes have been shown previously to be associated with cancerous tumors. CAV1 has been identified as a metastasis-related gene in prostate carcinoma (Yang et al., 1998) and in esophageal squamous cell cancer (Kato et al., 2002). Research done in the mid-1990s found evidence suggesting that there is a connection between creatinine kinase (CK) and

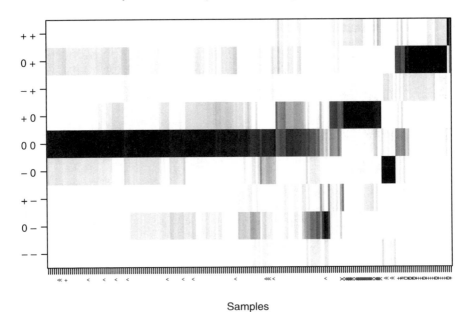

Samples

FIGURE 16.8. Molecular profiles based on genes CKB and CAV1.

malignancy and cancer. Specifically Walliman and Hemmer (1994) observed up-regulation of CK expression by adenoviral oncogenes.

In this analysis, we used the entire set of samples from the study, but we treated it as an unsupervised problem, using phenotype information only to provide ex-post evidence that the algorithm had mined interesting profiles. Because the adenos represent such a large proportion of the population of samples, the algorithm often associates the expression level in adenos to the baseline, as is the case for the two genes under investigation. This does not imply that the baseline level is normal biologically because we see that normal samples have higher expression levels in gene CKB than the adeno samples.

16.5 Discussion and Future Work

Unsupervised molecular classification is a challenging data analysis problem. Most approaches combine in different measure two general strategies. The first is to generate low-dimensional summaries of the gene expression information, such as distance matrices or lower dimensional projections, including principal components. The second is to identify, via visualization, expert opinion, or more formal tools, a manageable number of genes. The first approach has been more heavily exploited in the statistical literature, while the second is more common in successful biological work. One problem with the first approach is that molecular profiles are defined in a way that is specific to the overall set of genes measured on the array. An arraywise molecular profile from a custom array can be different in meaning from a molecular profile from a generic commercial array.

In this chapter, we discussed a way to reexpress continuous gene expression array data from any platform into expression measured on a probability scale using latent qualitative variables. This can support simple and interpretable dimension reduction, filtering, mining, and visualization of genes without requiring the use of hard-to-interpret projections. Also, extreme expression values are less influential than they would be in more standard analyses, where correlation and distance matrices tend to be sensitive to outlying observations. Finally, the transformation to the probability scale allows for gene expression data from different experiments to be studied using the same scale—in this case, the probability scale. Each experiment can be independently transformed to the probability scale using POE, and then the resulting probability matrices for the multiple experiments can be analyzed together. With current technologies, this is achievable when platforms are similar, but it is still challenging when combining oligonucleotide array results with cDNA arrays.

We also described a simple data-mining strategy specific for expression arrays in the probability scale that identifies multiple homogeneous sets of genes that show differential expression across samples. A close relative of our strategy is the "gene-shaving" algorithm (Hastie et al., 2000). Gene-shaving searches for groups of genes showing high variation across samples, high correlation of genes within a cluster, and high diversity of gene expression across groups. The same gene can belong to more than one group.

The sets of genes can then be used to create molecular profiles—the first step toward developing clinical classifications of different types of samples based purely on gene expression levels of a small set of genes. We emphasize the importance of model checking and data exploration by including data visualization tools that allow the user to explore the raw data, the transformed data, the fit of the Bayesian hierarchical model to each of the genes, the behavior of the gene sets (i.e., "groups") across samples, and the defined molecular profiles across samples.

Ongoing work is focusing on developing diagnostics of model fit. In the interim, users can assess fit on a gene-by-gene basis by measuring nonlinearity of the residuals of the regression of the normal q-q plot using $1 - \hat{p}_{ji}$ as the weights. Genes with large nonlinearity can be selected for further visualization and diagnostics. Another useful mining tool is the estimated gene self-consistency $q(j,j)$. Further ongoing work includes a supervised analog in which the mixture model is estimated using phenotypic information that is provided by the user. This allows a baseline to be defined using samples of known biology, such as normal samples in tumor classification.

Acknowledgments. The work of Parmigiani and Garrett is partly supported by NCI grants P50CA88843, DK-58757, and 5P30 CA06973-39. We would like to thank Jiang Hu for computing and programming assistance.

References

Alizadeh AA, Eisen MB, Davis RE, Ma C, Lossos IS, Rosenwald A, Boldrick JC, Sabet H, Tran T, Yu X, Powell JI, Yang L, Marti GE, Moore T, J. Hudson Jr J, Lu L, Lewis DB, Tibshirani R, Sherlock G, Chan WC, Greiner TC, Weisenburger DD, Armitage JO, Warnke R, Levy R, Wilson W, Grever MR, Byrd JC, Botstein D, Brown PO, Staudt LM (2000). Distinct types of diffuse large B-cell lymphoma identified by gene expression profiling. *Nature* 403:503–511.

Berger JO (1985). *Statistical Decision Theory and Bayesian Analysis, 2nd ed.* New York: Springer-Verlag.

Bhattacharjee A, Richards WG, Staunton J, Li C, Monti S, Vasa P, Ladd C, Beheshti J, Bueno R, Gillette M, Loda M, Weber G, Mark EJ, Lander ES, Wong W, Johnson BE, Golub TR, Sugarbaker DJ, Meyerson M (2001). Classification of human lung carcinomas by mRNA expression profiling reveals distinct adenocarcinoma subclasses. *Proceedings of the National Academy of Sciences USA* 98:13790–13795.

Bittner M, Meltzer P, Chen Y, Jiang Y, Seftor E, Hendrix M, Radmacher M, Simon R, Yakhini Z, Ben-Dor A, Sampas N, Dougherty E, Wang W, Marincola F, Gooden C, Lueders J, Glatfelter A, Pollock P, Carpten J, Gillanders E, Leja D, Dietrich K, Beaudry C, Berens M, Alberts D, Sondak V, Hayward N, Trent J (2000). Molecular classification of cutaneous malignant melanoma by gene expression profiling. *Nature* 406:536–540.

Clyde MA, Parmigiani G, Vidakovic B (1998). Multiple shrinkage and subset selection in wavelets. *Biometrika* 85:391–402.

Cowles MK, Carlin BP (1996). Markov chain Monte Carlo convergence diagnostics: A comparative review. *Journal of the American Statistical Association* 91:883–904.

Diebolt J, Robert CP (1994). Estimation of finite mixture distributions through Bayesian sampling. *Journal of the Royal Statistical Society, Series B* 56:363–375.

Duggan D, Bittner M, Chen Y, Meltzer P, Trent J (1999). Expression profiling using cDNA microarrays. *Nature Genetics* 21:10–14.

Eisen MB, Spellman PT, Brown PO, Botstein D (1998). Cluster analysis and display of genome-wide expression patterns. *Proceedings of the National Academy of Science, USA* 95:14863–14868.

Fraley C, Raftery AE (1998). How many clusters? Which clustering method? – Answers via model-based cluster analysis. *Computer Journal* 41:578–588.

Garber ME, Troyanskaya OG, Schluens K, Petersen S, Thaesler Z, Pacyna-Gengelbach M, van de Rijn M, Rosen GD, Perou CM, Whyte RI, Altman RB, Brown PO, Botstein D, Petersen I (2001). Diversity of gene expression in adenocarcinoma of the lung. *Proceedings of the National Academy of Sciences USA* 98:13784–13789.

George EI (1986). Minimax multiple shrinkage estimation. *The Annals of Statistics* 14:188–205.

Golub TR, Slonim DK, Tamayo P, Huard C, Gaasenbeek M, Mesirov JP, Coller H, Loh M, Downing JR, Caligiuri MA, Bloomfield CD, Lander ES (1999). Molecular classification of cancer: Class discovery and class prediction by gene expression monitoring. *Science* 286:531–537.

Hastie T, Tibshirani R, Eisen MB, Alizadeh A, Levy R, Staudt L, Chan WC, Botstein D, Brown P (2000). "Gene shaving" as a method for identifying distinct sets of genes with similar expression patterns. *Genome Biology* 1:research0003.1–research0003.21.

Kato K, Hida Y, Miyamoto M, Hashida H, Shinohara T, Itoh T, Okushiba S, Kondo S, Katoh H (2002). Overexpression of caveolin-1 in esophageal squamous cell carcinoma correlates with lymph node metastasis and pathologic stage. *Cancer* 94:929–933.

Lee ML, Kuo FC, Whitmore GA, Sklar J (2000). Importance of replication in microarray gene expression studies: Statistical methods and evidence from repetitive cDNA hybridizations. *Proceedings of the National Academy of Sciences USA* 97(18):9834–9839.

McLachlan GJ, Bean RW, D P (2002). A mixture model-based approach to the clustering of microarray expression data. *Bioinformatics* 18:413–422.

Parmigiani G, Garrett ES, Anbazhagan R, Gabrielson E (2002). A statistical framework for expression-based molecular classification in cancer. *Journal of the Royal Statistical Society, Series B*, to appear.

Quackenbush J (2001). Computational analysis of microarray data. *Nature Reviews Genetics* 2:418–427.

Rousseeuw P, Struyf A, Hubert M (1996). Clustering in an object-oriented environment. *Journal of Statistical Software* 1:1–30.

Walliman T, Hemmer W (1994). Creatine kinase in non-muscle tissues and cells. *Molecular Cell Biochemistry* 133–134:193–220.

West M, Turner D (1994). Deconvolution of mixtures in analysis of neural synaptic transmission. *The Statistician* 43:31–43.

Yang G, Truong L, Wheeler T, Park S, Nasu Y, Bangma M, Kattan P, Scardino P, Thompson T (1998). Elevated expression of caveolin is associated with prostate and breast cancer. *Clinical Cancer Research* 4:1873–1880.

Yeung K, Fraley C, Murua A, Raftery A, Ruzzo W (2001). Model-based clustering and data transformations for gene expression data. *Bioinformatics* 17:977–987.

17

Bayesian Decomposition

MICHAEL F. OCHS

Abstract

Gene chips and gene expression microarrays offer the opportunity to study biological systems on a genome-wide basis, exploring the full transcriptional response in an experiment or therapy. Because of the complexity of living organisms, these transcriptional responses are complex, with multiple, overlapping groups of genes being expressed in response to continuing internal and external stimuli. In order to use expression measurements to identify upstream modifications in signaling pathways, it is necessary to disentangle these overlapping responses. Bayesian Decomposition provides a method of identifying such overlap and correctly assigning genes to multiple groups, allowing easier identification of pathway modifications. Here the results of the application of Bayesian Decomposition to cell cycle data are shown.

17.1 Introduction

17.1.1 Role of Signaling and Metabolic Pathways

Biological systems such as cells are complex, with the ability to respond to the external environment and internal changes by modifying their behavior. These modifications can take place through changes in metabolic networks, driven by concentration gradients and their subsequent effect on chemical reactions, or through modification of proteins involved in signaling networks. Signaling networks, such as the MAPK cascades (Roberts et al., 2000; Sebolt-Leopold, 2000), generally modify the transcriptional program of a cell, resulting in changes in gene expression, although many other changes (e.g., phosphorylation and cleavage of proteins) are possible as well (Golemis et al., 2001). When transcriptional changes do occur, the newly transcribed genes will usually be processed into proteins that aid the cell in responding to environmental or internal changes.

Many human cancers can be traced back to errors in these signaling pathways (e.g., p53, ras, c-kit). Since cancer is the second-leading cause

388

of death in the western world (Alison and Sarraf, 1997), and since treatment is difficult due to the diverse nature of the underlying causes of the development of malignancies (Cooper, 1992; Macdonald and Ford, 1997), understanding the roots of cancer and the role of errors arising in signaling pathways is an important undertaking. Since many modifications to signaling pathways, including those resulting from cellular errors that can lead to the development of cancer, lead to changes in transcription and thus levels of gene expression, the nascent technologies for genome-wide monitoring of gene expression offer hope for improving our knowledge of the causes of cancer and of potential targets for therapies.

17.1.2   Gene Expression Microarrays

Gene expression array technology provides genome-wide measurements of mRNA levels in cells. Although studies have already shown that it is possible in some cases to identify disease states more accurately using mRNA expression profiles than can be done using classic pathology methods (Golub et al., 1999; Alizadeh et al., 2000; Zhang et al., 2001), the link between changes in expression and changes in signaling pathways is more difficult to deduce.

The popular methods for analyzing gene expression data divide into two general categories. In the first category, the algorithms use statistical methods to identify individual genes that are differentially regulated between multiple states (e.g., between treated and nontreated cells, between different types of cancer) (Claverie, 1999; Ideker et al., 2000; Kerr et al., 2000; Kerr et al., 2002). In the second category, the algorithms perform pattern recognition, often by application of matrix operations, since array data typically take the form of two-dimensional sets of numbers (e.g., expression levels for many genes at different conditions). These methods include principal component analysis (Alter et al., 2000), self-organizing maps (Tamayo et al., 1999), support vector machines (Brown et al., 2000), and clustering (Eisen et al., 1998; Heyer et al., 1999; Getz et al., 2000; Kerr and Churchill, 2001; Lukashin and Fuchs, 2001; Yeung et al., 2001), among others. However, all these methods still lack an ability to recover fundamental behavior, since each gene within the experiment can be assigned to only one coexpression group. Since the underlying biology dictates that many individual genes be coexpressed in multiple groups in response to different stimuli (Roberts et al., 2000), this inability to allow overlapping coexpression groups in the analysis leads inevitably to the loss of information related to behavior arising from multiple inputs, which is critical for understanding cellular behavior (Bittner et al., 2000).

In order to use a set of gene expression measurements as an indicator of a change to a signaling pathway, the coexpression related to that pathway must be extracted from the complex transcriptional response of the system under study. Because of individual variation, changes occurring due to stress, or other changes unrelated to the variables of interest, the transcriptional response can include numerous changes in gene expression unrelated to the specific pathway change. Thus, the goal for analysis of gene expression data with Bayesian Decomposition is to identify the overlapping groups of genes that vary together within the experiment, with the potential for these genes to provide insights into the pathways modified during the experiment. This requires identification of sets of coordinated expression changes that link genes together even in the presence of overlapping coregulation, followed by the assignment of these sets of genes to specific pathways. The behavior of such a set of genes as a whole then provides insight into the changes within the corresponding signaling pathway.

17.2   Methods

17.2.1   Matrix Decomposition

Transcriptional response is complex, with many genes having multiple potential signals leading to their transcription, so the coexpression groups will show considerable overlap. The analysis in this case can then be viewed mathematically as the identification of physiologically meaningful basis vectors for the data, with these basis vectors being nonorthogonal in the general case. The basis vectors describe the transcriptional response of a pathway or a group of pathways that are activated or deactivated simultaneously within the experiment. In addition, there can be background transcription unrelated to those signaling pathways affected during the experiment (e.g., such an experiment could be the administration of a potential therapeutic agent to a cell line). Such background transcription would include ongoing normal metabolic processes and cellular maintenance as well as possible cyclic changes due to normal growth. As such, one can expect that the set of expression changes of interest will comprise one or a few multiple patterns (i.e., basis vectors) present within the data. Figure 17.1 shows a diagram of the matrix decomposition performed by Bayesian Decomposition (BD), with the pattern matrix, P, containing the nonorthogonal basis vectors. The relationship of the matrices in Figure 17.1 is given by

$$X_{ji} = \sum_{p=1}^{N_p} F_{jp} P_{pi} + E_{ji} = M_{ji} + E_{ji}, \qquad (17.1)$$

where X_{ji} is the measured relative expression level for the jth gene in the ith condition, M_{ji} is the modeled relative expression level for the jth gene in the ith condition, F_{jp} is the amount of pattern p that is included in the model for gene j, P_{pi} is the fractional level of expression in pattern p under condition i (relative to the other conditions), and E_{ji} is the error or misfit of the model to the data for the expression of the jth gene in the ith

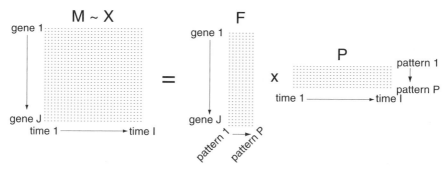

FIGURE 17.1. The matrices used in Bayesian Decomposition. The algorithm creates simultaneously the F and P matrices that define the model. These are combined by standard matrix multiplication to create the M matrix. The algorithm compares the M matrix to the data matrix X, since M is the data that the model would generate without noise.

condition, which under ideal circumstances is a random error. However, if the number of patterns is incorrectly estimated or if systematic error remains in the data, this term can include nonrandom components.

17.2.2   Markov Chain Monte Carlo

Since equation (17.1) is mathematically degenerate, allowing multiple, mathematically allowable solutions, a Markov chain Monte Carlo (MCMC) procedure (Metropolis et al., 1953; Hastings, 1970) is used to identify the F and P matrices that best explain the observed data within constraints provided by a Bayesian model. Specific examples of constraints are discussed below, but often even a positivity constraint alone is adequate to eliminate all but a single set of solutions (which differ only within a normally distributed uncertainty). In general, a single solution may still not exist; however, when the number of data points (elements of X) significantly exceeds the number of parameters, then a single solution (mean and standard deviation of each element in F and P) can often be recovered. At a maximum, the number of elements of F and P will be the number of parameters. However, if correlations can be identified a priori in the F and P matrices, then the number of parameters can be significantly lower. For example, in magnetic resonance relaxographic imaging (Labadie et al., 1994), each row of P must be a sum of inversion recovery curves, which allows substantial encoding of prior knowledge and a reduction in the number of free parameters, leading to a considerable improvement in the resolving power of the algorithm (Ochs et al., 2001). In the present work, the matrix F can be viewed as a flux through a given pathway, indicating the strength of a transduced signal as measured by the transcriptional response. Each element of F will be the relative level of expression for each gene related to that pathway.

17.2.3 Bayesian Framework

The Markov chain uses a Gibbs sampler that requires relative probability estimates between points in the posterior distribution (i.e., space of possible solutions). These are provided within a Bayesian framework (Besag, 1986; Besag and Green, 1993; Grenander and Miller, 1994). Using Bayes' equation, the probability at each point in the sampling space can be determined to be

$$p(Model|Data) = \frac{p(Data|Model)p(Model)}{p(Data)}, \tag{17.2}$$

where $p(Model|Data)$ is the conditional probability of the model given the data (the posterior), $p(Data)$ is the probability of the data (the evidence), $p(Data|Model)$ is the conditional probability of the data given the model (the likelihood), and $p(Model)$ is the probability of the model (the prior). Since the evidence usually acts as a scaling parameter, it can be ignored in this case, as the sampler requires only relative probabilities.

The relative probability between two points in the posterior is then determined by the likelihood, which is easily determined by comparing the model to the data, and the prior, which is the probability of the model independent of the data. Inserting the F and P matrices specifically generates the form of equation (17.2) used by Bayesian Decomposition (ignoring the scale factor),

$$p(F, P|X) \propto p(X|F, P)p(F, P). \tag{17.3}$$

The sampling from the posterior distribution and the encoding of the prior are done using a customized form of the Massive Inference (TM) Gibbs sampler from Maximum Entropy Data Consultants, Cambridge, England (Sibisi and Skilling, 1997). The prior, which can encode domain knowledge about the problem under study, is discussed below.

The likelihood can be recovered from the normalized χ^2 distribution, with the log likelihood, L, given by

$$L = -\sum_{j=1}^{J}\sum_{i=1}^{I}\left\{\frac{1}{2\sigma_{ji}^2}\left(X_{ji} - \sum_{p=1}^{N_p}F_{jp}P_{pi}\right)^2\right\}, \tag{17.4}$$

where σ_{ji} is the standard deviation for the expression measurement of the jth gene in the ith condition. Equation (17.4) assumes a normal error distribution. Although other likelihoods could be encoded, in our experience, a normal error model has been adequate. For any change to the model, the change in the log likelihood can be calculated by inserting the change in equation (17.4) and subtracting the result from equation (17.4). For a change δP at matrix element $[x, y]$ in the P matrix, for example, the change in the log likelihood is

$$\delta L(\delta P_{xy}) = -\sum_{j=1}^{J} \left\{ 2\Delta_{jy} F_{jx} \delta P_{xy} + \frac{1}{2\sigma_{jy}^2} F_{jx}^2 \delta P_{xy}^2 \right\}, \qquad (17.5)$$

with

$$\Delta_{ji} = \frac{1}{2\sigma_{ji}^2} \left\{ X_{ji} - \sum_{p}^{N_p} F_{jp} P_{pi} \right\} = \frac{1}{2\sigma_{ji}^2} E_{ji} \qquad (17.6)$$

defining the normalized mismatch between the model and data at each point. A similar equation applies for any change in the F matrix. In addition, the prior probability is included in determining the probability for the Markov chain step. In order to simplify the calculations, we do not allow simultaneous changes in F and P, since allowing such changes would require evaluation of terms involving $\delta F \delta P$. Note that barring such changes does not prevent the system from reaching any state and should have no effect on the final result since the sampler can move δP followed by δF and reach the same point. As long as detailed balance is maintained, the sampler still samples the space correctly.

17.2.4   The Prior Distribution

Much of the power of Bayesian techniques comes from the proper use of the prior during sampling. The search space for an analysis of a gene expression experiment comprises an enormous multidimensional space, and a prior effectively condenses this space by reducing the probability that the sampler will visit regions of zero or extremely low probability. Bayesian Decomposition encodes the prior by creating multiple domains, as shown in Figure 17.2. The first domain is an atomic domain that consists of two infinitely divisible lines (limited to 2^{32} points computationally), one that maps to the F matrix and one that maps to the P matrix. Atoms (point masses) are placed along these lines, with each atom being completely defined by its position and its amplitude. All possible steps in the sampler include prior probabilities with a distribution comprising a uniform spatial distribution and a logarithmic flux distribution (Sibisi and Skilling, 1997), together with the change in the likelihood, i.e., equation (17.5). This prior tends to permit atoms that are not forced to exist by the data to be removed during the Markov chain sampling, leading to a minimal structure in the model. The prior in this space also enforces positivity and additivity of atoms, which effectively reduces the search space for the sampler by a

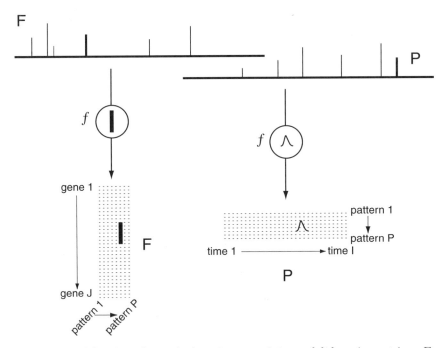

FIGURE 17.2. Mapping of atomic domain atoms into model domain matrices. For each domain, atoms are mapped to the model matrices (F and P) according to a mapping function. Here the mapping functions take an atom from the F atomic domain and distribute it over a number of elements in a column of the F matrix and take an atom from the P atomic domain and distribute it into a rise-and-fall curve in the P matrix.

factor of 2^N, where N is the number of dimensions in F and P. The second domain is the model domain (see Figure 17.2), essentially the F and P matrices.

The mapping of atoms into the model domain relies on prior knowledge encoded as convolution functions, as noted by Sibisi and Skilling (1997), where a distributed quantity Ψ can be constructed from a family of measures Φ_w using kernels K_w as

$$\Psi = \sum_w K_w \Phi_w. \tag{17.7}$$

This permits each atom to be spread to multiple matrix elements. Generally, the overall magnitude of the elements in the matrix equals the amplitude of the corresponding atom. For example, in gene expression analysis, the convolution function could encode the time behavior of mRNA species following initiation of transcription or known coregulation of genes (as in Figure 17.2). In this case, the integral of the curve in the P matrix would be equal to the amplitude of the atom, with its center in the matrix determined by the location of the atom in the atomic domain. Such convolution functions were used to model inversion recovery curves, allowing analysis in magnetic resonance relaxographic imaging (Ochs et al., 2001). In the work presented here, each atom maps to a single matrix element, so no additional information beyond positivity is included. The algorithm in this case is similar to nonnegative matrix factorization (Lee and Seung, 1999). No additional constraints are placed on the matrices, so the two matrices are independently generated.

An atomic prior offers significant advantages for the application of the Markov chain technique to physical data. Whereas equation (17.1) is similar in form to the standard equation of factor analysis (Lawley and Maxwell, 1971), the use of equation (17.7) allows Bayesian Decomposition to model an underlying physical process much more accurately than most implementations of Bayesian methods in factor analysis (see, for example, Utsugi and Kumagai, 2001). The modeling of the underlying processes also appears to resolve the rotation problem, as BD generally identifies the physically meaningful basis vectors, or factors, as opposed to rotated versions, as shown previously (Ochs et al., 1999; Ochs et al., 2001).

17.2.5   Summary Statistics

The sampler equilibrates for a user-specified amount of time and then takes samples from the posterior distribution. The samples comprise values for the matrix elements in the model domain, and standard summary statistics are generated from them.

Since the matrices permit exchange of amplitude between them without change of fit to the data, namely

$$\sum_p F_{jp} P_{pi} = \sum_p (k_p F_{jp}) \left(\frac{1}{k_p} P_{pi} \right), \tag{17.8}$$

normalization is required prior to generating the summary statistics. This is done by requiring each row of the P matrix to be normalized to total amplitude 1. The samples themselves are unconstrained; it is only when the summary statistics are generated that normalization is enforced.

The algorithm presently calculates the mean and variance at each matrix element, although more complex analyses of the samples could be performed. These statistics are included in the output files generated by Bayesian Decomposition. The program creates three output files upon completion. The first is the inference file (.inf), which includes the mean model and associated uncertainties. The second is the data file (.dat), which includes the input data (the X matrix), the input uncertainties (σ values), and the calculated mock data (the M matrix). The third is a movie file (.mov), which contains individual samples recorded during the Markov chain sampling process.

17.3 Software

17.3.1 Implementation

This implementation of Bayesian Decomposition for gene expression analysis directly maps each atom (i.e., element in the atomic domain) into a single matrix element and includes code that performs simulated annealing to minimize the possibility of the Markov chain becoming trapped in a local maximum in the posterior distribution. Simulated annealing is a method that derives its name from statistical mechanics and the relationship of optimization to the annealing of a physical system (Kirkpatrick et al., 1983; Geman and Geman, 1984). Essentially, the Markov chain is slowly "cooled," permitting the sampler to escape local maxima by effectively flattening the posterior distribution for early steps. The simulated annealing can take place in two ways. The method of annealing used here is simply to use an annealing schedule so that equation (17.3) is modified to

$$p(F, P|X) \propto p(X|F, P)^{1/T} p(F, P), \tag{17.9}$$

with the temperature slowly decreasing, which increases the effect of the data on the model. This annealing occurs during the equilibration or "burn-in" phase, when the sampler does not gather statistics but merely propagates the Markov chain to reach an area of high probability. The annealing stops when $1/T = 1$, so that equation (17.3) is recovered for sampling. In this case, the steps in annealing are uniform, so $\delta(1/T) = 1/S$, where S is the number of iterations prior to sampling (i.e., the burn-in time). The second method of annealing is to use multiple Markov chains and to allow comparisons of the likelihood between chains to determine the annealing schedule. This provides more coverage for the posterior distribution and also permits chains with low likelihoods to be replaced by chains with high likelihoods, increasing the coverage of regions of high probability. Equation (17.9) is still used for the annealing; however, the change in temperature is now determined by the behavior of the chains, with greater discrepancies between the likelihoods of the chains leading to slower lowering of the temperature. This method is not yet fully implemented for technical reasons related to the ordering of the rows of the P matrix between Markov chains.

17.3.2 Files and Installation

The Bayesian Decomposition executable for Linux, Solaris, or Macintosh OS X can be downloaded by academic users free of charge from the Fox Chase Cancer Center Bioinformatics Web site at `http://bioinformatics.fccc.edu/` (follow the proprietary software link on the left-hand side of the home page). You will be prompted while following links to download and to fill out a Material Transfer Agreement (MTA) form, which needs to be faxed back to the Fox Chase Cancer Center Technology Transfer Office. Once the Technology Transfer Office receives the form, a password will be e-mailed to the user to access the download Web page. In addition to the executable, some Java2 programs for viewing the output of Bayesian Decomposition are included. Detailed installation instructions as well as sample files are included so that the user can test the installation of the algorithm.

Presently, BD uses ASCII files for input and output. There are two input files for BD: a control file `ExpName.ctrl` that includes parameters concerning how the algorithm is run and a data file `ExpName.data` that includes the actual data. The control file provides the program with a root output name, the number of Markov chains to initiate (presently limited to 1), the number of iterations to run prior to sampling, the number of snapshots of the data to take during sampling, a random number seed, and parameters governing the expected number of atoms in both the F and P atomic domains. The data file includes the sizes of each of the matrices in equation (17.1), and thus the number of dimensions, N_P, in equation (17.1), as well as the normalized ratio and uncertainty for each point in the X matrix. Complete specifications are given in the downloadable files on the Web site.

As noted above, the program generates three output files, which include statistical measures of the matrices as well as the input data and mock data generated from the calculated mean matrices. The inference and data files are used by the Java output parser to create the serialized output file (`.bdo`) that serves as input to the visualization tool. The other Java2 program provided will load the serialized file, permit onscreen graphing, and allow the data to be saved as tab-delimited files for importing into spreadsheet programs. In addition, the users can directly read the ASCII files into their favorite visualization system (e.g., Matlab, IDL).

17.3.3   Issues in the Application of Bayesian Decomposition

There are a number of issues that should be kept in mind when applying Bayesian Decomposition to gene expression data. The first, which follows from equations (17.5) and (17.6), is that BD relies on uncertainty measurements at all data points when calculating the likelihood and change in likelihood during equilibration and sampling. The input data therefore need to include uncertainty estimates at each data point. Many public domain datasets do not include such measurements, so it is necessary to estimate the standard deviation for each data point. In our own work, we have found that in such cases the data variation tends to have a strong signal-dependent component, as reported elsewhere (Ideker et al., 2000), so that we use an estimate that varies with the ratio. An advantage of BD is that missing data do not cause a problem, as missing values can be estimated cavalierly as long as the uncertainty assigned to these values is large. The likelihood (from equation (17.5)) will then essentially ignore that value, as the standard deviation will dominate, so that the data point will not influence the model. In general, BD is not sensitive to errors in uncertainty estimates as long as they are not too large. However, misestimations will reduce the usefulness of the χ^2 values for interpreting overfitting of the model.

A continuing issue in the application of BD is the need to estimate the dimensionality of the model (i.e., N_P in equation (17.1)). Previous work in spectral imaging (Ochs et al., 1999) and relaxographic imaging (Ochs et al., 2001) used principal component analysis (PCA) to make estimates, as the principal components often showed indications of noise, with the true dimensionality usually within 1 or 2 of the PCA estimate. However, application of PCA to estimate dimensionality of gene expression data has not been successful in our experience. Instead, other information must be used. For the tutorial in Section 17.4, the knowledge of the cell cycle will be used. For complex datasets, such as the Rosetta compendium (Hughes et al., 2000), the correct number of dimensions remains unresolved (Bidaut et al., 2002). Since the actual dimensionality is generally unknown, BD should be run multiple times positing different dimensionalities.

A problem with all Markov chain Monte Carlo methods is the lack of convergence criteria (Besag et al., 1995), so there is no method available internally to the algorithm to validate the solution as the "best." In order to explore the possibility of other solutions fitting the data as well or better, BD should always be run multiple times at the same posited dimensionality with different random seeds (equivalent to different starting points for the Markov chains). If the same model is recovered each time, it is probably the most likely one. In addition, the snapshots created by BD and stored in the movie file can be viewed to make sure that the model is stable during the Markov chain process. In cases where multiple solutions fit the data equally well, these procedures can lead to identification of multiple models (Ochs et al., 1999).

17.4   Application of Bayesian Decomposition to Yeast Cell Cycle Data

17.4.1   Preparation of the Data

This tutorial will use the yeast cell cycle data from the Stanford cell cycle dataset (Spellman et al., 1998) and will focus on data from the cells synchronized through a temperature sensitive CDC28 mutant. The dataset should be downloaded from http://genome-www.stanford.edu/ through the SMD link. The data for the CDC28 series (0 to 160 minutes at 10 minute intervals) should be extracted, antilog-transformed, converted into the BD data file format (the file Test.data can be used as a template), and named MyExp.data. Estimates of the standard deviation at each point should be made, with 25% of signal providing the user with the ability to compare results to the published analysis (Moloshok et al., 2002). Missing data points can be set to 1.0, with the uncertainty set to 100.0. At this point, the dimensionality needs to be estimated, and use of six dimensions will match the publication. Originally, four through seven dimensions were explored; however, six dimensions was determined to reconstruct all cell cycle behaviors.

The control file provided (`Test.ctrl`) should be copied to `MyExp.ctrl` and modified to provide a unique output name (first line) and to set the number of iterations. Typically, at least 2000 iterations should be done before sampling; however, in complex, highly overlapping datasets, more may be needed. By performing multiple runs as discussed above, the adequacy of the equilibration time can be tested (i.e., all runs of BD should recover the same model within the uncertainties). Finally, the number of atoms in the F and P atomic domains should be estimated. Generally, these estimates only weakly affect the program, and if there is no convolution function, as is the case here, a good estimate is generally to use the number of elements in each respective matrix.

17.4.2 Running the Program

Bayesian Decomposition is run from the command line by giving the executable name together with the root form of the control and data files. For the control and data files named `MyExp.ctrl` and `MyExp.data`, the program is invoked with

```
./genpat MyExp,
```

where it is assumed that the executable is in the working directory. (Alternatively, for knowledgeable Unix users, the executable can be placed in a directory included in the PATH environment variable.) The program will echo to the terminal window the parameters for the run, including the input data size, the number of dimensions posited, and the number of iterations. Then the program will print lines during annealing at each tenth of the way until reaching the run temperature ($1/T = 1$). A typical line appears as

```
1/T = 0.10 of 1.00   2194 (702)    78503.06,
```

where the inverse temperature is given as a point out of 1.00, the first number (2194) gives the present number of atoms in the F atomic domain, the number in parentheses (702) gives the present number of atoms in the P atomic domain, and the final number is the χ^2 value for this sample. Depending on the number of iterations and speed of the computer, the appearance of the first equilibration output can take considerable time (about 10 minutes on a 700 MHz Pentium III machine running RedHat

Linux). Once equilibration has finished, sampling will begin. The progress of sampling is reported every 100 iterations, with a typical line being

```
Sample 1400 of   2000   4289 (351)      33665.26,
```

where the present sample number is shown, the total number of samples to be visited by the Markov chain is given, and the additional values match those given during equilibration. Upon completion, the program summarizes the process with a few final lines

```
Random seed was   2147483647
< Chisquared >  = 33545.66
< A Atoms >  =   4273.13
< P Atoms >  =   323.30
log[e]Prob(Data) = -21495.20
```

which give the random seed, the average χ^2 value of the samples, the average number of atoms in the samples (here A is used instead of F for the distributions), and the evidence, which is not used in this example. It should be noted that the average χ^2 is not the χ^2 of the mean model, but rather is an average of the χ^2 of the individual samples. One useful measure that can be made on the mean model returned by the algorithm is to check that its χ^2 value is lower than the average χ^2 of the samples. This helps to verify that the sampler has visited a reasonably large space in the posterior distribution.

The average number of atoms provides an indication of the number of parameters in the model. Essentially, each atom is defined by two values: a location and an amplitude. These are the only values that the algorithm is free to fix, so the number of parameters used to create the model can be viewed as twice the number of atoms. For data with well-measured noise with a normal distribution, this can provide an indication of whether the model is overfitting the data. However, for expression data this is not yet possible.

17.4.3 Visualizing the Output

As noted above, the program creates three output files upon completion. The results can either be visualized directly from the ASCII output files using a 4GL program such as Matlab or IDL or by using the Java visualization tool. Specific directions for using the visualization tool are provided in the downloadable files. Plots that are particularly useful are comparisons of the mock (M) and data (X) matrices to verify that the algorithm has created a model that reconstitutes the data. In addition, the residuals can provide indications of the need for more patterns (if a clear pattern remains in the residuals) or indications of which parts of the data are particularly poorly fit by the model. Comparisons can be made row-by-row through the

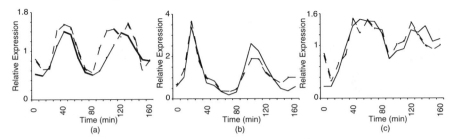

FIGURE 17.3. Comparison of the data and the model of the data for several genes with complex regulation. The original data (row of the X matrix) are shown as a dashed line, with the reconstruction (row of the M matrix) shown as a solid line. In each case, BD has successfully modeled the complex behavior as arising from a mixture of the simple behaviors shown in Figure 17.4. The genes shown are (a) ISR1, (b) RAD53, and (c) HSL7.

data matrix or for the full matrix if adequate visualization tools are available (work is ongoing to add this capability to the Java tool). Examples of some row comparisons are shown in Figure 17.3.

Next, the pattern matrix should be viewed. For data with expected coherent structure, the individual patterns should make physiological sense (for these time series data, for instance, we expect continuity). In Figure 17.4, the six patterns (P matrix) from these data are shown, and they

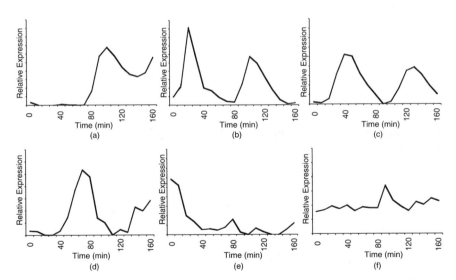

FIGURE 17.4. The patterns generated by Bayesian Decomposition analysis of the yeast cell cycle data. The first five patterns align with cell cycle phase, with (a) and (b) aligning with the G1 phase, (c) aligning with S/G2, (d) aligning with M, and (e) aligning with M/G1. The final pattern (f) is an oscillator; see the text for details.

align nicely with the cell cycle phase and show the expected rise-and-fall behavior of cell cycle regulated genes. The exception to this is the pattern in Figure 17.4(f), which shows constant expression with a superimposed oscillation. This pattern shows an approximately 40 minute oscillation, in agreement with previous measurements of a metabolic yeast oscillator (Moloshok et al., 2002).

For gene expression data, the visualization of the F matrix is often uninformative. Instead, it is useful to rank the genes by their assignment to the individual patterns, noting which genes are linked exclusively to a pattern and which genes are strongly represented in a pattern. Although many genes presently being spotted on microarrays are of unknown function, this ranking of genes can provide an indication of the signaling pathway associated with their expression.

17.4.4   Interpretation

The most difficult aspect of the analysis is interpretation of the patterns and distributions generated by Bayesian Decomposition. The cell cycle dataset was chosen for this work because it provides a good pedagogical example, with interpretation simplified by knowledge of the cell cycle progression in yeast. In more complex datasets, interpretation will be substantially more difficult. The patterns from BD analysis of the cell cycle data (Figure 17.4) can be viewed as an orderly progression through the cell cycle. The patterns have been ordered in Figure 17.4 to show this, with G1 cycles, an S/G2 cycle, an M cycle, and an M/G1 cycle. The two G1 cycles result from the absence of one pattern on exit from the arrest.

Interpretation of the distributions (F matrix) is problematic but can be done by creating a list of the genes most strongly represented within each pattern. Essentially, the genes attached to a pattern should indicate the purpose of the coregulation, such as a response to some internal or external stimulus with subsequent activation of one or more pathways. For these data, of course, the stimuli are the specific cyclin-dependent kinases and related proteins activating the steps through the cell cycle. Since the cell cycle is a well-understood process at this level of detail, interpretation of the patterns is straightforward without the need to analyze the genes in each pattern.

For most datasets, including those that are not time series data, interpretation will be more difficult. For example, data such as the Rosetta compendium do not produce patterns with correlations between points like a time series (Hughes et al., 2000). In these cases, interpretation is dependent on the identification of specific cellular behavior from the genes tied to a pattern (i.e., the significant values in the column of the F matrix related to the pattern). In recent work applying BD to the Rosetta Compendium of gene expression measurements in yeast deletion mutants (Hughes et al., 2000), BD identified patterns across the deletion mutants. For each pattern, the genes included were identified by their cellular role as recorded in the Proteome database (Costanzo et al., 2000; Costanzo et al., 2001). For one pattern, 23 of the top 50 scoring genes had known function, with 13 involved in the mating response and 5 involved in meiosis, allowing this pattern to be linked to the MAPK signaling pathway leading to the mating response (Posas et al., 1998). Validation of this conclusion was obtained by looking at the deletion mutants targeting the mating response pathway in the compendium, where all knockouts of proteins essential to the mating pathway lead to loss of this pattern, validating assignment of the pattern as the endpoint of activation of the mating pathway (Bidaut et al., 2002).

17.4.5   Advantages of Bayesian Decomposition

Bayesian Decomposition offers two primary advantages over widely used methods of gene expression data analysis. First, it allows genes to be assigned to multiple coregulation groups so that complex behavior can be correctly recovered, allowing expression such as that shown in Figure 17.3 to be correctly modeled. This aids in linking coregulation groups to pathways within the cell so that they provide an indirect measure of the behavior of the pathway. The refined coregulation groups should also improve the information available to systems aimed at building genetic network models (D'Haeseleer et al., 2000; Wessels et al., 2001).

The second primary advantage is the ability to introduce prior knowledge into the system through the use of the mapping functions between the atomic and model domains (see Figure 17.2). Although this has not been exploited here, the ability to encode information about known coregulation by linking genes in the columns of the F matrix or to encode coordinated behavior within a row in the P matrix (such as a time curve) has been demonstrated to be very powerful (Ochs et al., 2001). Work at encoding this prior knowledge is presently ongoing.

17.5   Discussion

Gene expression arrays provide insight into genome-wide expression in biological systems, allowing researchers to follow cellular processes during experiments or therapeutic treatments with unprecedented breadth. Such large-scale measurements can lead to improvements in our understanding of disease and in our ability to distinguish types of disease. Since the volume of data and complexity of biological systems make it extremely difficult to fully utilize these data using traditional biological approaches, mathematical methods of analysis are needed. However, these methods generally fail to incorporate mathematical descriptions matched to the underlying biology, which limits the knowledge that can be recovered from the data. Of particular importance to research on cancer and many other diseases is the failure of most methods to allow genes to be assigned to multiple coexpression groups, which the fundamental biology requires since many genes are multiply regulated, responding to many different stimuli. The signals from these stimuli often are transduced through the cell by signaling pathways, and these signaling pathways are often the site of protein mutations that lead to disease.

The goal of the application of Bayesian Decomposition to expression analysis is the identification of all genes involved in coexpression groups within an experiment, even when some or all of these genes belong to multiple such groups. The identification of these groups can allow successful linking of coexpression groups to specific signaling pathways so that expression changes can be used as indicators of activation or deactivation of a pathway. Since many therapeutics target specific pathways or members of pathways (e.g., STI-571 (Hirota et al., 1998; Hirota, 2001)), the recovery of such information is critical for understanding treatment response, treatment failure, and the development of resistance. The tutorial example given here shows the ability of the algorithm to identify coexpression and successfully model gene expression arising from multiple stimuli (e.g., at multiple points in the cell cycle) as demonstrated in Figure 17.3 and in previous work (Moloshok et al., 2002). Other work has demonstrated that such coexpression groups can be successfully linked to specific signaling pathways (Bidaut et al., 2002).

Bayesian Decomposition provides a unique insight into expression data; however, there are several areas where the algorithm can be improved. First, the typical analysis takes many hours of runtime, even on moderately powerful workstations. Second, the use of multiple Markov chains would allow the system to sample more of the posterior distribution. Third, the estimation of dimensionality remains a critical issue, especially for extremely complex datasets. Fourth, previous work has demonstrated that maximizing encoded prior information greatly improves the capabilities of the algorithm. We are presently addressing each of these issues.

In order to both use multiple Markov chains and to improve runtime, we are modifying the algorithm to run on a Beowulf cluster, with each chain running on a single node and with interchain comparison done using the message passing interface (MPI). These modifications require that the issue of the order of solutions within the matrices be handled (i.e., simultaneous interchange of two rows in P and the corresponding two columns in F does not change M). We are presently testing a system that uses autocorrelations to order the chains at the end of equilibration and prior to gathering of samples.

The issue of dimensionality is difficult since known correlated structures in the patterns do not exist for most datasets. This is not the case for time series analysis; however, time series data comprise only a small portion of the total set of gene expression experiments. We are presently pursuing methods of analysis aimed at estimating dimensionality in the data. It is important to note that this is a different question than asking the number of expression units (Manson McGuire and Church, 2000), as Bayesian Decomposition identifies a minimal set of basis vectors for the specific dataset. The number of basis vectors is unlikely to be identical to the number of groups of genes that together provide some cellular function but instead may be the number of groups of cellular functions that are activated simultaneously within the experiment. For example, many functions are activated for DNA synthesis in the S phase of the cell cycle; however, only one pattern is needed from BD, as they are activated simultaneously.

Prior knowledge in gene expression generally takes the form of known coexpression. We are presently modifying the algorithm to permit linking within the columns of the F matrix (as in Figure 17.2), equivalent to enforcing coexpression. This will not limit any pattern to a predetermined set of genes, as additional genes can be added, but it will require that when one gene in a set is changed then the set as a whole changes. These changes may be enforced within a single pattern or over all patterns. With these additions, Bayesian Decomposition should continue to provide an excellent tool for gene expression analysis.

Acknowledgments. This work was supported by the National Institutes of Health, National Cancer Institute (Grant CA06927 to R. Young, Pilot Grant to MFO under Grant CA83638 to R. Ozols), and the Pew Foundation.

References

Alison M, Sarraf C (1997). *Understanding Cancer.* Cambridge: Cambridge University Press.

Alizadeh AA, Eisen MB, Davis RE, Ma C, Lossos IS, Rosenwald A, Boldrick JC, Sabet H, Tran T, Yu X, Powell JI, Yang L, Marti GE, Moore T, Hudson JJ, Lu L, Lewis DB, Tibshirani R, Sherlock G, Chan WC, Greiner TC, Weisenburger DD, Armitage JO, Warnke R, Levy R, Wilson W, Grever MR, Byrd JC, Botstein D, Brown PO, Staudt LM (2000). Distinct types of diffuse large b-cell lymphoma identified by gene expression profiling. *Nature* 403(6769):503–511.

Alter O, Brown PO, Botstein D (2000). Singular value decomposition for genome-wide expression data processing and modeling. *Proceedings of the National Academy of Sciences USA* 97(18):10101–10106.

Besag J (1986). On the statistical analysis of dirty pictures. *Journal of the Royal Statistical Society B* 48:259–302.

Besag J, Green PJ (1993). Spatial statistics and bayesian computation. *Journal of the Royal Statistical Society B* 55:25–37.

Besag J, Green P, Higdon D, Mengersen K (1995). Bayesian computation and stochastic systems. *Statistical Science* 10(1):3–66.

Bidaut G, Moloshok TD, Grant JD, Manion FJ, Ochs MF (2002). Bayesian decomposition analysis of gene expression in yeast deletion mutants. In: K Johnson, S Lin (eds.), *Methods of Microarray Data Analysis II*, 105–122. Boston: Kluwer Academic.

Bittner M, Meltzer P, Chen Y, Jiang Y, Seftor E, Hendrix M, Radmacher M, Simon R, Yakhini Z, Ben-Dor A, Sampas N, Dougherty E, Wang E, Marincola F, Gooden C, Lueders J, Glatfelter A, Pollock P, Carpten J, Gillanders E, Leja D, Dietrich K, Beaudry C, Berens M, Alberts D, Sondak V (2000). Molecular classification of cutaneous malignant melanoma by gene expression profiling. *Nature* 406(6795):536–540.

Brown MP, Grundy WN, Lin D, Cristianini N, Sugnet CW, Furey TS, Ares MJ, Haussler D (2000). Knowledge-based analysis of microarray gene expression data by using support vector machines. *Proceedings of the National Academy of Sciences USA* 97(1):262–267.

Claverie JM (1999). Computational methods for the identification of differential and coordinated gene expression. *Human Molecular Genetics* 8(10):1821–1832.

Cooper GM (1992). *Elements of Human Cancer*. Boston: Jones and Bartlett Publishers.

Costanzo MC, Crawford ME, Hirschman JE, Kranz JE, Olsen P, Robertson LS, Skrzypek MS, Braun BR, Hopkins KL, Kondu P, Lengieza C, Lew-Smith JE, Tillberg M, Garrels JI (2001). Ypd, pombepd and wormpd: model organism volumes of the bioknowledge library, an integrated resource for protein information. *Nucleic Acids Research* 29(1):75–79.

Costanzo MC, Hogan JD, Cusick ME, Davis BP, Fancher AM, Hodges PE, Kondu P, Lengieza C, Lew-Smith JE, Lingner C, Roberg-Perez KJ, Tillberg M, Brooks JE, Garrels JI (2000). The yeast proteome database (ypd) and *caenorhabditis elegans* proteome database (wormpd): Comprehensive resources for the organization and comparison of model organism protein information. *Nucleic Acids Research* 28(1):73–76.

D'Haeseleer P, Liang S, Somogyi R (2000). Genetic network inference: from co-expression clustering to reverse engineering. *Bioinformatics* 16(8):707–726.

Eisen MB, Spellman PT, Brown PO, Botstein D (1998). Cluster analysis and display of genome-wide expression patterns. *Proceedings of the National Academy of Sciences USA* 95(25):14863–14868.

Geman S, Geman D (1984). Stochastic relaxation, Gibbs distributions, and the bayesian restoration of images. *IEEE Transactions on Pattern Analysis and Machine Intelligence* PAMI-6(6):721–741.

Getz G, Levine E, Domany E (2000). Coupled two-way clustering analysis of gene microarray data. *Proceedings of the National Academy of Sciences USA* 97(22):12079–12084.

Golemis EA, Ochs MF, Pugacheva EN (2001). Signal transduction driving technology driving signal transduction: Factors in the design of targeted therapies. *Journal of Cellular Biochemistry Supplement* 37:42–52.

Golub TR, Slonim DK, Tamayo P, Huard C, Gaasenbeek M, Mesirov JP, Coller H, Loh ML, Downing JR, Caligiuri MA, Bloomfield CD, Lander ES (1999).

Molecular classification of cancer: Class discovery and class prediction by gene expression monitoring. *Science* 286(5439):531–537.

Grenander U, Miller MI (1994). Representations of knowledge in complex systems. *Journal of the Royal Statistical Society B* 56:549–603.

Hastings W (1970). Monte Carlo sampling methods using Markov chains and their applications. *Biometrika* 57:97–109.

Heyer LJ, Kruglyak S, Yooseph S (1999). Exploring expression data: identification and analysis of coexpressed genes. *Genome Research* 9(11):1106–1115.

Hirota S (2001). Gastrointestinal stromal tumors: their origin and cause. *International Journal of Clinical Oncology* 6(1):1–5.

Hirota S, Isozaki K, Moriyama Y, Hashimoto K, Nishida T, Ishiguro S, Kawano K, Hanada M, Kurata A, Takeda M, Muhammad Tunio G, Matsuzawa Y, Kanakura Y, Shinomura Y, Kitamura Y (1998). Gain-of-function mutations of c-kit in human gastrointestinal stromal tumors. *Science* 279(5350):577–580.

Hughes TR, Marton MJ, Jones AR, Roberts CJ, Stoughton R, Armour CD, Bennett HA, Coffey E, Dai H, He YD, Kidd MJ, King AM, Meyer MR, Slade D, Lum PY, Stepaniants SB, Shoemaker DD, Gachotte D, Chakraburtty K, Simon J, Bard M, Friend SH (2000). Functional discovery via a compendium of expression profiles. *Cell* 102(1):109–126.

Ideker T, Thorsson V, Siegel AF, Hood LE (2000). Testing for differentially-expressed genes by maximum-likelihood analysis of microarray data. *Journal of Computational Biology* 7(6):805–817.

Kerr MK, Afshari CA, Bennett L, Bushel P, Martinez J, Walker NJ, Churchill GA (2002). Statistical analysis of a gene expression microarray experiment with replication. *Statistica Sinica* 12(1):203–218.

Kerr MK, Churchill GA (2001). Bootstrapping cluster analysis: assessing the reliability of conclusions from microarray experiments. *Proceedings of the National Academy of Sciences USA* 98(16):8961–8965.

Kerr MK, Martin M, Churchill GA (2000). Analysis of variance for gene expression microarray data. *Journal of Computational Biology* 7(6):819–837.

Kirkpatrick S, Gelatt CD, Vecchi MP (1983). Optimization by simulated annealing. *Science* 220:671–680.

Labadie C, Lee JH, Vetek G, Springer Jr. CS (1994). Relaxographic imaging. *Journal of Magnetic Resonance B* 105(2):99–112.

Lawley DN, Maxwell AE (1971). *Factor analysis as a statistical method.* New York: American Elsevier, 2nd edition.

Lee DD, Seung HS (1999). Learning the parts of objects by non-negative matrix factorization. *Nature* 401(6755):788–791.

Lukashin AV, Fuchs R (2001). Analysis of temporal gene expression profiles: Clustering by simulated annealing and determining the optimal number of clusters. *Bioinformatics* 17(5):405–414.

Macdonald F, Ford CHJ (1997). *Molecular Biology of Cancer.* Oxford: BIOS Scientific Publishers, Ltd.

Manson McGuire A, Church GM (2000). Predicting regulons and their cis-regulatory motifs by comparative genomics. *Nucleic Acids Research* 28(22):4523–4530.

Metropolis N, Rosenbluth A, Rosenbluth M, Teller A, Teller E (1953). Equations of state calculations by fast computing machines. *Journal of Chemical Physics* 21:1087–1091.

Moloshok TD, Klevecz RR, Grant JD, Manion FJ, Speier WI, Ochs MF (2002). Application of bayesian decomposition for analysing microarray data. *Bioinformatics* 18(4):566–575.

Ochs MF, Stoyanova RS, Arias-Mendoza F, Brown TR (1999). A new method for spectral decomposition using a bilinear bayesian approach. *Journal of Magnetic Resonance* 137(1):161–176.

Ochs MF, Stoyanova RS, Brown TR, Rooney WD, Springer Jr. CS (2001). A Bayesian Markov chain Monte Carlo solution of the bilinear problem. In: JT Rychert, GJ Erickson, CR Smith (eds.), *Bayesian Inference and Maximum Entropy Methods in Science and Engineering: 19th International Workshop*, 274–284. Melville: American Institute of Physics.

Posas F, Takekawa M, Saito H (1998). Signal transduction by map kinase cascades in budding yeast. *Current Opinion in Microbiology* 1(2):175–182.

Roberts CJ, Nelson B, Marton MJ, Stoughton R, Meyer MR, Bennett HA, He YD, Dai H, Walker WL, Hughes TR, Tyers M, Boone C, Friend SH (2000). Signaling and circuitry of multiple mapk pathways revealed by a matrix of global gene expression profiles. *Science* 287(5454):873–880.

Sebolt-Leopold JS (2000). Development of anticancer drugs targeting the map kinase pathway. *Oncogene* 19(56):6594–6599.

Sibisi S, Skilling J (1997). Prior distributions on measure space. *Journal of the Royal Statistical Society, B* 59(1):217–235.

Spellman PT, Sherlock G, Zhang MQ, Iyer VR, Anders K, Eisen MB, Brown PO, Botstein D, Futcher B (1998). Comprehensive identification of cell cycle-regulated genes of the yeast saccharomyces cerevisiae by microarray hybridization. *Molecular Biology of the Cell* 9(12):3273–3297.

Tamayo P, Slonim D, Mesirov J, Zhu Q, Kitareewan S, Dmitrovsky E, Lander ES, Golub TR (1999). Interpreting patterns of gene expression with self-organizing maps: Methods and application to hematopoietic differentiation. *Proceedings of the National Academy of Sciences USA* 96(6):2907–2912.

Utsugi A, Kumagai T (2001). Bayesian analysis of mixtures of factor analyzers. *Neural Computing* 13(5):993–1002.

Wessels LF, van Someren EP, Reinders MJ (2001). A comparison of genetic network models. *Pacific Symposium on Biocomputing* 508–519.

Yeung KY, Haynor DR, Ruzzo WL (2001). Validating clustering for gene expression data. *Bioinformatics* 17(4):309–318.

Zhang H, Yu CY, Singer B, Xiong M (2001). Recursive partitioning for tumor classification with gene expression microarray data. *Proceedings of the National Academy of Sciences USA* 98(12):6730–6735.

18

Bayesian Clustering of Gene Expression Dynamics

PAOLA SEBASTIANI
MARCO RAMONI
ISAAC S. KOHANE

Abstract

This chapter presents a Bayesian method for model-based clustering of gene expression dynamics and a program implementing it. The method represents gene expression dynamics as autoregressive equations and uses an agglomerative procedure to search for the most probable set of clusters, given the available data. The main contributions of this approach are the ability to take into account the dynamic nature of gene expression time series during clustering and an automated, principled way to decide when two series are different enough to belong to different clusters. The reliance of this method on an explicit statistical representation of gene expression dynamics makes it possible to use standard statistical techniques to assess the goodness of fit of the resulting model and validate the underlying assumptions. A set of gene expression time series, collected to study the response of human fibroblasts to serum, is used to illustrate the properties of the method and the functionality of the program.

18.1 Introduction

Microarray technology (Lockhart et al., 1996; Schena et al., 1995) enables investigators to simultaneously measure the expression level of the genome of entire organisms under a particular condition and is reshaping molecular biology. The promise of this technology is the ability to observe the entire genome in action and, in so doing, uncover its underlying expression mechanisms. Cluster analysis is today one of the favorite unsupervised learning approaches to identify these mechanisms (Alter et al., 2000; Butte et al., 2000; Eisen et al., 1998; Tamayo et al., 1999). Although different, these clustering algorithms share the general strategy of grouping together genes according to the similarity of their behavior across different experimental conditions or different samples. The intuition behind this approach is that genes acting together belong to similar, or at least related, functional categories. Cluster analysis has become widely popular in molecular biology and has been successfully applied to the genome-wide discovery and characterization of the regulatory mechanisms of several biological processes and organisms (Golub et al., 1999; Iyer et al., 1999; Lossos et al., 2000; Wen et al., 1998).

Several applications of genome-wide clustering methods focus on the temporal profiling of gene expression patterns. Temporal profiling offers the possibility of observing the cellular mechanisms in action and tries to break down the genome into sets of genes involved in the same, or at least related, processes. In these experiments, different experimental conditions correspond to the observation of the genome at a particular time point during the temporal evolution of some biological process. In these cases, standard clustering methods cannot be used any longer because they typically rest on the assumption that the set of observations for each gene are *independent and identically distributed* (iid). Although this assumption holds when expression measures are taken from independent biological samples, such as different subjects or different experimental conditions, it is no longer valid when the observations are realizations of a time series, where each observation may depend on prior ones (e.g., Box and Jenkins, 1976; West and Harrison, 1997). Standard similarity measures currently used for clustering gene expression data, such as correlation or Euclidean distance, are invariant with respect to the order of observations: if the temporal order of a pair of series is permuted, their correlation or Euclidean distance will not change. Biomedical informatics investigators over the past decade have demonstrated the risks incurred by disregarding the dependency among observations in the analysis of time series (Haimowitz et al., 1995; Shahar et al., 1992). Not surprisingly, the functional genomic literature is becoming increasingly aware of the specificity of temporal profiles of gene expression data as well as of their fundamental importance in unraveling the functional relationships between genes (Aach and Church, 2001; Holter et al., 2001; Reis et al., 2001).

A second critical problem of clustering approaches to gene expression data is the arbitrary nature of the actual partitioning process. The method described here automatically identifies the number of clusters and partitions the gene expression time series in different groups on the basis of the principled measure of the posterior probability of the clustering model. In this way, it allows the investigator to assess whether the experimental data convey enough evidence to support the conclusion that the behavior of a set of genes is significantly different from the behavior of another set of genes. This feature is particularly important because decades of cognitive science research have shown that the human eye tends to overfit observations by selectively discounting variance and "seeing" patterns in randomness (see Gilovich et al., 1985, Kahneman et al., 1982; Tversky and Kahne-

man, 1974). By contrast, a recognized advantage of a Bayesian approach
to model selection, such the one adopted in this chapter, is the ability to
automatically constrain model complexity (MacKay, 1992; Tenenbaum and
Griffiths, 2001) and to provide appropriate measures of uncertainty.

We describe here a Bayesian model-based clustering method (Ramoni et
al., 2002) to profile gene expression time series that explicitly takes into
account the dynamic nature of temporal gene expression data. This method
is a specialized version of a more general class of methods called *Bayesian
Clustering by Dynamics* (BCD) (Ramoni, 2002) that have been applied to
a variety of time series data, ranging from cognitive robotics (Ramoni et al.,
2000) to official statistics (Sebastiani and Ramon, 2001). The main novelty
of BCD is the concept of similarity: two time series are similar when they are
generated by the same stochastic process. With this concept of similarity,
the Bayesian approach to the task of clustering a set of time series consists
of searching the *most probable* set of processes generating the observed time
series. The method presented here models temporal gene expression profiles
by autoregressive equations (Box and Jenkins, 1976) and groups together
the profiles with the highest posterior probability of being generated by the
same process. Although this chapter will adopt autoregressive equations
to model the dynamic of gene expression time series, the method here
presented can easily incorporate other representations, such as polynomial
trend models (West and Harrison, 1997). Another important character of
the method here presented is its reliance on an explicit statistical model of
gene expression dynamics. This reliance makes it possible to use standard
statistical techniques to assess the goodness of fit of the resulting model
and validate the underlying assumptions. This method is implemented in
a computer program, called CAGED (Cluster Analysis of Gene Expression
Dynamics). This chapter will first describe the theoretical framework and
the clustering method, then summarize the functionalities of the computer
program implementing this method, and finally illustrate the use of the
program on a publicly available database of gene expression dynamics.

18.2   Methods

The design of a microarray experiment exploring the temporal behavior of
a set of J genes usually consists of a set of n microarrays, each measuring
the gene expression level x_{jt} of a set of genes at a time point t. We regard
the expression values of a single gene across these measurements as a *time
series* $S_j = \{x_{j1}, \ldots x_{jt}, \ldots x_{jn}\}$ and the entire experiment as a set of
J time series $S = \{S_1, S_2, \ldots, S_J\}$ generated by an unknown number of
stochastic processes. The task here consists of merging these expression
profiles into groups (*clusters*) so that each cluster groups the time series
generated by the same process. Our method searches for the most probable
set of processes responsible for the observed gene expression time series.

Our clustering method has two components: a stochastic description of a set of clusters, from which we derive a probabilistic scoring metric to rank the different ways of combining gene expression profiles, and a heuristic search procedure to efficiently explore the space of these combinations.

18.2.1   Modeling Time

CAGED takes a Bayesian approach to clustering and searches for the most probable set of processes responsible for the observed data. To do so, CAGED looks for the set of clusters (i.e., ways of combining genes on the basis of their expression values along time) with maximum posterior probability. The critical point here is that the expression measurements of each gene along time are not independent and identically distributed. CAGED represents this dependency using autoregressive equations. More formally, a stationary time series $S_j = \{x_{j1}, \ldots x_{jt}, \ldots x_{jn}\}$ of continuous values follows an autoregressive model of order p, say AR(p), if the value of the series at time $t > p$ is a linear function of the values observed in the previous p steps. We can describe this model in matrix form as

$$x_j = F_j \beta_j + \epsilon_j, \tag{18.1}$$

where x_j is the vector $(x_{j(p+1)}, \ldots, x_{jn})^T$, F_j is the $(n - p) \times q$ regression matrix whose tth row is $(1, x_{j(t-1)}, \ldots, x_{j(t-p)})$, for $t > p$, and $q = p + 1$. The elements of the vector $\beta_j = \{\beta_{j0}, \beta_{j1}, \ldots, \beta_{jp}\}$ are the autoregressive coefficients, and $\epsilon_j = (\epsilon_{j(p+1)}, \ldots, \epsilon_{jn})^T$ is a vector of uncorrelated errors that we assume to be normally distributed, with expected value $E(\epsilon_{jt}) = 0$ and variance $V(\epsilon_{jt}) = \sigma_j^2$, for any t. The value p is the autoregressive order and specifies that, at each time point t, x_{jt} is independent of the past history before p, given the previous p steps. The time series is stationary if it is invariant by temporal translations. Formally, stationarity requires that the coefficients β_{jk} be such that the roots of the polynomial $f(u) = 1 - \sum_{k=1}^{p} \beta_{jk} u^h$ have moduli greater than unity. The model in equation (18.1) represents the evolution of the process around its mean μ_j, which is related to the β_j coefficients by the equation $\mu_j = \beta_{j0}/(1 - \sum_{k=1}^{p} \beta_{jk})$. In particular, μ_j is well-defined as long as $\sum_{k=1}^{p} \beta_{jk} \neq 1$. When the autoregressive order $p = 0$, the series S_j becomes a sample of independent observations from a normal distribution with mean $\mu_j = \beta_{j0}$ and variance σ_j^2.

Given a time series S_j, we wish to estimate the parameters β_j and σ_j^2 from the data. The Bayesian estimation of β_j and σ_j^2 consists of updating their prior distribution into a posterior distribution by Bayes' Theorem, so, with $f(\beta_j, \sigma_j^2)$ denoting the prior density, we need to compute the posterior density

$$f(\beta_j, \sigma_j^2 | x_j, p) = \frac{f(x_j | \beta_j, \sigma_j^2, p) f(\beta_j, \sigma_j^2)}{f(x_j | p)}.$$

The likelihood function $f(x_j|\beta, \tau, p)$ is

$$f(x_j|\beta_j, \sigma_j^2) = \sqrt{\frac{\sigma_j^{2n}}{(2\pi)^n}} \exp\left(-\frac{(x_j - F_j\beta_j)^T(x_j - F_j\beta_j)}{2\sigma_j^2}\right) \qquad (18.2)$$

and it is in fact a function of the first p values of the series, but we omit the explicit dependence for simplicity of notation. As prior distributions of β_j and σ_j^2, we assume the family of improper distributions with density $f(\beta_j, \sigma_j^2) \propto \sigma_j^{2\gamma}$ for $\sigma_j^2 > 0$ and $\gamma \geq 0$. When $\gamma = 0$, this formula represents the uniform prior, when $\gamma = 1$ it is the typical reference prior, and when $\gamma = 2$, it becomes the so-called Jeffreys prior (Bernardo and Smith, 1994). The quantity $f(x_j|p)$ is the averaged likelihood function in which β_j and σ_j^2 are integrated out. Suppose that F_j is of full rank, and define

$$\hat{\beta}_j = (F_j^T F_j)^{-1} F_j^T x_j \qquad (18.3)$$

$$\text{RSS}_j = x_j^T(I_n - F_j(F_j^T F_j)^{-1}F_j^T)x_j. \qquad (18.4)$$

Then, the quantity $f(x_j|p)$ is

$$f(x_j|p) = \frac{\left(\frac{\text{RSS}}{2}\right)^{(q+\gamma-n)/2} \Gamma\left(\frac{n-q-\gamma}{2}\right)}{(2\pi)^{(n-q)/2} \det(F_j^T F_j)^{1/2}}, \qquad (18.5)$$

where q is the dimension of β_j. The posterior distribution of σ_j^2 and β_j is Normal Inverse Gamma, with

$$\beta_j|x_j, \sigma_j^2 \sim N(\hat{\beta}_j, [\tau(F_j^T F_j)]^{-1}), \qquad (18.6)$$

$$1/\sigma_j^2|x_j \sim \text{Gamma}\left(\frac{\text{RSS}_j}{2}, \frac{n-q-\gamma}{2}\right), \qquad (18.7)$$

and we define the density function of a $\text{Gamma}(a, b)$ by $f(\tau) = a^b\Gamma(b)^{-1}\tau^{b-1}\exp(-\tau a)$. The posterior distributions (18.6) and (18.7) are proper as long as F_j is of full rank and $n > q + \gamma$.

18.2.2   Probabilistic Scoring Metric

We are actually interested in finding the clustering model with the highest posterior probability given the observed gene expression time series. Since all clustering models are compared with respect to the same data, and we assume uniform prior distributions, the posterior probability of a clustering model becomes proportional to its *marginal likelihood*. We describe a set of c clusters of gene expression time series as a statistical model M_c consisting of c autoregressive models with coefficients β_k and variance σ_k^2. Each cluster C_k groups the time series data of J_k genes that are jointly modeled as

$$x_k = F_k \beta_k + \epsilon_k,$$

where the vector x_k and the matrix F_k are defined by stacking the J_k vectors x_{kj} and regression matrices F_{kj}, one for each time series, as

$$x_k = \begin{pmatrix} x_{k1} \\ \vdots \\ x_{kJ_k} \end{pmatrix} \qquad F_k = \begin{pmatrix} F_{k1} \\ \vdots \\ F_{kJ_k} \end{pmatrix}.$$

Note that we now label the vectors x_j assigned to the same cluster C_k with the double subscript kJ, and k denotes the cluster membership, so that $\sum_k J_k = J$, and J is the total number of genes. The vector ϵ_k is the vector of uncorrelated errors with zero expected value and constant variance σ_k^2. In principle, given a set of possible clustering models, the task is to rank them according to their posterior probabilities. The posterior probability of each clustering model M_c is:

$$P(M_c|x) \propto P(M_c) f(x|M_c),$$

where $P(M_c)$ is the prior probability of M_c, x consists of all the time series data $\{x_k\}$, and the quantity $f(x|M_c)$ is the marginal likelihood. The marginal likelihood $f(x|M_c)$ is the solution of the integral

$$\int f(x|\theta) f(\theta|M_c) d\theta,$$

where θ is the vector of parameters specifying the clustering model M_c, $f(\theta|M_c)$ is its prior density, and $f(x|\theta)$ is the overall likelihood function. By independence of the series assigned to different clusters, the overall likelihood function is

$$f(x|\theta) = \prod_{k=1}^{c} p_k^{m_k} f(x_k|F_k, \beta_k, \sigma_k^2),$$

where p_k is the marginal probability that a time series is assigned to the cluster C_k. We assume independent uniform prior distributions on the model parameters β_k, σ_k^2 and a symmetric Dirichlet distribution on the parameters p_k, with hyperparameters $\alpha_k \propto p_k$ and overall precision $\alpha = \sum_k \alpha_k$. By independence of the time series conditional on the cluster membership, and parameter independence, the marginal likelihood $f(x|M_c)$ can be computed as

$$f(x|M_c) = \frac{\Gamma(\alpha)}{\Gamma(\alpha + m)} \prod_{k=1}^{c} \frac{\Gamma(\alpha_k + m_k)}{\Gamma(\alpha_k)} \frac{\left(\frac{\text{RSS}_k}{2}\right)^{(q+\gamma-n_k)/2} \Gamma\left(\frac{n_k-q-\gamma}{2}\right)}{(2\pi)^{(n_k-q)/2} \det(F_k^T F_k)^{1/2}}$$

$$(18.8)$$

where n_k is the dimension of the vector x_k, and $\text{RSS}_k = x_k^T(I_n - F_k(F_k^T F_k)^{-1}F_k^T)x_k$ is the residual sum of squares in cluster C_k. When all clustering models are a priori equally likely, the posterior probability $P(M_c|x)$ is proportional to the marginal likelihood $f(x|M_c)$, which becomes our probabilistic scoring metric.

18.2.3 Heuristic Search

The Bayesian approach to the clustering task is to choose the model Mc with maximum posterior probability. As the number of clustering models grows exponentially with the number of time series, we use an agglomerative, finite-horizon search strategy that iteratively merges time series into clusters. The procedure starts by assuming that each of the J observed time series is generated by a different process. Thus, the initial model M_J consists of J clusters, one for each time series, with score $f(x|M_J)$. The next step is the computation of the marginal likelihood of the $J(J-1)$ models in which two of the J series are merged into one cluster. The model M_{J-1} with maximal marginal likelihood is chosen and, if $f(x|M_J) \geq f(x|M_{J-1})$, no merging is accepted and the procedure stops. If $f(x|M_J) < f(x|M_{J-1})$, the merging is accepted, a cluster C_k merging the two time series is created, and the procedure is repeated on the new set of $J-1$ clusters, consisting of the remaining $J-2$ time series and the cluster C_k.

Although the agglomerative strategy makes the search process feasible, the computational effort can still be extremely demanding when the number J of time series is large. To further reduce this effort, we use a heuristic strategy based on a measure of similarity between time series. The intuition behind this strategy is that the merging of two similar time series has better chances of increasing the marginal likelihood of the model. The heuristic search starts by computing the $J(J-1)$ pairwise similarity measures of the J time series and selects the model M_{J-1} in which the two closest time series are merged into one cluster. If $f(x|M_{J-1}) > f(x|M_J)$, the merging is accepted, the two time series are merged into a single cluster, an *average profile* of this cluster is computed by averaging the two observed time series, and the procedure is repeated on the new set of $J-1$ time series containing the new cluster profile. If this merging is rejected, the procedure is repeated on pairs of time series with decreasing degrees of similarity until an acceptable merging is found. If no acceptable merging is found, the procedure stops. Note that the decision to merge two clusters is actually made on the basis of the posterior probability of the model and that the similarity measure is only used to improve efficiency and to limit the risk of falling into local maxima.

CAGED includes several similarity measures to assess the similarity of two time series, both model-free, such as Euclidean distance, correlation, and lagcorrelation, and model-based, such as the symmetric Kullback-

Liebler distance. This distance is computed for every pair of parameter vectors β_k, β_j, using the normal distribution of each β_k, conditional on the cluster variance σ_k^2. The variance is then replaced by the posterior estimate. For a clustering model specifying c clusters C_k, with matrices F_k and data x_k, the conditional posterior distribution of $\beta_k | x_k, \sigma_k^2$ is $N(\hat{\beta}_k, \sigma_k^2[(F_k^T F_k)]^{-1})$, and the symmetric Kullback-Liebler divergence between conditional distributions of β_k, β_j is

$$d_{kj} = \int f(\beta | x_k, F_k, \sigma_k^2) \log \frac{f(\beta | x_k, F_k, \sigma_k^2)}{f(\beta | x_j, F_j, \sigma_j^2)} \, d\beta$$

$$+ \int f(\beta | x_j, F_j, \sigma_j^2) \log \frac{f(\beta | x_j, F_j, \sigma_j^2)}{f(\beta | x_k, F_k, \sigma_k^2)} \, d\beta,$$

where β denotes the generic integration variable, and $f(\beta | x_k, F_k, \sigma_k^2)$ is the density function of a distribution $N(\hat{\beta}_k, \sigma_k^2[(F_k^T F_k)]^{-1})$.

Model-free distances are calculated on the raw data. Since the method uses these similarity measures as heuristic tools rather than scoring metrics, we can actually assess the efficiency of each of these measures to drive the search process toward the model with maximum posterior probability. In this respect, the Euclidean distance of two time series $S_i = \{x_{i1}, \ldots, x_{1n}\}$ and $S_j = \{x_{j1}, \ldots, x_{jn}\}$, computed as

$$D_e(S_i, S_j) = \sqrt{\sum_{t=1}^{n} (x_{it} - x_{jt})^2},$$

performs best on the short time series of our dataset. This finding is consistent with the results of (Golub et al., 1999) claiming a better overall performance of Euclidean distance in standard hierarchical clustering of gene expression profiles.

18.2.4 Statistical Diagnostics

Standard statistical diagnostics are used as independent assessment measures of the cluster model found by the heuristic search. Once the procedure terminates, the coefficients β_k of the AR(p) model associated with each cluster C_k are estimated as $\hat{\beta}_k = (F_k^T F_k)^{-1} F_k^T x_k$, while $\hat{\sigma}_k^2 = \text{RSS}_k / (n_k - q - \delta)$ is the estimate of the within-cluster variance σ_k^2. The parameter estimates can be used to compute the fitted values for the series in each cluster as $\hat{x}_{kj} = F_{kj}\hat{\beta}_k$, from which we compute the standardized residuals $r_{kj} = (x_{kj} - \hat{x}_{kj})/\hat{\sigma}_k$. If AR$(p)$ models provide an accurate approximation of the processes generating the time series, the standardized residuals should behave like a random sample from a standard normal distribution. A normal probability plot, or the residuals histogram per cluster, is used to assess normality. Departures from normality cast doubt on the autoregressive assumption, so that some data transformation, such as a logarithmic transformation, may be needed. Plots of fitted versus observed values and of fitted values versus standardized residuals in each cluster provide further diagnostics. To choose the best autoregressive order, we repeat the clustering for $p = 0, 1, \ldots, w$, for some preset w, by using the same p for every clustering model, and compute a goodness-of-fit score defined as

$$s = c(q + \gamma) - (1 + \log(2\pi)) \sum_k n_k + \sum n_k \log(n_k - q - \gamma) - \sum n_k \log(\text{RSS}_k)$$

where c is the number of clusters, n_k is the size of the vector x_k in C_k, $q = p + 1$, p is the autoregressive order, and RSS_k is the residual sum of squares of cluster C_k. This score is derived by averaging the log-scores cumulated by the series assigned to each clusters. The resulting score trades off model complexity, measured by the quantity $cq + \sum_k n_k \log(n_k - q)$, with lack of fit, measured by the quantity $\sum n_k \log(\text{RSS}_k)$, and it generalizes the well-known AIC goodness-of-fit criterion of Akaike (1973) to a set of autoregressive models. We then choose the clustering model with the autoregressive order p that maximizes this goodness-of-fit score.

18.3   Software

The method described in the previous section is implemented in a computer program called CAGED (Cluster Analysis of Gene Expression Dynamics). The program runs under the various versions of Microsoft Windows, and the graphic user interface is implemented as a *Wizard interface*. The Wizard interface is composed of succeeding screens (see Figure 18.1, see color insert) that guide the user through the steps of analyzing a database of gene expression dynamics. This section describes the use of the program screen-by-screen.

18.3.1   Screen 0: Welcome Screen

When the program is started, a Welcome screen appears, containing a welcome message and a summary of the End-User License. The bottom of the screen contains six buttons, which will remain present in all subsequent screens. The buttons are, from left to right, an About button, containing some basic information about the program; an Help button, evoking an online help file system; a Cancel button, to quit the program at any time; a Back button, to move backward in the succession of screens; a Next button, to move forward; and a Finish button, which will become active at the end of the analysis process. By hitting the Next button, the user is taken to the next screen.

18.3.2 Screen 1: Getting Started

The second screen, shown in Figure 18.1a is divided into three parts. The first part allows the user to load a database for the analysis. CAGED expects files in ASCII tab-delimited format. The file format follows the rules of most statistical genomics analysis programs: gene expression time series are reported along the rows of the database. The first column contains a description of the gene, the second column an accession number to an Internet-available database (such as GenBank, UniGene, or an Affymetrix accession number), and the following columns report the expression values of such genes across time; that is, each column reports the expression value of a gene in a particular microarray. The first row, except for the first two columns, reports a label denoting the experimental conditions and, in this case, it is expected to report a time stamp. Below the loading dialog, the user is given the option to convert the absolute expression values of each gene into ratios between each time point value and the value recorded at the first time point. This option can be particularly useful when the investigator is interested in the relative dynamics of each gene expression time series rather than the absolute values. For microarray platforms measuring relative gene expression values, such as cDNA or oligonucleotide microarrays, this is usually the preferred option. For platforms measuring absolute expression values, such as SAGE, the absolute expression values can be meaningful by themselves, and this option is usually left unchosen. The user can also decide to filter out gene expression time series where a gene does not show at least one value change higher than a user-defined threshold. The second section of the screen allows the user to load a previously saved analysis session. In this case, the button Next will take the user directly to Screen 3 (Cluster Model). The last section of the screen allows the user to open a Web site on the current machine to serve the program over the Internet.

18.3.3 Screen 2: Analysis

If the user has chosen to load a database for analysis, the Next button will display the Analysis screen, shown in Figure 18.1b. This screen allows the user to choose the statistical hypotheses to run the analysis, the distance to guide the heuristic search, and some optional data transformations. Statistical hypotheses are encoded by some model parameters in the Modeling panel in the top left-hand corner of the screen. Here, the user can set the *order* p in equation 18.1, representing the length of the memory of the model. The user can also set the prior precision α and the parameter γ used to compute marginal likelihood (equation 18.5) and the cluster score (Equation (18.8)). These three parameters are usually dealt with using sensitivity analysis: the user will run various analyses using different settings of the parameters (typically, 0, 1, and 2 for all three) to check the robustness of the clustering model with respect to these assumptions. The last parameter in the panel sets a threshold on the Bayes factor: the ratio between the marginal likelihood of two alternative models. Setting this parameter to δ will tell the program to merge two gene expression time series or two clusters if the marginal likelihood of the model resulting from such a merging is at least δ times larger than the marginal likelihood of the model in which the two gene expression time series or clusters are kept separated. This screen also offers the user the opportunity to choose the distance used by the heuristic search process and described in the previous section. Since this measure is meant to simply guide the search process, the best distance will be the one leading to the clustering model with the largest marginal likelihood. The program also offers the possibility of imposing some optional transformations over the data, ranging from the common logarithmic transformation to some standard power transformations.

18.3.4   Screen 3: Cluster Model

When the user hits the Next button from the Analysis screen, the program will run the clustering process described in the previous section. The results of this analysis are displayed in the Cluster Model screen, shown in Figure 18.1c. The model is primarily described by a set of plots collecting the gene time series members of each cluster. The members and the basic statistical properties of each cluster can be viewed by clicking the button below each plot. The screen also offers the display of a dendrogram with a binary tree representing the clustering order of the gene expression time series. The nodes of the tree in this representation report the Bayes factor of the merging (i.e., how many times the marginal likelihood of the model is increased by the merging of the two subtrees with respect to the model in which the two subtrees are kept separated). General properties of the clustering model are displayed by the Property window, which also contains a validation program to list which repeated genes fall in the same cluster. An example of such a tree is given in the dendrogram shown in Figure 18.1e.

18.3.5   Screen 4: Pack and Go!

The last screen of the program, shown in Figure 18.1d, allows the user to save the results of the analysis in two formats. The first option saves the analysis results in CAGED format so that the results will be loadable and viewable through the program. The second option generates a complete report, including images and statistical diagnostics, in HTML format, which can be posted on the World Wide Web or loaded in some word-processing program. An optional dialog allows the user to insert a URL template to link the Accession numbers in the second column of the input data with the appropriate Internet resource database, such as GenBank, UniGene, or an Affymetrix accession numbers repository.

18.4   Application

This section illustrates the properties of this method and the use of CAGED using a dataset of gene expression dynamics. Iyer et al. (1999) report the results of a study of the temporal deployment of the transcriptional program underlying the response of human fibroblasts to serum. The study uses two-dye cDNA microarrays to measure the changes of expression levels of 8613 human genes over 24 hours at unequally spaced time points. The actual data described in the study comprise a selection of 517 genes whose expression level changed in response to serum stimulation. At the time of their original publication, 238 genes were unknown expressed sequence tags (ESTs). We relabeled the dataset using the most recent Uni-Gene database (http://www.ncbi.nlm.nih.gov/unigene), and 45 genes were left unknown. The UniGene classification was used to identify repeated genes in the dataset. We found that 20 genes appeared at least twice in the dataset and were not known to be part of the same UniGene cluster at the time of the original report.

18.4.1   Analysis

The analysis of a database of gene expression dynamics typically involves more than one run of the program. Statistical diagnostics play the fundamental role of assessing the best-fitting model and, in general, validating the soundness of the conclusions. After the database is loaded in Screen 1, with no filter on the minimum required change, Screen 2 contains all the parameters to set in order to explore different statistical hypotheses. The data in question are actually ratios of two conditions, so we choose to log-transform the data in order to treat symmetrically positive and negative fold changes.

   For this analysis, we choose a uniform prior ($\gamma = 0$) and a minimal prior precision ($\alpha = 1$), the default values in the Modeling panel in the top left-hand corner of Screen 2. Also, because the time points were not equally spaced, we assumed that the spacing of the time points was irrelevant. In other words, intervals of different lags were taken to be equally informative about the underlying process. We ran the clustering algorithm with four autoregressive orders $p = 0, 1, 2, 3$ and the similarity measures available in the Distance panel. Since the role of the distance is simply to guide the search process, we can assess the best working distance by checking the *Marginal Likelihood* of the resulting model in the Properties window in Screen 3. For all values of p, Euclidean distance gave the best results (i.e., the model with highest marginal likelihood). At this point, we must choose the best *Model Order* p, the first parameter in the Modeling panel. The number of clusters found for $p = 0, 1, 2, 3$ varied between 4 ($p = 0, 1$) and 3 ($p = 2, 3$). To choose a clustering model among these four, we used the goodness-of-fit score called *Autoregressive Score* in the Properties window of Screen 3. The scores for the four models were, for increasing p, 10130.78, 13187.15, 11980.38, and 11031.12, and the model with order $p = 1$ was therefore selected. This model merges the 517 gene time series into four clusters of 3, 216, 293, and 5 time series, with estimates of the autoregressive coefficients and within-cluster variance $\hat{\beta}_{10} = 0.518, \hat{\beta}_{11} = 0.708, \hat{\sigma}_1^2 = 0.606$ in cluster 1, $\hat{\beta}_{20} = 0.136, \hat{\beta}_{21} = 0.776, \hat{\sigma}_2^2 = 0.166$ in cluster 2, $\hat{\beta}_{30} = -0.132, \hat{\beta}_{31} = 0.722, \hat{\sigma}_3^2 = 0.091$ in cluster 3, and $\hat{\beta}_{40} = -0.661, \hat{\beta}_{41} = 0.328, \hat{\sigma}_4^2 = 0.207$ in cluster 4.

18.4.2   Statistical Diagnostics

In the selected model, merging any of these clusters decreases the posterior probability of the clustering model by at least 10.05 times, thus providing *strong* evidence in favor of their separation (Kass and Raftery, 1995). The symmetry of the standardized residuals in Figure 18.2, together with the lack of any significant patterns in the scatterplot of the fitted values versus the standardized residuals and the closeness of fitted and observed values, suggests that AR(1) models provide a good approximation of the processes generating these time series. This impression is further reinforced by the averages of the fitted time series in each cluster, shown in Figure 18.2, which closely follow their respective cluster average profiles. In CAGED, the button *Show Residuals* on Screen 4 shows the residuals plots and their basic statistics for each cluster.

18.4.3   Understanding the Model

The most evident difference between the model in Figure 18.3 (see color insert) and the model obtained in the original article by visual inspection (Iyer et al., 1999) is the number of clusters: our method detects four distinct clusters, characterized by the autoregressive models described above, while hierarchical clustering merges all 517 genes into a single cluster and leaves it to the investigator to identify subgroups by visual inspection. For example, Iyer et al. (1999) identify, by visual inspection, eight subgroups of genes— labeled A, B, C,..., I, J—from eight large contiguous patches of color. With the exception of a few genes, our cluster 2 merges the subgroups of time series labeled as D, E, F, G, H, I, and J, and cluster 3 merges the majority of time series assigned to subgroups A, B, and C. Interestingly enough, the members of subgroups A, B, and C differ, on average, by one single temporal value and, similarly, members of groups D and G differ by a single temporal value, as well as F, H, J, and I. The assignment of time series to different groups on the basis of one temporal point could be a consequence of the fact that the human eye tends to overfit.

Across the four clusters, both average profiles and averages of the fitted time series appear to capture different dynamics. Our cluster 1 collects the temporal patterns of three genes—interleukin 8, prostaglandin-

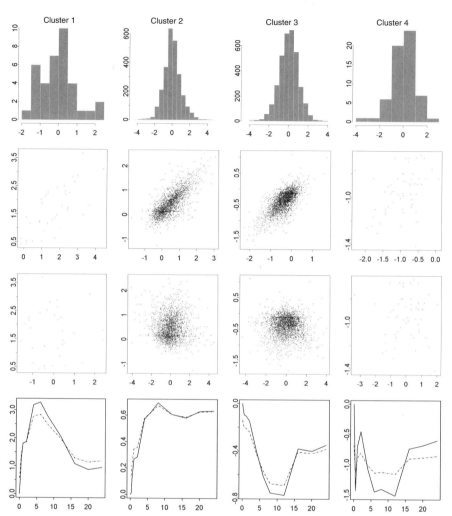

FIGURE 18.2. Diagnostic plots for the clustering model identified by the method when the autoregressive order is $p = 1$. The first row reports histograms of standardized residuals. The second row reports the scatterplot of fitted values versus observed values. The third row shows the scatterplot of fitted values versus standardized residuals. The fourth row displays the four cluster average profiles (continuous lines) computed as averages of the observed time series in each cluster and the averages of the fitted time series in each cluster (dashed lines). In these plots, the x-axis reports time in hours.

endoperoxide synthase 2, and interleukin 6 (interferon beta 2). These time series were assigned by Iyer et al. (1999) to the subgroups F, I, and J, respectively. Cluster 4 collects the time series of five genes—receptor tyrosine kinase-like orphan receptor, TRABID protein, death-associated protein kinase, DKFZP586G1122 protein, and transcription termination factor-like protein. Three of these time series were assigned by Iyer et al. (1999) to the A and B subgroups. These two smaller clusters—clusters 1 and 4—are particularly noteworthy because they illustrate how our method automatically identifies islands of particular expression profiles. The first of these two clusters merges cytokines involved in the processes of the inflammatory response and chemotaxis and the signal transduction and cell-cell signaling underlying these processes. The cluster includes interleukin 8, interleukin 6, and prostaglandin-endoperoxide synthase 2, which catalyzes the rate-limiting step in the formation of inflammatory prostaglandins. The second small cluster includes genes that are known to be involved in the cell-death/apoptosis processes. It includes kinases and several transcription factors reported to be involved in these processes. The cluster includes receptor tyrosine kinase-like orphan receptor 2, TRAF-binding protein domain, and death-associated protein kinase. The cluster also includes the transcription termination factor-like protein, which plays a central role in the control of rRNA and mRNA synthesis in mammalian mitochondria (Fernandez-Silva et al., 1997), and DKFZP586G1122 protein, which has an unknown function but has strong homology with murine zinc finger protein Hzf expressed in hematopoiesis.

The number of clusters found by our algorithm is directly inferred from the data, which also provide evidence in favor of a temporal dependency of the observations: the goodness-of-fit score of the AR(0) clustering model, where the observations are assumed to be marginally independent, is lower than the goodness-of-fit score of the AR(1) clustering model, which assumes that each observation depends on its immediate predecessor. The allocation of the 20 repeated genes in the dataset seems to support our claim that identifying subgroups of genes by visual inspection may overfit the data: with the exception of the two repeats of the DKFZP566O1646 protein, our model assigns each group of repeated genes to the same cluster, whereas four of the repeated genes are assigned to different subgroups in Iyer et al. (1999). Details are shown in Table 18.1. The risks of overfitting by visual inspection can be easily appreciated by looking at the color patterns in Figure 18.3. As the dendrogram is built, genes with highly similar temporal profiles are merged first, thus producing subtrees with similar patterns of colors. However, according to our analysis, the data do not provide enough evidence to conclude that such subtrees contain time series generated by different processes, and they are therefore merged into a single cluster.

An example of this phenomenon is shown in detail in the dendrogram displayed in Figure 18.1e, which enlarges part of the dendrogram in Figure 18.3 around the breaking point between cluster 1 and cluster 2. The first 28 time series from the top of the image, which appear to be more homogeneous by visual inspection, are merged in a subtree showing that, at each step of the iterative procedure, merging the time series induces a model more likely than the model determined by not merging them. Sim-

TABLE 18.1. Assignment of gene repeats to subgroups by Iyer et al. Iyer et al. (1999) (column 2) and by our method (column 3). The first column reports the UniGene name of the repeated genes. Subgroups in column 2 are identified by A to J letters, with NA denoting a placement outside the eight clusters identified by the authors.

Gene name	Group	Cluster
serum/glucocorticoid regulated kinase	J, J	2, 2
pre-B-cell colony-enhancing factor	J, NA	2, 2
myeloid cell leukemia sequence 1	J, J	2, 2
serine proteinase inhibitor	I, I	2, 2
stromal cell-derived factor 1	NA, H	2, 2
neurotrimin	H, H	2, 2
dual specificity phosphatase 6	F, F	2, 2
v-ets avian erythroblastosis virus E26	F, F	2, 2
ESTs	H, H	2, 2
DKFZP566O1646 protein	B, A	2, 3
stearoyl-CoA desaturase	C, C, C	3, 3, 3
pregnancy-associated plasma protein A	C, C	3, 3
DEAD/H box polypeptide 17	B, B	3, 3
KIAA0923 protein	B, B, B, B	3, 3, 3, 3
WW Domain-Containing Gene	B, B	3, 3
Bardet-Biedl syndrome 2	B, B	3, 3
calcium/calmodulin-dependent protein kinase	B, B	3, 3
Tax1 (human T-cell leukemia virus type I)	A, B	3, 3
AD036 protein	A, A	3, 3
DKFZp586I1823	A, A	3, 3

ilarly, the next 18 time series are merged in a subtree labeled, at the top, by a Bayes factor of 10, in logarithmic scale. These two subtrees are then merged in a larger tree, with a Bayes factor of $\exp(33)$, meaning that the model in which the two subtrees are merged together is $\exp(33)$ times more likely than the model in which these subtrees are taken as two separate clusters. The dramatic change of the color patterns two time series below the end of this tree marks the beginning of cluster 2.

18.5 Conclusions

The analysis of gene expression data collected along time is at the basis of critical applications of microarray technology. This contribution addresses a fundamental property of temporal data—their directed dependency along time—in the context of cluster analysis. We have represented the dependency of temporal observations as autoregressive equations and have taken a Bayesian approach to the problem of selecting the number and members of clusters. To explore the exponential number of possible clustering models, we have devised a heuristic search procedure based on pairwise distances to guide the search process. In this way, our method retains the important visualization capability of traditional distance-based clustering and acquires a principled measure to decide when two time series are different enough to belong to different clusters. It is worth noting that the measure adopted here, the posterior probability of the clustering model, takes into account all the available data, and such a global measure also offers a principled way to decide whether the available evidence is sufficient to support an empirical claim. Our analysis shows that sometimes the available evidence is not sufficient to support the claim that two time series are generated by two distinct processes. Figure 18.3 shows contiguous patches of colors, but the posterior probability of the model does not support the claim that these subgroups are sufficiently distinct to be viewed as distinct processes. This finding has interesting implications for experiment design and sample size determination, because it allows the analyst to assess whether the available information is sufficient to support significant differentiations among gene profiles and, if necessary, collect more data. A third feature of the method presented here is the reliance of the clustering process on an explicit statistical model. Contrary to other approaches (Holter et al., 2001), our method builds the clustering model using the parametric content of the statistical model rather than providing statistical content to an established clustering model. This stochastic content allows us to use standard statistical techniques to validate the goodness of fit of the clustering model. Although the biological validation of microarray experiments plays a critical role in the development of modern functional genomics, practical considerations often limit this validation to few genes, while the claims and the scope of a microarray experiment involve thousands. A proper use of available statistical diagnostics provides analytical tools to independently assess the global validity of a clustering model.

Autoregressive equations are very simple representations of process dynamics and they rely on the assumption that the modeled time series are stationary. Our reason to choose this representation is its simplicity: since the time series of gene expression experiments are typically very short, more sophisticated representations could be prone to overfitting. Stationarity conditions can be checked using the method here described but, both in the data analyzed here and in our general experience, the clustering process seems to be largely unaffected by the presence of nonstationary time series. In principle, however, other representations can be integrated within the Bayesian framework described in this chapter. The forthcoming version of CAGED will include, besides autoregressive models, polynomial trend models to tackle the problem of shorter time series and the explicit dependency on time, and state-space models to handle comparative experiments along time and multiple arrays.

Acknowledgments. The authors thank Stefano Monti (Whitehead Institute) and Alberto Riva (Harvard Medical School) for their insightful comments on an early draft of this chapter. This research was supported in part by the National Science Foundation (Bioengineering and Environmental Systems Division—Biotechnology) under contract ECS-0120309.

References

Aach J, Church GM (2001). Aligning gene expression time series with time warping algorithms. *Bioinformatics*, 17:495–508.

Akaike H (1973). Information theory and an extension of the maximum likelihood principle. In *2nd International Symposium on Information Theory*, pp 267–281, Budapest, Hu. Kiado.

Alter O, Brown PO, Botstein D (2000). Singular value decomposition for genome-wide expression data processing and modeling. *Proceedings of the National Academy of Sciences USA*, 97:10101–10106.

Bernardo JM, Smith AFM (1994). *Bayesian Theory*. Wiley, New York, NY.

Box GEP, Jenkins GM (1976). *Time Series Analysis: Forecasting and Control*. Holden-Day, San Francisco, CA.

Butte AJ, Tamayo P, Slonim D, Golub TR, Kohane IS (2000). Discovering functional relationships between rna expression and chemotherapeutic susceptibility using relevance networks. *Proceedings of the National Academy of Sciences USA*, 97:12182–12186.

Eisen M, Spellman P, Brown P, Botstein D (1998). Cluster analysis and display of genome-wide expression patterns. *Proceedings of the National Academy of Sciences USA*, 95:14863–14868.

Fernandez-Silva P, Martinez-Azorin F, Micol V, Attardi G (1997). The human mitochondrial transcription termination factor (mterf) is a multizipper protein but binds to DNA as a monomer, with evidence pointing to intramolecular leucine zipper interactions. *Embo J*, 16(5):1066–79.

Gilovich T, Vallone R, Tversky A (1985). The hot hand in basketball: On the misperception of random sequences. *Cognitive Psychology*, 17:295–314.

Golub R, Slonim DK, Tamayo P, Huard C, Gaasenbeek M, Mesirov JP, Loh IML Coller H, Downing JR, Caligiuri MA, Bloomfield CD, Lander ES (1999). Molecular classification of cancer: Class discovery and class prediction by gene expression monitoring. *Science*, 286:531–537.

Haimowitz IJ, Le PP, Kohane IS (1995) Clinical monitoring using regression-based trend templates. *Artif Intell Med*, 7(4):471–472.

Holter NS, Maritan A, Cieplak M, Fedoroff NV, Banavar JR (2001). Dynamic modeling of gene expression data. *Proceedings of the National Academy of Sciences USA*, 98(4):1693–1698.

Iyer VR, Eisen MB, Ross DT, Schuler T, Moore G, Lee JM, Trent JC, Staudt LM, Hudson J, Boguski MS, Lashkari D, Shalon D, Botstein D, Brown PO (1999). The transcriptional program in the response of human fibroblasts to serum. *Science*, 283:83–7.

Kahneman D, Slovic P, Tversky A (1982). *Judgment under Uncertainty: Hueristic and Biases.* Cambridge University Press, New York, NY.

Kass RE, Raftery A (1995). Bayes factors. *Journal of the American Statistical Association*, 90:773–795.

Lockhart DJ, Dong H, Byrne MC, Follettie MT, Gallo MV, Chee MS, Mittmann M, Wang C, Kobayashi M, Horton H, Brown EL (1996). Expression monitoring by hybridization to high-density oligonucleotide arrays. *Natural Biotechnology*, 14:1675–1680.

Lossos IS, Alizadeh AA, Eisen MB, Chan WC, Brown PO, Botstein D, Staudt LM, Levy R (2000). Ongoing immunoglobulin somatic mutation in germinal center b cell-like but not in activated B cell-like diffuse large cell lymphomas. *Proceedings of the National Academy of Sciences USA*, 97(18):10209–10213.

MacKay DJC (1992). Bayesian interpolation. *Neural Computing*, 4:415–447.

Ramoni M, Sebastiani P, Cohen PR (2000). Multivariate clustering by dynamics. In *Proceedings of the 2000 National Conference on Artificial Intelligence (AAAI-2000)*, pp 633–638, San Francisco, CA. Morgan Kaufmann.

Ramoni M, Sebastiani P, Cohen PR (2002). Bayesian clustering by dynamics. *Mach Learn*, 47(1):91–121.

Ramoni M, Sebastiani P, Kohane IS (2002). Cluster analysis of gene expression dynamics. *Proceedings of the National Academy of Sciences USA*, 99(14):9121–6.

Reis BY, Butte AS, Kohane IS (2001). Extracting knowledge from dynamics in gene expression. *J Biomed Inform*, 34(1):15–27.

Schena M, Shalon D, Davis RW, Brown PO (1995). Quantitative monitoring of gene expression patterns with a complementary DNA microarray. *Science*, 270:467–70.

Sebastiani P, Ramoni M (2001). Common trends in European school populations. *Research in Official Statistics*, 4(1):169–183.

Shahar Y, Tu S, Musen M (1992). Knowledge acquisition for temporal abstraction mechanisms. *Knowl Acquis*, 1(4):217–236.

Tamayo P, Slonim D, Mesirov J, Zhu Q, Kitareewan S, Dmitrovsky E, Lander ES, Golub TR (1999). Interpreting patterns of gene expression with self-organizing maps: Methods and application to hematopoietic differentiation. *Proceedings of the National Academy of Sciences USA*, 96:2907–2912.

Tenenbaum JB, Griffiths TL (2001). Generalization, similarity, bayesian inference. *Behavi Brain Sci*, 24(3).

Tversky A, Kahneman D (1974). Judgment under uncertainty: Heuristics and biases. *Science*, 185:1124–1131.

Wen X, Fuhrman S, Michaels GS, Carr DB, Smith S, Barker JL, Somogyi R (1998). Large-scale temporal gene expression mapping of central nervous system development. *Proceedings of the National Academy of Sciences USA*, 95:334–339.

West M, Harrison J (1997). *Bayesian Forecasting and Dynamic Models.* Springer, New York, NY.

19

Relevance Networks: A First Step Toward Finding Genetic Regulatory Networks Within Microarray Data

ATUL J. BUTTE
ISAAC S. KOHANE

Abstract

An increasing number of methodologies are available for finding functional genomic clusters in RNA expression data. In this chapter, we describe a technique, termed relevance networks, that computes comprehensive pairwise measures of similarity for all genes in such a dataset. Associations with high positive or negative measures are saved and displayed in a graph-network-type diagram. Advantages of this method over others include: (1) negative associations (e.g., those from tumor suppressing genes) are shown: (2) disparate data types can be included (i.e., clinical, expression, and phenotypic); and (3) multiple connections are allowed (e.g., a transcription factor may be responsible for regulating the expression of multiple other genes). Java-based software is available for academic use to construct relevance networks, and operation of the software is also explained in this chapter.

19.1   Introduction

Most algorithms used in the analysis of large RNA expression datasets fall into one of two categories: *supervised* methods (finding predictors of conditions so that unknowns can be accurately labeled) and *unsupervised* methods (finding characteristics of a dataset without a priori labels). Instead of learning the best way to predict a "correct answer," unsupervised algorithms find useful or interesting patterns within a dataset. Obviously, the same dataset can be analyzed using both supervised and unsupervised methods, and these yield very different results. There are several applications of unsupervised methods in functional genomics, such as finding genes with interesting properties without looking for a specific pattern, or determining groups of genes or samples with similar patterns of gene expression. One category of unsupervised methods in functional genomics that we will focus on in this chapter is that of *network determination*, which includes the use of Boolean networks (Liang et al., 1998; Szallasi and Liang, 1998; Wuensche, 1998), Bayesian networks (Friedman et al., 2000), and relevance networks (Butte and Kohane, 1999, 2000; Butte et al., 2000).

In the reincarnated deconstructionist movement currently known as *systems biology*, one hopes to be able to model all the components in biology, ascertain networks and pathways linking these components, perturb the systems in vitro and in silico, and update the networks with new information (Milburn, 2001). Microarrays, one of the first comprehensive genomic measurement tools to be available to the majority of biologists, are felt to be crucial in the determination of the expression state of cells, and knowing the pattern of gene expression across many cells is just one step toward systems biology.

However, despite all the progress made in functional genomics since the mid-1990s, there are still major areas of bioinformatics analysis that are currently undeveloped. One goal in functional genomics is to be able to ascertain biological regulatory pathways from microarray datasets. There have been a few attempts to try to reconstruct pathways from gene expression measurements (DeRisi et al., 1997; Friedman et al., 2000), inspired by the initial success of rediscovering steps of the glycolytic pathway from measurements of substrate levels (Arkin et al., 1997).

Relevance networks offer a method for constructing networks of similarity, with the principal advantages being the ability to (1) include features of more than one data type, (2) represent multiple connections between features, and (3) capture negative as well as positive correlations. As in dendrogram construction, this algorithm begins by evaluating the similarity of features by comprehensively comparing all features with each other in a pairwise manner over the same cases. Several dissimilarity measures have been used in this methodology, including mutual information and the correlation coefficient. Those pairwise associations that do not exceed a threshold measure are then ignored, and the remaining associations are visualized as a graph, with features as nodes and associations as edges.

19.1.1 Advantages of Relevance Networks

We will now discuss some advantages of using relevance networks. First, many alternative clustering and visualization algorithms require imputation to handle missing data. For example, without imputation, it is not obvious where to place a gene in a multidimensional space when even a single gene expression measurement is inaccurately measured or missing. Relevance networks handle missing data without the additional assumptions required by imputation by ignoring any case in the calculation of a pairwise dissimilarity measure that is missing either of the two measurements.

Second, relevance networks naturally handle negative interactions. As an example of a negative interaction in biology, *p53* is a tumor suppressor gene in that increased levels of *p53* are known to be associated with decreased expression of other genes. The concept of negative interaction is clearly different from the concept of no interaction. Since Euclidean distance ranges from zero through the positive numbers, where zero is the strongest score, there is no representation for a negative interaction. In a multidimensional space, *p53* and the genes under its control would not be close together, and clustering techniques using Euclidean distance would not cluster these genes together. Dendrograms, as commonly computed with correlation coefficients, also miss negatively correlated interactions; it is not clear how to simultaneously visualize both positive and negative associations in the same tree. Without taking into account negative interactions, the behavior of tumor suppressor genes and other negative transcriptional factors will be ignored. Because relevance networks use correlation coefficients and mutual information as dissimilarity measures, the methodology takes into account negative interactions between genes.

Third, the process of constructing a dendrogram always attempts to connect all leaves, and there is no rapid method for determining the stronger links (i.e., the more believable ones) compared to the weaker ones. In essence, relevance networks provide a dial for "believability." One can quickly construct relevance networks at a high threshold, and then, if more novel hypotheses are needed, the dial can be lowered gradually to introduce slightly weaker links.

Fourth, although biological functional clusters likely have variable numbers of genes in them, dendrograms connect all clusters into a single structure. A visual inspection is often needed to determine where to cut the tree apart. Relevance networks create multiple networks containing varying numbers of genes.

Fifth, while dendrograms effectively cluster data of a single data type, mixing phenotypic measurements with expression measurements will produce trees with the leaves of phenotypic measurements scattered throughout the tree, which may not be useful. Instead, two-dimensional dendrograms, where phenotypic and expression measurements are separately clustered, are used to mix these two data types (Ross et al., 2000). Relevance networks can naturally mix phenotypic measurements with expression measurements, and direct hypotheses and links are provided between different types of data (Butte et al., 2000).

Sixth, features can only be positioned in a single place within a dendrogram. Each gene is directly connected to the tree with only one stem. In reality, a transcription factor may be responsible for regulating the expression of multiple other genes, but a dendrogram will link that transcription factor only with the one gene it most closely resembles in expression pattern. Relevance networks clearly show that if a gene is closely linked to few or many other genes, then each link is shown separately. This is also important when a gene is similar to two different groups of genes or when a pharmaceutical agent displays activity similar to two different classes of compounds (Butte et al., 2000).

Finally, trees are constructed with only a single dissimilarity measure. A single dendrogram can only cluster genes based on correlation coefficient or Euclidean distance, but not both. As mentioned above, relevance networks can mix multiple types of dissimilarity measures. For example, gene A and gene B may be linked because of high correlation coefficient, but gene B and gene C may be linked because of a high mutual information. Relevance networks can be constructed to include both types of links simultaneously.

19.2   Methodology

19.2.1   Formal Definition of Relevance Networks

Relevance networks are one of a growing number of unsupervised learning methodologies being used in functional genomics, with the principal advantages being the ability to (1) include features of more than one data type, (2) represent multiple connections between features, and (3) capture negative as well as positive correlations.

In this technique, one evaluates the similarity of features by comprehensively comparing all features with each other in a pairwise manner over the same set of cases. Strictly speaking, relevance networks are defined and implemented as a graph

$$G = \{g_1, g_2, \ldots, g_n, \{e_{(1,1)}, e_{(1,2)}, \ldots, e_{(1,m)}\}, \{e_{(2,1)},$$
$$e_{(2,2)}, \ldots, e_{(2,m)}\}, \ldots, \{e_{(p,1)}, e_{(p,2)}, \ldots, e_{(p,m)}\}\}$$

where n nodes $(g_1, g_2, \ldots, g_n)$ are connected by p sets of m edges, each holding a score

$$(\{e_{(1,1)}, e_{(1,2)}, \ldots, e_{(1,m)}\}, \{e_{(2,1)}, e_{(2,2)}, \ldots,$$
$$e_{(2,m)}\}, \ldots, \{e_{(p,1)}, e_{(p,2)}, \ldots, e_{(p,m)}\}),$$

where

$$m = \frac{n^2 - n}{2}.$$

In other words, each set of p edges completely connects the n nodes, and each pair of nodes is connected by a single edge with a score. In practice, each set of p edges represents a different dissimilarity measure (such as Euclidean distance, correlation coefficient, or mutual information), and several of these may be simultaneously applied to the same set of n nodes. When used with microarray data, genes are represented as nodes, and edges are labeled with a real-valued score, which represents the strength of association between two genes.

Relevance networks are viewed by specifying a subset

$$G_s(G, f_1, f_2, \ldots, f_p, t_1, t_2, \ldots, t_p),$$

where $t_1 \ldots t_p$ are values (considered as thresholds), and $f_1 \ldots f_p$ are functions applying the threshold to each set $1 \ldots p$ of edges. In other words, each of the p sets of edges in G has a threshold t and a function f for applying that threshold to each of the edges in that set. Only those edges in G where the function returns true are kept in the subset G_s. Typically, if the edges $\{e_{(i,1)}, e_{(i,2)}, \ldots, e_{(i,m)}\}$ contain values between -1.0 and 1.0, then t_i is set to a value between 0 and 1, and f_i returns true for $e_{(i,j)}$ if $|e_{(i,j)}| >= t_i$ for all j between 1 and m, where $|e_{(i,j)}|$ means the absolute value of $e_{(i,j)}$. When applied to microarray data, this translates into a biological hypothesis that those edges assigned a higher positive value or lower negative value are more likely to represent hypotheses of a biological relationship. Using a threshold serves to break apart the completely connected network graph into a set of smaller graphs. The resultant relevance networks are displayed in a graphical manner similar to Figure 19.1.

The choice of dissimilarity measures used to calculate the scores is arbitrary, and, as described below, implementations of relevance networks have used correlation coefficients and mutual information to score the similarity of patterns of features (Butte et al., 2000). Relevance networks are interpreted by translating the scores, relations, networks, and the dissimilarity measure into a set of hypotheses, but, in general, those pairs of features with the highest scores of association correspond to hypotheses of interaction that can be tested.

19.2.2   Finding Regulatory Networks in Phenotypic Data

We initially chose clinical laboratory results as a source of patient phenotypic data. Bayesian networks (Heckerman, 1996) have traditionally been used to model conditional probabilities between variables in the medical domain. However, there are three reasons why using Bayesian networks is difficult for unsupervised learning in genomics. First, computing the structure of a Bayesian network with over 10,000 features without any prior assignment is currently computationally intractable. Second, updating and maintaining the conditional probabilities in networks with cycles is difficult, and biological networks with cycles should not necessarily be excluded. Third, computing and updating conditional probabilities is problematic when continuous variables are used instead of discrete values (for instance, 100, 210, or 345 mg/dL versus "high" or "normal"). In addition to these

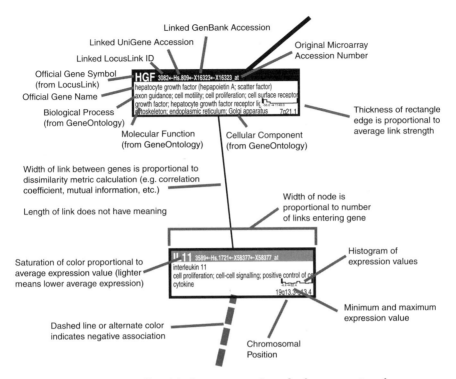

FIGURE 19.1. Graphical representation of relevance networks.

three problems, information on the specific relatedness between variables may be lost when these variables are forced into discrete values.

Our first goal was to exploit existing electronic databases for unsupervised medical knowledge discovery without a prior model or information. We wanted to identify candidate models and systems of these putative relationships for further exploration. Specifically, we wanted to ascertain the relationships between laboratory tests, to see if an unsupervised technique could discover the physiologic, mathematical, and other classes of relationships between types of tests.

We took 410,514 distinct laboratory tests (features) with numeric results belonging to 798 different types of tests and measured on 5158 patients at 28,566 patient time points (cases). We used relevance networks to analyze these large datasets without a priori assumptions. As described above, the algorithm starts with large datasets and proceeds in an unsupervised manner to find networks of similar features (in this case, phenotypic measures).

After eliminating those pairwise associations with r^2 between -0.50 and 0.50 and eliminating those associations constructed with fewer than 15 data points, 31 networks were found, shown in Figure 19.2. At this threshold, most networks were easily validated with "known laws of physiology and pathophysiology." However, potentially novel links were also found, in-

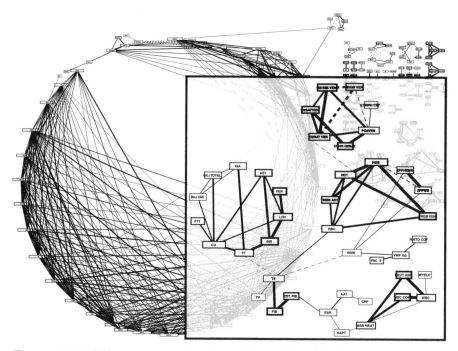

FIGURE 19.2. Relevance networks formed from clinical, laboratory data, with threshold r^2 at 0.5 and n at 15. Thirty-one networks were formed. The largest network connects multiple measurements from pulmonary function testing. The inset describes two positive associations from thyroxine (T4) with prothrombin time (PT) and red cell count (RBC) and a negative association with red cell distribution width (RDW).

cluding a positive association between serum transferrin level and calcium levels, a negative association between CD19-positive cells (a receptor on B-cells) and CD4-positive cells (associated with the T-cell antigen receptor), and a positive association between von Willebrand factor levels and red cell distribution width. More importantly, we found that changing the threshold r^2 allows specificity in the accuracy and consistency of the generation of a range of networks with varying degrees of belief.

19.2.3   Using Entropy and Mutual Information to Evaluate Gene–Gene Associations

Our next goal was to use this method to take large datasets of RNA expression measured under varying conditions and generate networks of hypotheses of gene–gene interactions. We compute the *entropy* of gene expression patterns and the *mutual information* between RNA expression patterns for each pair of genes. The entropy of an RNA expression pattern is a measure of the information content in that pattern and is calculated using equation (19.1)

$$H(A) = -\sum_{i=1}^{n} p(x_i) \log 2(p(x_i)),$$ (19.1)

where $\log 2$ is base 2 logarithm (Shannon and Weaver, 1949). Higher entropy for a gene means that its expression levels are more randomly distributed.

Gene expression is measured on a continuous scale, yet equation (19.1) shows how entropy is computed using discrete probabilities. To calculate entropy, we use a histogram or binning technique. We first calculate the range of values for each gene separately and then divide that range into n subranges. In equation (19.1), $p(x_i)$ equals the proportion of measurements in subrange x_i. As n approaches infinity, the histogram will more accurately model the probability density function (PDF) for the gene.

Ideally, n would be set to a specific appropriate value for each gene. For example, if a particular gene's range of expression level is known to have only two functional states, say "on" and "off," n for that gene could be set to 2. Previous work in applying mutual information to gene expression measurements has operated in this way by assigning only two allowable states for all genes and quantizing continuous expression measurements into these two states (Michaels et al., 1998). By assuming only binary values for each gene, one can represent a state of expression of all genes as a single Boolean vector. State transitions can then be modeled as a Boolean network, and logical rules can be intuited from these networks (Liang et al., 1998).

However, many genes are known to interact with other genes and proteins in a dose-response type of manner, where the specific amount of a gene affects downstream processes on a continuous scale. Most importantly, the specific number of functional states is currently unknown for the majority of genes. For our computations, we arbitrarily set $n = 10$ for all genes, although we acknowledge that changing this piece of a priori information can have a large impact on the ordering of gene–gene interactions by mutual information (Bishop, 1995).

The mutual information is a measure of the additional information known about one expression pattern when given another, as shown in equation (19.2).

$$MI(A, B) = H(A) - H(A \mid B).$$ (19.2)

Equation (19.2) can be restated as equation (19.3). Mutual information can be calculated by subtracting the entropy of the joint RNA expression patterns from the individual gene entropies.

$$MI(A, B) = H(A) + H(B) - H(A, B).$$ (19.3)

Mutual information at zero means that the joint distribution of expression values holds no more information than the genes considered separately. Higher mutual information between two genes means that one gene is non-randomly associated with the other. In this way, mutual information can be used as a dissimilarity measure between two genes related to their degree of independence. We hypothesize that the higher mutual information is between two genes, the more likely it is that they have a biological relationship.

Based on permutation testing, we chose a threshold mutual information (TMI) and displayed only those genes that were linked to others with MI higher than the threshold. At a TMI of 1.3, we found 22 relevance networks, the vast majority of which were validated through finding the proposed hypotheses as previously demonstrated in the biomedical literature. The largest network clustered 143 genes, and two branches of this network are shown in Figure 19.3. Here, *eft1*, an elongation factor, was linked to *ssb2*, a heat shock protein associated with translating ribosomes, which was linked to *yef3*, another elongation factor; *sah1*, s-adenosyl-l-homocysteine hydrolase, a cytoplasmic adenosine-binding protein; and *rpl4a*, one of two genes encoding ribosomal protein L4.

When used as a hypothesis-generation tool, these results can help direct biologists toward specific experiments to validate and explain the gene–gene interactions. Clearly stated, one hypothesis is that elongation factors,

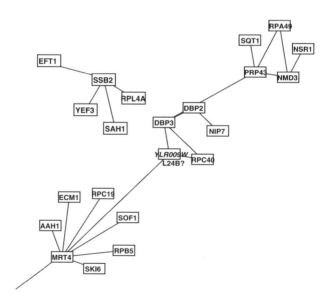

FIGURE 19.3. Two branches enlarged from a relevance network. A PDF version of this figure is available at http://www.smi.stanford.edu/projects/helix/psb00/butte.pdf, where the reader can pan and zoom using the online figure.

such as *eft1*, may be linked to *yef3* and specifically to the ribosome through *rpl4a*. A biologist could then conduct experiments using site-directed mutagenesis on *rpl4a* to determine whether elongation was impaired and could even generate a new set of relevance networks to determine the changes resulting from such a genetic modification.

We found two advantages to using MI in relevance networks. First, using MI is more general than using correlation coefficients to model the relationship between genes, and thus complex relationships between genes can potentially be modeled. For example, if one gene acts as a transcription factor only when it is expressed at a midrange level, then the scatterplot between this transcription factor and other genes might more closely resemble a bell curve rather than a linear model and might have a low correlation coefficient. MI does not require an a priori choice of any particular model.

A second advantage is that expression levels can then be modeled to include measurement noise. As mentioned earlier in this proposal, RNA expression levels are not exactly replicated when experiments are repeated. This noise in RNA expression level measurement can come from many sources: intrachip defects, variation within a single lot of chips, variation within an experiment, and biological variation for a particular gene. Instead of calculating discrete MI, one could instead use a Parzan density function to model the joint distribution and calculate continuous MI (Shannon and Weaver, 1949). This is important because as more is learned about the noise and reproducibility of expression level measurements, this methodology could be used to represent gene expression levels as a distribution instead of just a single point and still find functional patterns.

19.3 Applications

19.3.1 Finding Pharmacogenomic Regulatory Networks

Our next focus was to apply relevance networks to a pharmacogenomic dataset, generating hypotheses of putative functional relationships between pairs of genes and pharmaceuticals (Butte et al., 2000). Specifically, we used baseline RNA expression levels measured from the NCI60, a set of 60 human cancer cell lines used by the National Cancer Institute Developmental Therapeutics Program to screen anticancer agents since 1989 (Weinstein et al., 1997). We joined the gene expression levels to a database with measures of cancer susceptibility to anticancer agents to see how the baseline RNA expression levels in the cell lines correlated with the inhibition of growth of these same cell lines to thousands of anticancer agents. To be clear, RNA expression levels were measured without any exposure to anticancer agents.

Using this methodology, relevance networks were constructed from the 11,692 features (baseline expression of 6701 genes and measures of sus-

ceptibility to 4991 anticancer agents) in the 60 NCI60 cell lines. There
were 68,345,586 pairwise comparisons between features, of which roughly
22 million relationships were between a pair of genes, 12 million relationships were between two anticancer agents, and 33 million relationships were
between a gene and an anticancer agent. The relevance networks formed
from associations with r^2 outside -0.80 to 0.80 are shown in Figure 19.4.

At this threshold, only one network contains an association between
a gene expression and a measure of anticancer agent susceptibility. The
association is between the gene coding for lymphocyte cytosolic protein-1
(*LCP1*, *pp65*, or L-plastin, UniGene Hs.198260), and the anticancer agent
NSC 624044 (4-thiazolidinecarboxylic acid, 3-[[6-[2-oxo-2-(phenylthio)
ethyl]-3-cyclohexen-1-yl]acetyl]-2 thioxo-, methyl ester, [1R-[1α(R*),6α]]-
(9CI)). *LCP1* is an actin-binding protein involved in leukocyte adhesion
(Jones et al., 1998) whose regulation is steroid-hormone-receptor-dependent
(Zheng et al., 1997). A specific role for L-plastin in tumorogenicity has
been postulated; low-level expression of L-plastin is thought to occur in
most human cancer cell lines (Park et al., 1994). Prostate carcinoma in-

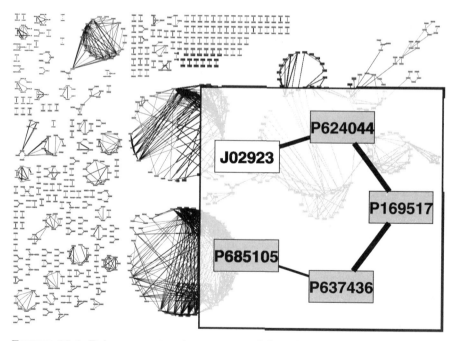

FIGURE 19.4. Relevance networks constructed from baseline gene expression in
60 cancer cell lines joined with susceptibility of the same cell lines to anticancer
agents. The pairs of features (anticancer agents in shaded boxes, genes in white
boxes) with r^2 greater than ±0.80 were drawn with line thickness proportional
to r^2. The inset shows the association between *LCP1* expression (J02923) and
susceptibility to a thiazolidinecarboxylic acid (P624044).

vasion is decreased when levels of L-plastin are suppressed (Zheng et al., 1999). Expression of T-plastin, a related gene, is increased in cisplatin-resistant cell lines (Hisano et al., 1996). Although there is no known relationship between this specific anticancer agent and gene in the biomedical literature, other thiazolidine carboxylic acid derivatives are known to inhibit tumor cell growth (Prevost et al., 1999). Specifically stated, the generated hypothesis is that increasing *LCP1* makes a tumor cell more susceptible to anticancer agent NSC 624044 and possibly other pharmaceuticals.

We felt it was important to test the significance of this discovered association in a statistical and quantitative manner. Using permutations of the data, we calculated 100 distributions of pairwise correlations and were able to highlight only those associations and clusters that were statistically significant in the original data, meaning those demonstrated to be unlikely to be due to random chance. Although hypotheses representing true biological relationships may exist in associations with weaker strength, we felt they could not be statistically distinguished from random noise. It is possible that if additional experiments were performed or cell lines collected to "exercise" the expression space, the strength of these weaker associations could be enhanced.

19.3.2 Setting the Threshold

With so many pairwise comparisons being performed, it may not be immediately clear where one sets the threshold for visualizing the relevance networks. One way to answer this question is to first find the distribution of dissimilarity measure scores (i.e., correlation coefficient measurements), then randomly permute the dataset (shuffling the measurements for each gene or drug), then recompute the dissimilarity measure scores using the permuted datasets. For instance, if one randomly permutes the dataset 100 times and finds that the highest dissimilarity measure score is x, then one could set the threshold to be greater than or equal to x. However, an important point to raise is that the threshold does not have to be a fixed measurement but instead can be "dialed" up and down. In other words, one can set the threshold to be arbitrarily high to start with and the resulting relevance networks can be viewed. If the associations are strong *but are already known and no longer worthy of pursuit*, the threshold can be "dialed down" slightly until new genes and connections are added, creating potentially novel hypotheses. Of course, one must be careful not to drop the threshold so low as to allow potentially noisy and inaccurate associations. The permutation testing can help with this.

In Figure 19.5, we see in the pharmacogenomics experiment described above that after 100 random permutations we were still unable to find a pair of features with a correlation coefficient over 0.80 or under -0.80. Thus, we would use 0.80 as a lower bound for the threshold.

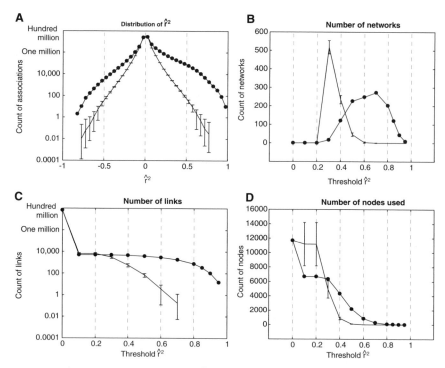

FIGURE 19.5. (A) Distribution of $\hat{r}^2$ calculated from approximately 70 million pairwise comparisons from the pharmacogenomics experiment described in the text, with counts plotted on a log axis. The $\hat{r}^2$ is equal to the r^2, but keeps the same sign (positive or negative) as r. Random permutation was unable to create an association with $\hat{r}^2$ at or over 0.80 or lower than -0.85. (B) As the threshold $\hat{r}^2$ is dropped from 1.0 to 0.0, the number of networks increases and then decreases (closed circles), with the maximum number of networks created with a threshold $\hat{r}^2$ at 0.70. This number of networks is significantly different from the number of networks seen when the dataset was permuted. (C) Significantly more associations are incorporated into networks at higher threshold $\hat{r}^2$ (closed circles) than compared to permuted controls. (D) Similarly, the number of nodes incorporated into networks (closed circles) is significantly higher than controls at higher threshold $\hat{r}^2$.

19.4   Software

A Java-based software package to compute and display relevance networks, called RelNet, is publicly available at no charge from the Children's Hospital Informatics Program at http://book.chip.org for academic use only. Commercial users can obtain Pathlinx software from Xpogen, Inc. at http://www.xpogen.com that calculates relevance networks and displays them in an interactive manner. We will describe the use of the publicly available RelNet software here.

Since it is Java-based, RelNet runs on any platform and under any operating system for which Java is available, including Windows, Linux, and Macintosh. If not already done, the Java runtime environment first needs to be installed; the necessary files can be found at `http://www.javasoft.com`. RelNet requires a computer on the Internet and works best with 300 MB or more RAM.

Users start RelNet with the command

```
java -Xmx500m -jar RelNet.jar
```

The first parameter, `-Xmx500m`, instructs the Java runtime environment to allow the software to use up to 500 MB of RAM (instead of the default 64 MB).

Input data files must be stored as tab-delimited text files, but to ease usage, the exact format for these text files is quite liberal. Each text file can hold the results of one or many microarrays. If the text file holds data from more than one microarray, each microarray should be placed in its own column and a tab-delimited header row (with the same number of tabs as the data rows) should be provided just before the data with the name of each microarray sample. At least one of the columns must contain a gene symbol. At the time of writing, if GenBank, UniGene, LocusLink, Incyte, Agilent, or Affymetrix accession codes are used, gene names, symbols, and details are automatically found over the Internet and added to the relevance network diagrams. Finally, nontab-delimited lines of text are allowed before and after the tab-delimited portion, and these are ignored.

RelNet has only three menu commands. First, all the input microarray files must be opened using the Pairwise:Open Data File command. Second, the comprehensive pairwise dissimilarity measures must be calculated across all genes using the Pairwise:Compute command. This command outputs a large calculation file. Finally, the calculation file must be processed to generate the relevance networks using the Network:Construct command.

First, choose Open Data File from the Pairwise menu and open each input microarray data file. Each file is displayed in a window similar to Figure 19.6 after opening. Column 1 shows each column from the tab-delimited text file. If a header row was found, column names from the header row are used. Column 2 shows the number of unique values found in the column. For example, in Figure 19.6, 12582 indicates that over twelve thousand unique values were found in the "Name" column. Column 3 shows the first value found for that column. Continuing our example in Figure 19.6, the first entry found in the "Name" column was "AFFX-MurIL2_at." Finally, Column 4 indicates the assigned data type for each column.

In reading the tab-delimited text files, RelNet is concerned with only four types of columns: "Sample Name," "Gene Name," "Amount," and "Absolute Call." RelNet uses many heuristics to assign roles to columns automatically; this is usually performed correctly, but the roles for specific columns may need to be changed. "Sample Name" indicates a column that

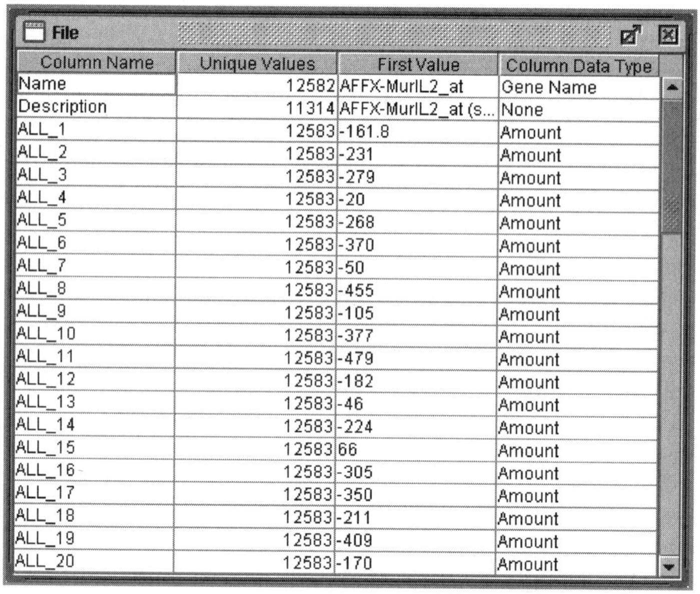

Column Name	Unique Values	First Value	Column Data Type
Name	12582	AFFX-MurIL2_at	Gene Name
Description	11314	AFFX-MurIL2_at (s...	None
ALL_1	12583	-161.8	Amount
ALL_2	12583	-231	Amount
ALL_3	12583	-279	Amount
ALL_4	12583	-20	Amount
ALL_5	12583	-268	Amount
ALL_6	12583	-370	Amount
ALL_7	12583	-50	Amount
ALL_8	12583	-455	Amount
ALL_9	12583	-105	Amount
ALL_10	12583	-377	Amount
ALL_11	12583	-479	Amount
ALL_12	12583	-182	Amount
ALL_13	12583	-46	Amount
ALL_14	12583	-224	Amount
ALL_15	12583	66	Amount
ALL_16	12583	-305	Amount
ALL_17	12583	-350	Amount
ALL_18	12583	-211	Amount
ALL_19	12583	-409	Amount
ALL_20	12583	-170	Amount

FIGURE 19.6. Input tab-delimited data files are automatically processed to determine which columns contain gene and sample information, expression levels, and absolute calls.

contains the name or symbol of a sample, and "Gene Name" indicates a gene; both of these are allowed on only a single column each, and both are optional. If columns specifying sample or gene names are missing, then temporary names are created. "Amount" indicates a numeric column containing a relevant gene expression measurement. This type of column may be labeled "Average Difference" or "AvgDiff" in Affymetrix data files or may be called "Ratio" or "Log Cy3/Cy5 Ratio" for spotted array data files. More than one "Amount" column is allowed if the results of more than one microarray are being stored in the file. Finally, "Absolute Call" delineates columns containing Affymetrix absolute calls; these are text columns containing the single letters "A," "P," or "M." These columns may only follow an "Amount" column and are optional.

After all the relevant files that are desired for inclusion in the relevance network analysis have been opened, choose Compute from the Pairwise menu. The options for this command are shown in Figure 19.7. If the columns marked "Amount" may be missing values, click the first option. If negative or zero expression measurements should be ignored before performing pairwise comparisons, click the second option. To have negative expression measurements (not unusual in data files from Affymetrix Microarray Analysis Software version 4.0) treated as zero, use the third op-

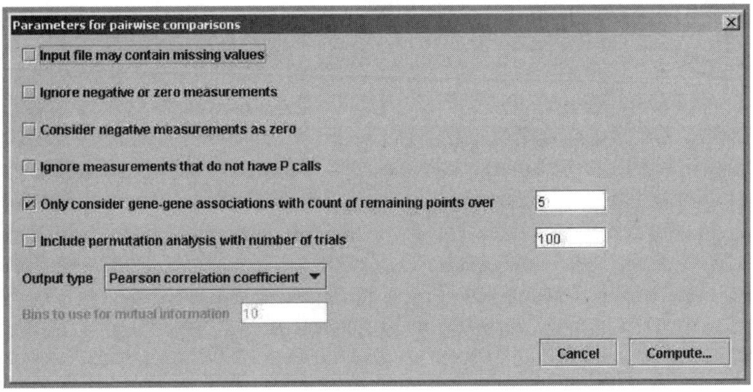

FIGURE 19.7. Parameters to be set for comprehensive pairwise calculations of similarity between genes.

tion. When using Affymetrix files containing columns marked "Absolute Call," click the fourth option to have non-"P" values ignored.

Using any of the first four options may eliminate expression measurements when a pairwise association is being calculated. As an example, two genes may have initially been measured on six microarrays, but after some values have been eliminated using the first four options, there may be only three expression measurements left for the calculation of a dissimilarity measure. The fifth option allows one to eliminate the calculation of a dissimilarity measure when too few points are left. For example, comparing a correlation coefficient with only three points is not at all ideal since it then becomes relatively easy to find pairs of genes with a high correlation coefficient. This value should be set to 4 or higher.

Use the sixth option to specify for permutations to be performed. When selected, comprehensive pairwise associations will be calculated on any number of permuted datasets in addition to the original dataset. The distribution of original and permuted dissimilarity measures, similar to Figure 19.5a, appears later with the output relevance networks.

The final option is to choose the dissimilarity measure to use in the calculation of pairwise associations using a pop-up menu. The choices available include Pearson correlation coefficient and mutual information. The use of Pearson correlation coefficients in constructing relevance networks was described in Butte et al. (2000), and the use of mutual information was described in Butte and Kohane (2000). If mutual information is chosen, an additional option appears allowing the number of bins to use in quantizing the expression measurements to be set; typically, this is set to 5 or higher.

After clicking the Compute button and naming the new output file, RelNet computes the pairwise associations. After this has completed, choose Construct from the Network menu to create the actual relevance networks.

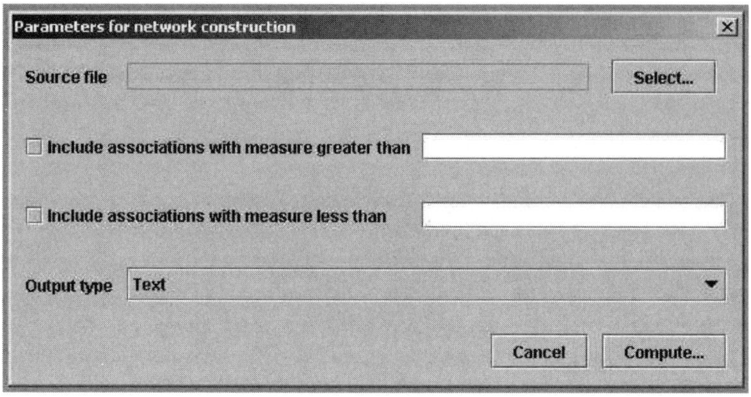

FIGURE 19.8. Parameters to be set for construction of relevance networks.

The options for this command are shown in Figure 19.8. First, click on Select to find and select the output file created from the previous command. RelNet can create relevance networks from a single output file. Clicking on the appropriate checkboxes will allow associations greater than and less than specified thresholds to be included in the output relevance networks.

Finally, the output file type can be specified using the pop-up menu; choices include text, Encapsulated PostScript (EPS), Portable Document Format (PDF), and GMF (a graph layout file supported by Tom Sawyer Software). If PDF format is selected, the output relevance networks will appear similar to Figure 19.1, following several initial pages containing histograms documenting the distribution of dissimilarity measures of the original and permuted datasets and a table of contents listing each network.

Acknowledgments. Portions of this chapter were originally published in Atul Butte's Masters Thesis, "Mutual Information Relevance Networks: Functional Genomic Networks Built From Pair-wise Entropy Measurements."

Atul Butte was supported in part by the National Library of Medicine grant "Research Training in Health Informatics," 5T15 LM07092-07 and R01 LM06587-01; the National Heart, Lung, and Blood Institute, 1U01HL066582-01; the National Institute of Diabetes and Digestive and Kidney Diseases, 1U24DK058739-01; National Institute of Neurological Disorders and Stroke, 1P01NS040828-01A1; National Institute of Allergy and Infectious Diseases, 1R01AI050987-01; the Merck/Massachusetts Institute of Technology Graduate Research Fellowship; the Lawson Wilkins Pediatric Endocrine Society Award; the Endocrine Fellows Foundation; and the Genentech Center for Clinical Research and Education.

Isaac Kohane has been funded in part through the generosity of the John F. and Virginia B. Taplin Award of the Harvard–MIT Division of Health

Sciences and Technology as well as through funding by the NIH, including Grant N01 LM-9-3536, "Personal Internetworked Notary and Guardian," from the National Library of Medicine; Grants HL066582-01 and HL-99-24 through the Program for Genomic Applications of the National Heart, Lung, and Blood Institute; Grant U24 DK058739, "NIDDK Biotechnology Center," by the National Institute of Diabetes and Digestive and Kidney Diseases; Grant 1R21 NS41764-01, "Functional Genomic Analysis of the Developing Cerebellum," by the National Institute of Neurological Disorders and Stroke; Grant U01 CA091429-01, "Shared Pathology Informatics Network," of the National Cancer Institute; and Grant P01 NS 40828-01, "Gene Expression in Normal and Diseased Muscle During Development," by the National Institute of Neurological Disorders and Stroke.

References

Arkin A, Shen P, Ross J (1997). A test case of correlation metric construction of a reaction pathway from measurements. *Science*, 277:1275–1279.

Bishop CM (1995). *Neural Networks for Pattern Recognition*. Clarendon Press: Oxford.

Butte AJ, Kohane IS (1999). Unsupervised knowledge discovery in medical databases using relevance networks. In: Lorenzi N (Ed), Fall Symposium, American Medical Informatics Association, pp. 711–715. Hanley and Belfus: Washington, DC.

Butte AJ, Kohane IS (2000). Mutual information relevance networks: functional genomic clustering using pairwise entropy measurements. *Pacific Symposium on Biocomputing*, 418–429.

Butte AJ, Tamayo P, Slonim D, Golub TR, Kohane IS (2000). Discovering functional relationships between RNA expression and chemotherapeutic susceptibility using relevance networks. *Proceedings of the National Academy of Sciences USA*, 97:12182–12186.

DeRisi JL, Iyer VR, Brown PO (1997). Exploring the metabolic and genetic control of gene expression on a genomic scale. *Science*, 278:680–686.

Friedman N, Linial M, Nachman I, Pe'er D (2000). Using Bayesian networks to analyze expression data. *Journal of Computational Biology*, 7:601–620.

Heckerman D (1996). Bayesian networks for knowledge discovery. In: Fayyad UM, Piatetsky-Shapiro G, Symth P, Uthurusamy R (Eds), *Advances in Knowledge Discovery and Data Mining* pp 273–305. The MIT Press: Cambridge, MA.

Hisano T, Ono M, Nakayama M, Naito S, Kuwano M, Wada M (1996). Increased expression of T-plastin gene in cisplatin-resistant human cancer cells: Identification by mRNA differential display. *FEBS Letters*, 397:101–107.

Jones SL, Wang J, Turck CW, Brown EJ (1998). A role for the actin-bundling protein L-plastin in the regulation of leukocyte integrin function. *Proceedings of the National Academy of Sciences USA*, 95:9331–9336.

Liang S, Fuhrman S, Somogyi R (1998). Reveal, a general reverse engineering algorithm for inference of genetic network architectures. *Pacific Symposium on Biocomputing*, 18–29.

Michaels GS, Carr DB, Askenazi M, Fuhrman S, Wen X, Somogyi R (1998). Cluster analysis and data visualization of large-scale gene expression data. *Pacific Symposium on Biocomputing*, 42–53.

Milburn J (2001). Beyond the genome: Turning data into knowledge. *Drug Discovery Today* 6:881–883.

Park T, Chen ZP, Leavitt J (1994). Activation of the leukocyte plastin gene occurs in most human cancer cells. *Cancer Research*, 54:1775–1781.

Prevost GP, Pradines A, Viossat I, Brezak MC, Miquel K, Lonchampt MO, Kasprzyk P (1999). Inhibition of human tumor cell growth in vitro and in vivo by a specific inhibitor of human farnesyltransferase: BIM-46068. *International Journal of Cancer*, 83:283–287.

Ross DT, Scherf U, Eisen MB, Perou CM, Rees C, Spellman P, Iyer V (2000). Systematic variation in gene expression patterns in human cancer cell lines. *Nature Genetics*, 24:227–235.

Shannon CE, Weaver W (1949). *The Mathematical Theory of Communication*. University of Illinois Press: Chicago.

Szallasi Z, Liang S (1998). Modeling the normal and neoplastic cell cycle with "realistic Boolean genetic networks": Their application for understanding carcinogenesis and assessing therapeutic strategies. *Pacific Symposium on Biocomputing*, 66–76.

Weinstein JN, Myers TG, O'Connor PM, Friend SH, Fornace AJ Jr, Kohn KW, Fojo T (1997). An information-intensive approach to the molecular pharmacology of cancer. *Science*, 275:343–349.

Wuensche A (1998). Genomic regulation modeled as a network with basins of attraction. *Pacific Symposium on Biocomputing*, 89–102.

Zheng J, Rudra-Ganguly N, Miller GJ, Moffatt KA, Cote RJ, Roy-Burman P (1997). Steroid hormone induction and expression patterns of L-plastin in normal and carcinomatous prostate tissues. *American Journal of Pathology*, 150:2009–2018.

Zheng J, Rudra-Ganguly N, Powell WC, Roy-Burman P (1999). Suppression of prostate carcinoma cell invasion by expression of antisense L-plastin gene. *American Journal of Pathology*, 155:115–122.

Index

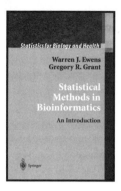
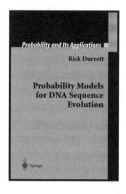
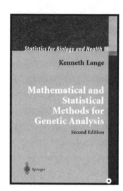